Introduction To Soil Mechanics and Shallow Foundations Design

Introduction To Soil Mechanics and Shallow Foundations Design

SAMUEL E. FRENCH, Ph.D., P.E.

School of Engineering Technology and Engineering
University of Tennessee at Martin

Prentice Hall, *Englewood Cliffs, New Jersey 07632*

Library of Congress Cataloging-in-Publication Data

French, Samuel E.
Introduction to soil mechanics and shallow foundations design
Samuel French.
p. cm.
Includes index.
ISBN 0-13-497454-9
1. Foundations—Design and construction. 2. Soil mechanics.
I. Title.
TA775.F736 1989
624.1'5136—dc19

Editorial/production supervision and
interior design: Lillian Glennon
Cover design: Wanda Lubelska Design
Manufacturing buyer: Robert Anderson

Printed in the United States of America
10 9 8 7 6 5 4 3 2 1

ISBN 0-13-497454-9

Prentice-Hall International (UK) Limited, *London*
Prentice-Hall of Australia Pty. Limited, *Sydney*
Prentice-Hall Canada Inc., *Toronto*
Prentice-Hall Hispanoamericana, S.A., *Mexico*
Prentice-Hall of India Private Limited, *New Delhi*
Prentice-Hall of Japan, Inc., *Tokyo*
Simon & Schuster Asia Pte. Ltd., *Singapore*
Editora Prentice-Hall do Brasil, Ltda., *Rio de Janeiro*

Contents

Preface

The study of foundation design is invariably treated as two separate and distinct topics, the study of shallow foundations and the study of deep foundations. The concepts, approaches, and practices involved in these two topics are quite different; it is a very natural choice to study them separately. This book deals only with the first topic, the design of shallow foundations. The elementary level of this text recognizes that the design of routine shallow foundations can be presented readily at the undergraduate level, while the study of deep foundations is ordinarily presented in more advanced textbooks, quite often at the graduate level.

A necessary part of foundation design is a recognition of the many ways that soil can respond to foundation loadings. Too, the interactive response between soil and structure throughout the life of the structure must be recognized and considered. Consequently, an introductory study of soil as a structural material becomes essential to a study of shallow foundations. Such an introductory study is included herein, limited for the sake of clarity to those topics that directly influence foundation design; this text is not intended to stand as an independent reference in soils.

The design of shallow foundations necessarily involves two disciplines, soil mechanics and structural mechanics. Their interface occurs at the founding line, where the concrete footing meets the load-bearing soil. There is a tradition in the industry to keep these two disciplines separate, if only for the sake of payment vouchers; it is the stated purpose of this book to blend them.

Specialists in soil mechanics are somewhat inclined to view foundation design as a soils problem with implications in structures. Specialists in structures are similarly inclined to view foundation design as a structures problem with implications in soils. Depending on the particular site and the particular structure, the problem could of course be either. In this text, only the most common case is considered, where foundation design is an interwoven indistinct mixture of the two specialties.

It is intended that this book will set forth a brief but balanced presentation from each of the two specialties. From soils, it includes a brief treatment of relevant soil properties and soil mechanics. From structures, it includes a condensed summary of building loads and design of reinforced concrete. The focus, however, is the mixture of these two specialties at the founding line, where the particular structure is matched to the particular soil conditions.

For shallow foundations, the matching of simple structures to the soil conditions is relatively straightforward. By comparison, the analysis of deep foundations can often be very complex, time consuming, and tedious. For both shallow and deep foundations, however, as the complexity of the analysis increases the accuracy of the analysis quite often becomes subject to serious question. In this text, only the more simple, provable cases are presented.

This book is intended for use by persons other than structural engineers. It is presumed that the user has taken the usual preparatory courses in statics and strength of materials and is fully familiar with concepts of stress and strain, to include the Mohr's circle solution for state of stress at a point. The flexure formula and the shear equation are the starting points for the chapter on concrete design; it is assumed that the user is familiar with their use.

The level of the presentation is directed at technologists, architects, construction pushers, and others who are not primarily engineers but who are sometimes called on to design foundations for simple structures. Such people are also occasionally called on to predict the foundation design for larger structures, sometimes even before the structural engineer has been engaged. Such practices allow initial earthwork to proceed while the final project design is under way, a common occurrence in today's construction industry.

The book could also serve, however, as a very elementary reference in foundation design for structural engineers. The information presented is sound; it is just intended for a less technical audience. A much broader study of soil mechanics, in addition to a detailed study of deep foundations, would be necessary for more sophisticated engineering designers.

The approach used in this book has been highly simplified where such simplification does not diminish accuracy. However, there should be no delusions about high levels of accuracy in foundation design, where at best, concrete properties are accurate to two figures and soil properties are accurate to one. Elaborate methods of analysis requiring high levels of accuracy in the material properties are avoided.

The Imperial system of weights and measures (inches, feet, pounds, slugs) is used exclusively in this text. In keeping with widespread practice, units of kips (1000 lb) per square foot as well as pounds per square foot are used for pressures in soils. There are occasional references to SI units where significant features or comparison between the two systems merit special attention.

It is intended that this book be used without additional references. Design tables for reinforced concrete sections are included in the back of the book. If a project is encountered where more elaborate tables or design aids are needed, then the project is beyond the scope of this book and a more comprehensive approach should be used.

Samuel E. French
Martin, Tennessee
1989

Introduction To Soil Mechanics and Shallow Foundations Design

1 Applications

By far the most common structural foundation in today's construction industry is the shallow foundation. Other types of foundations, such as piles, piers, caissons, and similar deep foundations, are used primarily for major structures, not for the ordinary buildings that constitute the overwhelming majority of all construction. This book is devoted to the study of shallow foundations for these small buildings; over the centuries, such foundations have been proven to be economical, serviceable, and reliable.

As with any structural system, shallow foundation systems have limitations in their use. Limitations arise from the quality of the soil, the magnitude of the loads, the configuration of the structure, the type of materials used, and any number of other conditions that can occur with each project. Many times, these limitations can be overcome, sometimes at little cost, sometimes at great cost. Wherever shallow foundations can be made to work, however, they will usually be more economical by a wide margin than the nearest alternative.

In succeeding chapters, the more common conditions that affect the use of shallow foundations will be introduced and analyzed. Because there are so many special circumstances that could occur, an introductory volume such as this is necessarily limited to the most common applications. These common types of foundations, structures, and soils are summarized in this chapter.

COMMON TYPES AND CONFIGURATIONS OF FOUNDATIONS

The most comon types of shallow foundations fall into three descriptions:

1. *Spread footings* supporting a concentrated load, such as a loading delivered by a column (Fig. 1-1a)
2. *Strip footings* supporting a line load, such as a loading delivered by a bearing wall (Fig. 1-1b)
3. *Grade beams* supporting a repetitive series of concentrated loads, such as a loading delivered by a line of several columns (Fig. 1-1c)

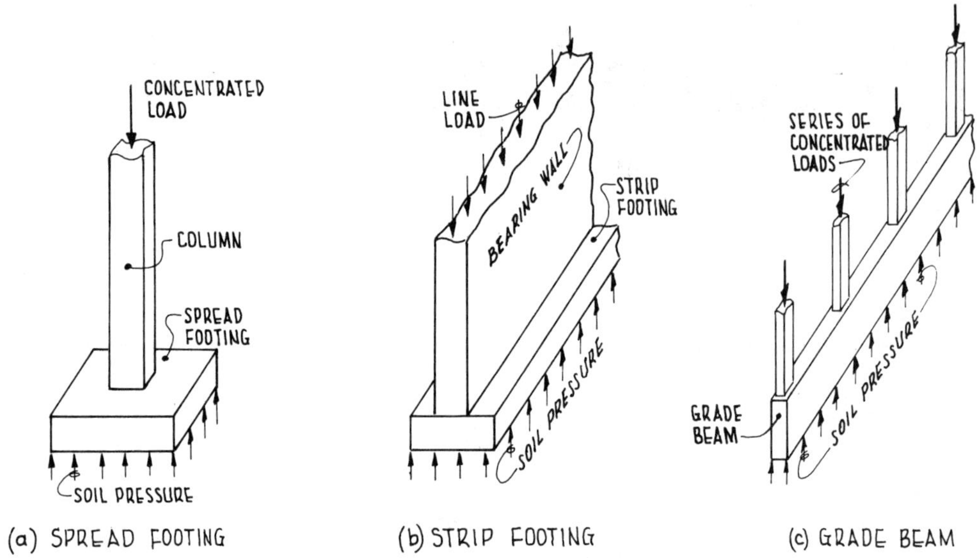

Figure 1-1 Typical shallow foundations.

All three of these types fulfill their function by distributing a concentration of load over a larger bearing area. The footings are said to be *founded* on the soil at the *founding line*. The area under the footing that bears on the soil is descriptively called the *contact area*.

The isolated spread footing shown in Fig. 1-1a should always be centered under the column. Loads occurring off center can produce very high peaks in soil bearing pressure and are to be avoided. Where a line of two or more columns are placed on one footing, the required symmetry is obviously reduced to only one axis.

Strip footings distribute line loads outward as indicated in Fig. 1-1b. The need for symmetry is obvious. There is no stress in a strip footing in its long direction; the wall itself must be designed to sustain any variations in load that occur along its length.

A grade beam is a continuous beam subject to flexure longitudinally, loaded by the line of columns it supports. It is supported at the contact area by the upward soil pressure. The loading system of the grade beam in Fig. 1-1c may look more familiar if the figure is viewed upside down.

The contact area of shallow foundations is kept a specified distance below surface level. That distance is called *depth of founding* and is denoted by D_f. Typical measurements of D_f are shown in Fig. 1-2. A minimum allowable depth of founding is usually set by the building code. The minimum depth is set in temperate climates by considerations of such things as frost penetration, in the tropics by such things as water erosion, and in the desert by such things as wind erosion. Minimum depth of founding is rarely less than 16 in. and may be considerably more depending on local conditions.

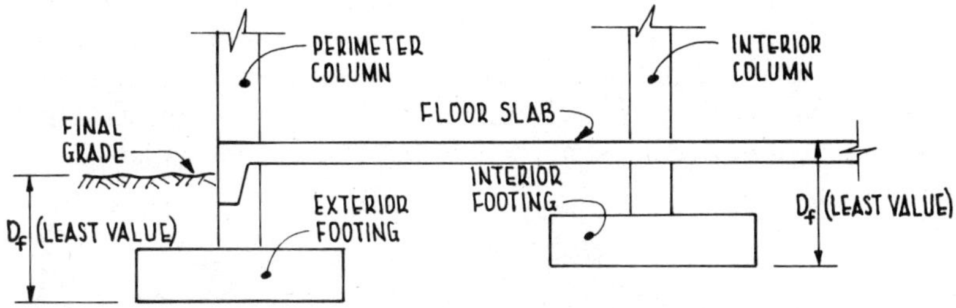

Figure 1-2 Depth of founding.

The maximum depth of founding that might be used and still have the footing classified as a "shallow" foundation is quite arbitrary. Significant changes begin to occur in many soils when the depth exceeds about 6 ft, due to both the higher confining pressures and more stable moisture contents. In addition, a practical limit for unbraced or open excavations is also about 6 ft. Due to such considerations, this depth of about 6 ft is taken here as a reasonable but arbitrary limit of D_f for shallow foundations.

The contact area of a shallow foundation is kept horizontal. Where a bearing wall is to be located along a sloping grade, both wall and footing are stepped to follow the grade as shown in Fig. 1-3. The wall and footing are stepped wherever needed to keep the depth of founding greater than the minimum allowable depth.

Presumably, the area of a shallow foundation could be made any size required. There are practical limits, however. When dimensions become too small,

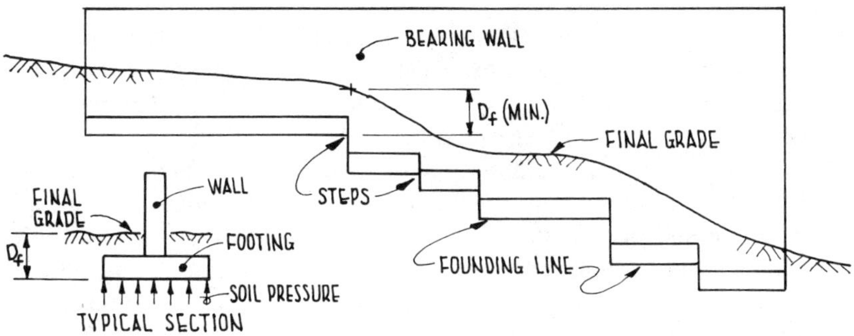

Figure 1-3 Stepped footing.

stress concentrations can cause problems. When dimensions become too large, simple economics suggest that other types of foundations be investigated. A minimum width of 2 ft and a maximum width of 10 ft are commonly observed.

COMMON TYPES AND CONFIGURATIONS OF BUILDINGS

One of the more common structural systems used for small buildings is the braced frame. Two typical examples of a braced frame system are shown in Fig. 1-4. The structural frame of Fig. 1-4a is usually built of steel. The shearwall bracing of Fig. 1-4b is usually built of concrete, although masonry shearwalls are quite common. In both systems, vertical loads are carried by columns; lateral loads (wind and earthquake) are carried by a separate bracing system. The floors or roof are carried by framed girders. Columns carry primarily vertical loads, but some small moments occur due both to the eccentricities of live load and to the unequal spans.

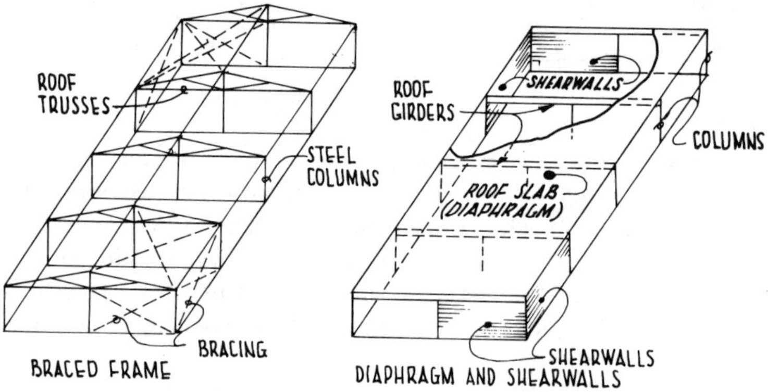

Figure 1-4 Braced frames.

Loads delivered to the footing in a braced frame are usually only the vertical loads. Due to the rigidity of the bracing system, there is little or no sidesway. There are few deformations that would produce significant rotations at the footings.

The lateral load in a braced frame affects only a few footings, those directly supporting the bracing system. The lateral load at these footings is treated as a separate load and is resisted separately, usually by friction but sometimes by lateral bearing on the surrounding soil.

Another very common type of structural system is the rigid frame, shown in Fig. 1-5. In this system, the wind and earthquake loads are carried by bending of the columns. As a result of the bending, the footings may be subject to rotations whenever the columns are rigidly fixed to the footings. Loads delivered to the soil by a rigid frame include the vertical load and the horizontal load as well as the effects of the rotations. All these effects must, of course, be considered in the foundation design.

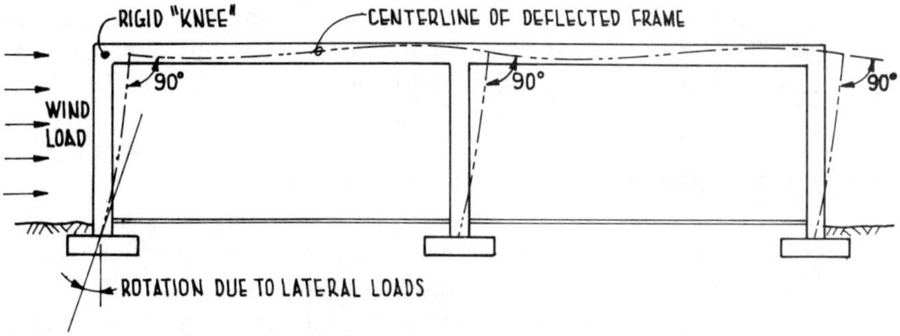

Figure 1-5 Rigid frame.

In common applications of the two systems, braced frames are more likely to be used for low buildings and rigid frames for higher buildings. There are no sharp distinctions, however. One of the most common single-story structures in use is the prefabricated steel building, which is a rigid frame in one direction and a braced frame in the other (Fig. 1-6).

Footing loads for these common types of structural systems can be carried by shallow foundations as long as the height of the building is not too great. For steel buildings, a height of some six to eight stories can usually be carried by shallow foundations. For concrete buildings only about four or five stories should be expected. For buildings higher than these, other types of foundations would probably be feasible. The limitation on height is simply one of economics. As height increases, column loads increase and footing sizes increase. When the total area of footings exceeds about one-third of the building footprint, other types of foundations could be more economical.

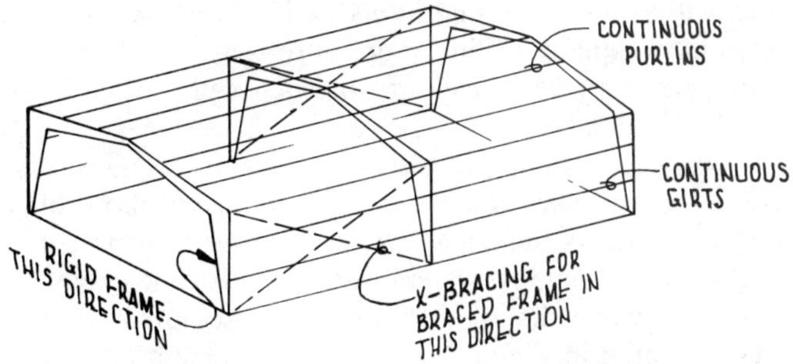

Figure 1-6 Typical prefabricated steel building.

With the suggested height limits, column loads in routine structures should not be more than about 150 tons. At a maximum soil pressure of about 5000 psf, the corresponding footing width is then about 8 ft, consistent with the general limits recommended earlier.

In many small buildings, a heightened first floor is a common architectural feature. In such designs, the columns at the first floor are quite long, usually around 16 ft. Where the extra length could affect the footing design, that length is included in the subsequent analyses.

Walls or columns cantilevered out of a spread footing are not considered here. The "flagpole" analysis for such cases is readily apparent but such cantilevered columns are not used in buildings.

COMMON SOIL PRESSURES AND SETTLEMENTS

For the sake of comparison, soils having an allowable bearing pressure greater than about 6000 psf would be classified as one of the better foundation materials. It would be unusual for a soil having that much strength to pose serious problems with settlements. For lower bearing strengths, 5000 psf and less, both strength and settlement are likely to be of concern.

Unfortunately, it is the soils that have an allowable bearing capacity below about 5000 psf that are most likely to be encountered where shallow foundations are located. At these shallow depths, the soils have been somewhat subject to weathering, water percolation, chemical action, desiccation, and other disturbances that reduce strength. Even with capacities of less than 5000 psf, however, it is usually more economical to use these shallow soils than to try to go deeper.

The lowest reasonable value that may be used for allowable soil pressure is not limited except by economics; at low pressures, footing sizes can simply be too large to be practical. In general, a soil having an allowable bearing pressure below about 1000 psf should not be used to support ordinary shallow foundations. For such soils, a mat or raft foundation (a very large flexible "plate") might be considered if the foundation must be kept shallow. Alternatively, piles or other deep foundations may be appropriate.

The soils of interest here are ordinary soils having an allowable bearing pressure within the foregoing limits of 1000 to 5000 psf. These soils will be assumed to have equally ordinary pressure–settlement characteristics (modulus of subgrade reaction). The response of such soils to load can be expected to be reasonably predictable.

As soil strength improves in these soils, rigidity also improves. Consequently, when a weaker soil is subjected to its low allowable bearing pressure and a stronger soil to its higher allowable bearing pressure, it can be expected that both will settle by roughly the same amount. As one may suppose, such things are borne in mind when the allowable bearing pressures are being established.

As indicated in Fig. 1-7, pressure–settlement curves for most soils are not usually linear (straight-line) where pressures are so high that they approach ultimate strength. Back at working levels, however, small variations are very nearly linear. Within the lower levels of pressures used in foundation design, pressure–settlement relations are assumed herein to be linear (i.e., when pressures at service levels increase by 50%, settlements increase by 50%).

There are soils that respond to load quite differently than in the ways just described. Means to identify these unconventional types of soils are discussed in Chapters 3 and 4. Wherever such problem soils exist, it is usually very common knowledge locally to those in the building industry; consequently, it is very easy

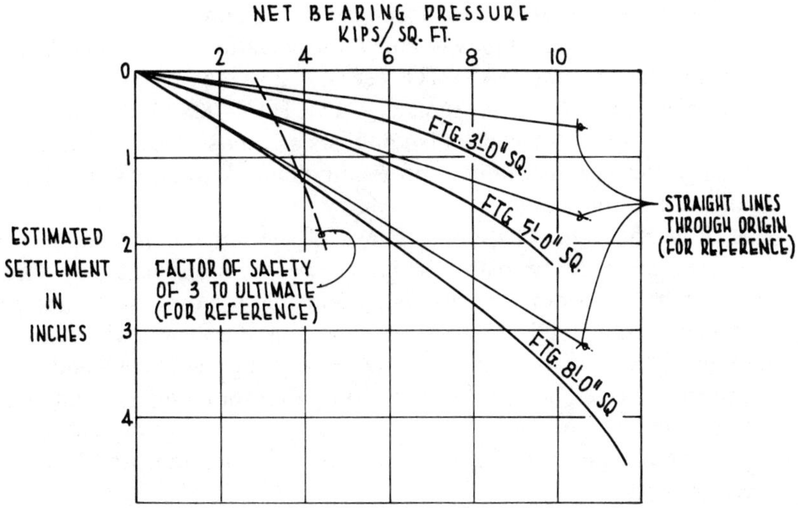

Figure 1-7 Typical pressure-settlement curves.

to determine if one should seek help. The design of foundations in such problem soils is best left to experienced professionals.

A common example of such a problem soil is a clay soil that has been heavily precompressed at some point in its geologic past. Such a clay could be subject to swelling when exposed to water. Such clays are called *active* (or *expansive*) clays and require special consideration when used as a foundation material. They are specifically excluded from the "conventional" soils treated in this book, although a brief discussion of foundations on these precompressed clays is presented in Chapter 10.

STANDARD TEST SPECIFICATIONS

It is a feature of an engineering soils investigation that certain tests will be run on the soil to determine its engineering properties. Such tests have become standardized over the years and specifications have been written prescribing exactly the procedures, equipment, quantities, and timing (if appropriate) to be used in conducting such tests.

The more common standard soils tests that are referenced in this book are contained in the specifications of the American Society for Testing and Materials (ASTM), Volume 19. For identification of a specific test, the ASTM designation for the test is a number, such as ASTM D698-78, in which the number following the dash indicates the year of the latest review or revision. When an ASTM test number appears in this book without such a number, it should be understood that the most recent revision is intended.

A second set of soil test specifications in common use is that of the American Association of State Highway and Transportation Officials (AASHTO). The designation of a specific AASHTO test is again given as a number, followed by the year of the latest revision, such as AASHTO T-180-74. In addition to test specifications for soils, AASHTO also publishes extensive sets of design specifications, construction specifications, and recommended practices that have also become standards in the industry.

For concrete, the specifications of the American Concrete Institute (ACI) are used exclusively in this book. A specification issued by ACI will bear the number of the committee that is responsible for originating and maintaining the specification, as well as a number indicating the year of the latest revisions. For example, ACI 318-83 is the specification "Building Code Requirements for Reinforced Concrete" as reported by committee number 318 and issued by ACI in 1983.

Lesser-known sources of tests and specifications are described as they are encountered.

USEFUL APPROXIMATIONS

It will be noted repeatedly in subsequent discussions that the engineering properties of soil might be subject to considerable error. Even when determined very accurately in a lab test, these properties can vary significantly from point to point within the actual soil stratum. As a consequence, calculations should not be expected to be accurate to more than one figure. In such computations, reasonable approximations can be made without disturbing overall accuracy. A typical example of such an approximation is the overall moment of inertia, $I = bh^3/12$, for concrete columns, where I is the moment of inertia, b the width of the cross section, and h the height. Another is the overall section modulus, $S = bh^2/6$, for columns or bearing walls in flexure, where S is the section modulus.

Unit weights for soil, γ, and concrete, γ_C, are somewhat different; for soil, γ = 90 to 125 pcf, and for concrete, γ_C = 145 pcf. In soil mechanics there is usually no distinction drawn due to this difference in unit weights; the weight of a concrete footing is commonly assumed to be equal to the weight of the soil it displaces.

2

Engineering Properties of Soils

In contemporary construction, soil is simply another construction material. The soil, however, is acquired along with the building site and in most cases must be used "as is." Engineers do not have the latitude to specify the engineering properties of the soil in the way that they do for steel, concrete, or timber.

As a construction material, soil is especially notable for its variability. Even within a single building site, there can be enough variation in soil properties that the foundations at one side of the building site may have to be designed differently from the foundations at the other side. A detailed study of the various types of soil and their engineering properties is, of course, the first step in learning to accommodate such variability.

Soil is classified into four broad groupings: gravels, sands, silts, and clays. These four classifications are developed in considerable detail later in this chapter. Highly organic soils may be suitable for agricultural production, but they are not suitable for carrying foundation loads; organic soils will probably have to be removed from the foundation area during site preparation and replaced by inorganic soils.

Gravels and coarse sands respond to loads in essentially the same way; gravels are therefore not discussed herein as a separate foundation material.

Subsequent discussions concerning foundations on sand should be understood to include gravels.

Sands are familiar to anyone who has ever been to the beach or the desert. Generically, sands are called "cohesionless" soils to describe their total lack of tensile strength. The tendency of sand particles to cling together when wet is only an apparent cohesion; it is not a permanent tensile strength.

Clays are familiar to anyone who has ever seen or used modeling clay. Its sticky, almost greasy texture is evidence of its tensile strength. All the strength of clay comes from its tensile strength; clays are generically called "cohesive" soils.

Silts are less distinctive than either sands or clays and are not as readily identified. Some silts are cohesive, some are cohesionless. In appearance, dry silt looks like a very fine, almost dusty sand.

The foregoing four groupings can be collected into two major soil types according to the way they respond to load. The first major soil type consists of the cohesionless soils: gravels, sands, and cohesionless silts. In all succeeding discussions, a general reference to sands should be understood to include gravels and cohesionless silts as well as sands; these three soils will respond to load in the same general way. The second major soil type consists of the cohesive soils: clays and cohesive silts. As before, a general reference to clays in succeeding discussions should be understood to include cohesive silts as well as clays; these two soils respond to load in the same way. There is a vast difference, however, between the way sands respond to load and the way clays respond to load.

RESPONSE OF A SOIL TO FOUNDATION LOADS

Sands and clays are thus the two extremes in soil types that may be encountered in the design of shallow foundations. Even at this early stage in the discussion, it is well to point out the distinct differences between the ways that these two major soil types respond to load. In many respects they are extreme opposites. Most soils, however, are neither pure sand nor pure clay but are mixtures of these types of soil; it will be seen later that the response to load by such mixtures will fall at a predictable point between the two extremes.

A comparison of ways in which sands and clays respond to load is shown in Table 2-1. Item 1 for footings on sands compares to item 1 for footings in clays. All other item numbers are similarly comparable. It should be immediately apparent from the comparisons of Table 2-1 that "soil" is not a single material. Its properties can be vastly different depending on the relative amounts of clay or sand in the mixture. At the two extremes shown in Table 2-1, either pure sands or pure clays, the two soils often respond to load in diametrically opposite ways.

It should be evident from the comparisons of Table 2-1 that it is essential to identify the foundation material before beginning the structural design. Further, the field investigation of the actual soil conditions should pinpoint problems with

TABLE 2-1 COMPARISON OF RESPONSE TO LOAD

Footings on sand or cohesionless silts	Footings on clay or cohesive silts
1. Sand strength increases with confining pressure, whether due to overburden or to surface loads.	1. Clay strength is relatively constant regardless of magnitude of confining pressure.
2. Sand strength is due entirely to friction, measured by the angle of internal friction ϕ.	2. Clay strength is due entirely to cohesion (tensile strength).
3. Sand strength is relatively insensitive to footing shape.	3. Clay strength is influenced considerably (up to 20%) by footing shape.
4. Sand strength increases markedly (doubled or tripled) with depth of burial of footing.	4. Clay strength is relatively insensitive to depth of burial of footing.
5. Loss of strength in sand is significant if overburden (confining pressure) is removed or eroded away.	5. Little loss of strength is caused by removal of overburden.
6. Strength of a dry sand is cut in half when the sand is submerged.	6. Strength of clay soils is relatively unaffected by submergence.
7. Settlements in sands occur soon after application of load, measured in weeks or months.	7. Settlements in clay occur *very* slowly following application of load, measured in months or years.
8. Settlements in sand can occur under relatively short-term loads.	8. Settlements in clays are relatively unaffected by short-term loads.
9. Settlements in a dry sand are essentially doubled if the sand is submerged (or if water table rises).	9. Settlements in clays are markedly affected (but not doubled) if the clay is submerged.
10. Deposits of sand are best compacted by vibration and submergence, with some pressure.	10. Deposits of clay are best compacted by long-term surcharge pressure.

the surface drainage or the water table, since the presence of water can profoundly affect the response to load, also noted in Table 2-1. Proceeding on assumptions or on past practices could require expensive (and embarrassing) redesign at a later date.

Table 2-1 is presented again in Chapter 3. Each item of the table is discussed there and verified. It is included here only to point out the sharp differences in the ways the two extremes in soil types can respond to loads.

Soil may be used as a structural material in many different ways; supporting the foundations of a building is only one of its many applications as a structural material. Highway embankments, dams, cuts and fills, tunneling, and countless other construction activities place soil under a load of one kind or another. Its response to such loads is included in the field of study called soil mechanics. In this book the discussions will be limited to the response of soil to the loads imposed on it by shallow foundations. It is, of course, only a very small part of the study of soil mechanics.

GEOLOGIC ORIGINS OF SOILS

It is a feature of earth's geologic activity that the rocks forming the earth's mantle are continually being pushed out of the earth onto the surface, where they are then exposed to earthquakes, glaciation, freeze–thaw, water erosion, chemical attack, waterborne abrasion, wind erosion, and other forms of weathering. Under such attack, the rocks are broken progressively into huge boulders, these into smaller boulders, these into cobbles, these into pebbles, and finally, the pebbles are reduced to grains. Deposits of these assorted particles of rock and rock grains are called *soils*.

There is at present no word in engineering geology that is used specifically to describe the ever-continuing reduction of rock into smaller and smaller particles. For the sake of simplicity, it is termed *degradation* in subsequent discussions. Degradation of rock into soil grains is not something that happened only in some vague geologic past. It is a continuing process, as active today as it was a million years ago.

As the size of the particles become progressively smaller, the particles become progressively easier to transport. Transported first by such things as glaciation and avalanche, the particles are reduced to smaller and smaller sizes as they are subjected successively to rushing floods, then, when smaller, by whitewater rapids, later by rapidly moving streams, eventually by muddy meandering rivers, and finally, when small enough, to wind and duststorms. At any stage, the particles might be deposited in a recognizable stratum for a few thousand years before something happens to cause them to be picked up, transported, and deposited in some new location along with particles transported similarly from a hundred other locations.

At every stage of deposition and stratification, water is the ever-present medium of erosion and transport. Water deep within the ground carries acids and bases that chemically attack the particles even when they are buried thousands of feet deep. At the surface, rain, snow, and sleet combine to erode, freeze, thaw, and further reduce chunks of rock into smaller and smaller pieces.

Whenever exposed, the soil particles become subject to organic attack by vegetation, carbon dioxide, and atmospheric acids, changing them chemically into other compounds or even other minerals. Picked up again, redeposited, and exposed again and again, the particles might undergo thousands of years of changes before they come to rest for a few hundred years in relative quiet. It is at one of these quiet periods that the foundation engineer is given a stratum of these particles on which to place a foundation.

Because nature works on such a huge scale, the strata are usually (but not always) so large that the foundation can be located entirely on one stratum. But underlying this stratum could be another stratum having vastly different engineering properties, and underlying this, yet another. The material to be used to support a building foundation is thus a heterogeneous mixture of minerals coming from countless sources over the breadth of a continent, randomly deposited, irregularly stratified, and absolutely inconsistent. It is, in short, soil.

With such a description, defining the engineering properties of a material as variable as soil would seem to be hopeless. Within recent years, however, real progress has been made in defining the engineering properties of soils. Although these properties are more appropriate within large brackets rather than to refined exact details, the response of most soils to a bearing load can now be predicted with some degree of confidence.

Insofar as the engineering properties of a soil are concerned, the mechanism of geologic transport, deposition, burial, and exposure is one of the more important influences. With few exceptions, all soils have been transported to their present locations from somewhere else. They have been blended, disturbed, chemically modified, reblended, mixed, crushed, restratified, picked up and redeposited, sometimes loaded by thousands of feet of overburden, and finally exposed when the overburden was eroded away. The mechanism by which the stratum was finally deposited and later exposed will be seen to be of profound importance when the soil is used as a foundation material.

A second major influence is the groundwater. The location of the water table, the amount of its rise and fall throughout the year, and the chemistry of the groundwater can profoundly affect both the type of foundation and the ability of the soil to carry the foundation. In some circumstances, however, the water entering the soil from the surface can be far more important than that in the water table below; this point is discussed further in later chapters.

A third major influence is the residue of vegetation. Even in the most barren deposits of soil, a few plants will manage to survive. The residue from their eventual death and decay will provide a somewhat more hospitable environment for the next generation of plants. The detritus of these plants in turn will deepen the fragile layer of organic material, providing a yet more fertile ground for other plants, and so on. Eventually, a gradient of organic material is developed, with the high organic content at the surface diminishing steadily with depth. As always, water is the primary vehicle for transporting the organic material downward, aided in this case by disturbances from root penetration.

These combined effects of deposition, water percolation, vegetation, and other influences on the soil eventually produce a typical distribution of soil and organic matter called a *soil profile* by agronomists. Such a profile, divided into soil *horizons,* is shown schematically in Fig. 2-1 for a temperate climate. The typical thicknesses shown in Fig. 2-1 are intended only as a broad indication; the thicknesses of the various horizons would be significantly larger for the rich soils of the valleys and bottomlands and much shallower for the sparse soils of the mountain slopes.

The presence of humus or organic material in a soil can cause serious changes in engineering properties as the organic material continues to decay. Due to this variability, engineered foundations must be placed well below the organic agricultural soils and into the parent soils of the C horizon. The engineering properties of the soils in the C horizon are reasonably constant and are not subject to change due to the decay of organic material. All further discussions in this book are limited to these inorganic (or minimally organic) soils of the C horizon.

Although soils have long been classified into broad groupings, the engineering properties of various classifications of soils cannot be determined once and

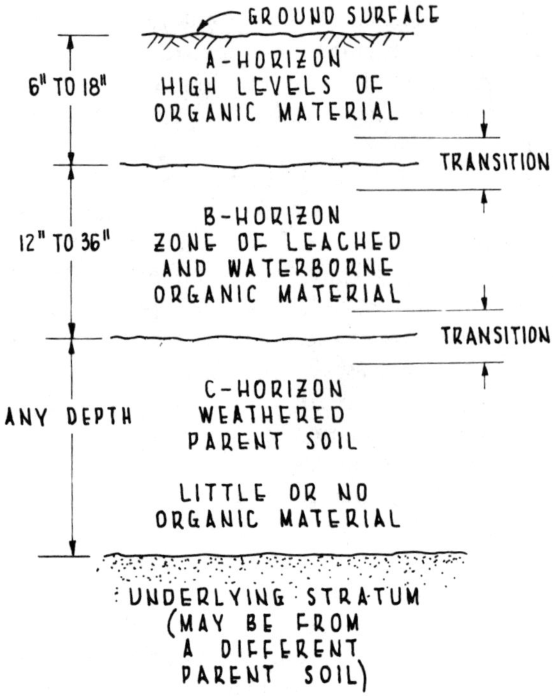

Figure 2-1 Soil horizons.

used thereafter as constants; a soil in Wyoming classified as clay might have vastly different properties from a soil in Georgia classified as clay. There are, however, certain characteristics that are common to all soils regardless of origin or chemistry. Such characteristics can be used as a basis for soil classifications. One such characteristic is the size and shape of the individual particles that comprise the soil.

GRAIN SIZE AND DISTRIBUTION

The size, shape, and distribution of the particles that make up a soil have long been used to describe a soil, first by agrarians and later by engineers. Several methods of screening or sieving to determine the sizes of the particles have been developed; all have some degree of merit. The system presented here is the one currently most popular with engineers, that prescribed by the American Society for Testing and Materials under its designation ASTM D422.

A sketch of a typical granular soil structure is shown in Fig. 2-2. The larger particles create void spaces which are filled by smaller particles, the remaining voids are in turn filled with yet smaller particles, and so on. When the size and gradation of the soil particles are such that the final amount of void space is minimal and the soil approaches its absolute maximum density, the soil is said to be *well graded*.

Soil particles have been classified into general groupings according to their

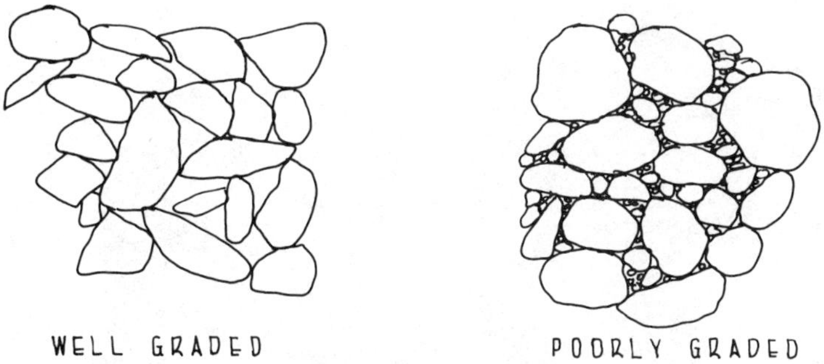

Figure 2-2 Typical soil structure.

size. From Fig. 2-3, for example, particle sizes between 0.075 and 0.425 mm are classified as fine sand. These particle sizes will pass through a standard sieve having 40 wires per inch but will be retained on a standard sieve having 200 wires per inch. Other soil classifications are shown in Fig. 2-3.

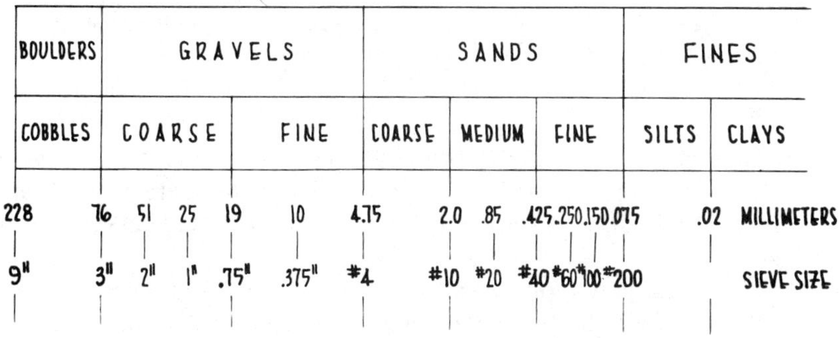

Figure 2-3 Particle and sieve sizes.

Standard sieves are prescribed by ASTM D422, in which the test method is standardized; a photograph of some standard sieves is shown in Fig. 2-4. Although many sieve sizes are available, the more common sizes are those listed in Fig. 2-3. The clear opening is given in inches down to $\frac{3}{8}$ in.; thereafter the sieve size is given as the number of wires per inch.

Over the years, it has been found that certain sieve sizes fall at natural break points between soil classifications; those sieve sizes will almost always be included in the sieve analysis of a soil. Typical of such "natural" break sizes are the 3-in. screen (separates cobbles from gravels), the No. 4 screen (separates gravels from sands), the No. 40 screen (separates coarse sands from fine sands), and the No. 200 screen (separates sands and gravels from silts and clays). By far the most significant of these natural break points is the No. 200 sieve size, which defines the separation between "fine-grained" soils (silts and clays) and the "coarse-grained" soils (sands and gravels).

In a standard sieve analysis as prescribed by ASTM D422, a soil sample is first weighed, then it is passed through the set of standard sieves. The total weight

Figure 2-4 Typical sieves. Nesting sieve (courtesy of Humboldt Mfg. Co.) Nesting sieves in a shaker (Courtesy of Soiltest, Inc.)

passing the 3-in. sieve is recorded, then successively, the total weight passing the 1 in., $\frac{1}{2}$ in., No. 4, No. 20, and so on. The results are plotted on semilog paper producing a curve as shown in Fig. 2-5; the resulting curves are called gradation curves.

Several gradation curves are shown in Fig. 2-5 for comparison. Curve C is a *well-graded* granular soil, having grain sizes well distributed throughout the normal range of grain sizes. The gentle slope and curvature of curve C indicates that

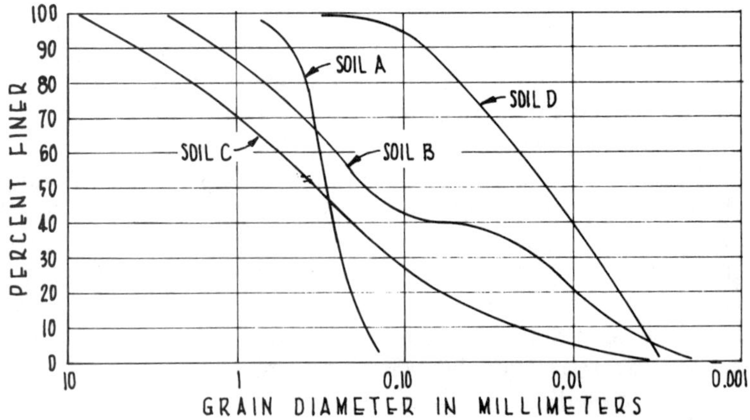

Figure 2-5 Typical gradation curves.

the soil is well graded, having no excessive amounts of particles in one size nor any gaps in another size. Curve A in Fig. 2-5 indicates a *uniformly* graded soil, having most of its particles grouped together in one narrow range of sizes. There is little distribution of sizes. Beach sands that have had their finer sizes removed by water erosion will typically produce a gradation curve similar to curve A. Curve B in Fig. 2-5 indicates a *gap-graded* soil in which a complete range of sizes is missing. Such soils can come about through marine deposition, where the soil has been deposited either by fast-moving water or by sluggish water, with nothing in between. Sometimes, blending a soil of this type with a uniformly graded soil such as that of curve A will produce a better gradation.

In discussing coarse-grained soils such as those shown in curves A, B, and C, it is often useful to designate grain sizes at certain points of their gradation curves. As an example, the designation D_{60} is used to denote the sieve size which allows 60% of the material to pass; 60% of the particles are therefore smaller than the D_{60} sieve size. Similarly, 30% of the particles are smaller than the D_{30} sieve size and 10% of the particles are smaller than the D_{10} size. The D_{10} sieve size is often called the *effective size* of a coarse-grained soil. It is used when studying the permeability of the soil.

The sieve sizes are also used to describe general properties of coarse-grained soils. For example, the *uniformity coefficient,* C_u, is a measure of the uniformity of soil particles and is computed by

$$C_u = \frac{D_{60}}{D_{10}} \tag{2-1}$$

Similarly, the smoothness of the curve is indicated by the *coefficient of curvature,* C_z, and is computed by

$$C_z = \frac{(D_{30})^2}{D_{60}D_{10}} \tag{2-2}$$

The coefficients C_u and C_z are empirically developed, but they afford a means to determine whether or not a coarse-grained soil is well graded. For gravels, if $C_u > 4$ and $1 < C_z < 3$, the gravel is well graded. For sands, if $C_u > 6$ and $1 < C_z < 3$, the sand is well graded. Any coarse-grained soil not meeting these limits is considered to be poorly graded.

The soil shown by curve D in Fig. 2-5 is a fine-grained soil in which more than half the soil passes the No. 200 sieve. For such fine soils, descriptions such as "well graded" have little meaning. A different property is obviously necessary when dealing with soils having extremely small particle sizes. That property is plasticity; it is presented in the next section.

PLASTICITY AND ATTERBERG LIMITS

Fine-grained soils (more than 50% passing the No. 200 sieve) are very sensitive to the amount of water in their void spaces, or pores. They may even change state as their water content increases, going gradually from a solid state to a semisolid state to a plastic state and eventually to a liquid. Coarse-grained soils undergo no such change of state.

An example of this phenomenon in fine-grained soils occurs in modeling clay. At the preferred water content, modeling clay is in a plastic state, readily molded into various shapes. As it dries, it becomes solid, even brittle. Adding water to the dried clay will restore its plasticity, but if too much water is added, the clay will pass beyond plasticity into a soupy or liquid state.

Four distinct states—solid, semisolid, plastic, and liquid—can be distinguished and identified by the amount of water contained in a fine-grained soil; a general description of these states is given in Table 2-2. The change from a solid state to a semisolid state occurs at the *shrinkage limit,* from a semisolid state to a plastic state at the *plastic limit* and from a plastic state to a liquid state at the *liquid limit*. As a means to remove the element of personal opinion or bias in establishing these limits, standard tests have been developed which are now generally accepted in the practice; these tests are described in subsequent discussions.

TABLE 2-2. PHYSICAL STATES OF CLAY SOIL

State	Limits	Comparative description
Solid		Rock candy
	Shrinkage limit	
Semisolid		Firm Swiss cheese
	Plastic limit	
Plastic		Stiff putty or modeling clay
	Liquid limit	
Liquid		Thick pea soup

The points where a soil changes state are commonly called the *Atterberg limits* [1] of the soil, being given the name of the scientist who introduced them. They are determined for that portion of a soil that passes the No. 40 sieve; some fine sands are therefore included in the standard tests. Only the plastic limit and the liquid limit are presented here. The third Atterberg limit, the shrinkage limit, has fallen from use and will not be considered further.

Atterberg limits are stated in terms of the water content. The *water content* of a soil, w, is a ratio of weights; it is defined as the weight of water in the soil, W_W, divided by the weight of solids, W_S. It is stated as whole numbers of percent without the percent sign.

$$w = \frac{W_W}{W_S} \times 100 \tag{2-3}$$

A water content of 20 therefore means that the water in the soil sample weighs 20% as much as the soil particles (*not* 20% of the total weight).

The *plastic limit* of a soil, w_P, is the value of the water content at the point when the soil passes from a solid state to a plastic state. The test for the plastic limit consists of rolling the sample manually into a string $\frac{1}{8}$ in. in diameter. When the water content is such that the string just starts to crumble when it is at $\frac{1}{8}$ in. diameter, the soil is at its plastic limit.

The *liquid limit* of a soil, w_L, is the value of the water content at the point when the soil passes from a plastic state to a liquid state. The test for liquid limit requires a simple mechanical device, shown in Fig. 2-6. As the crank is turned on the device, the cup is raised exactly 10 mm and is allowed to fall back onto the base of gravity, creating a ''blow'' on the bottom of the cup. The number of such blows that are required to cause movement in a soil sample is a measure of the liquidity of the soil. To determine the liquid limit, a sample of moist soil is placed in the cup of the liquid limit device shown in Fig. 2-6 and a groove is formed in the sample with a standard grooving tool. The crank is then turned, causing the cup to strike against the base, falling exactly 10 mm at each blow. When the water content of the soil is such that the groove closes at 25 blows, the soil is said to be at its liquid limit.

Figure 2-6 Liquid limit device. (Courtesy of Humboldt Mfg. Co.)

A third parameter, the *plasticity index* (PI), is used in distinguishing whether a soil is a silt or a clay. The plasticity index is the range of water content over which the soil is considered to be plastic, that is, the liquid limit minus the plastic limit:

$$PI = w_L - w_P \tag{2-4}$$

The soil is considered to be plastic throughout this range of water contents.

The determination as to whether a soil is a silt or clay is based on its plasticity. The plasticity chart for fine-grained soils is shown in Fig. 2-7. It was developed in 1948 by Arthur Casagrande [2] and has survived to date with few changes. To determine whether a soil is a silt or clay, enter the plasticity chart with the PI and the liquid limit of the soil and find the point thus defined. If the point falls above the A-line (the ''activity'' line), the soil is a clay; if below the A-line, it is a silt.

A further description can be made based on the magnitude of the liquid limit. For plastic soils having a liquid limit in excess of 50, the soil is described as having high plasticity (or high compressibility); if less than 50 it is said to have low plasticity (or low compressibility). The further distinction between low and medium plasticity shown in the plasticity chart of Fig. 2-7 is not widely used in

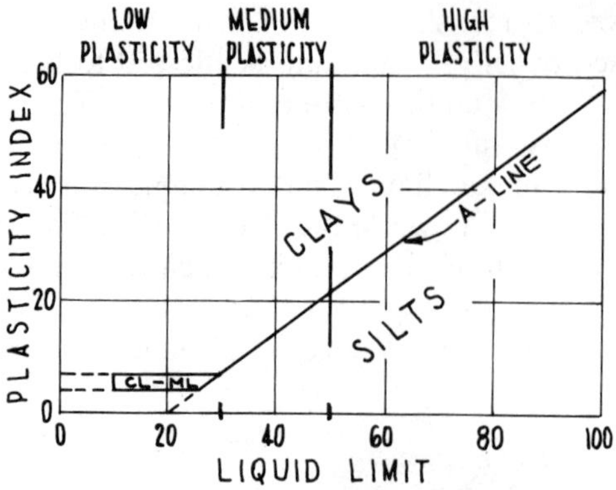

Figure 2-7 Plasticity chart.

foundation design and is ignored here; if the liquid limit is below 50, the soil is considered to be of low plasticity.

The PI can also be used as a quick indicator of the susceptibility of a clay to swelling (or expansion) when it is exposed to water. As a general indicator, a clay having a PI of 25 or more should be regarded as being potentially subject to swell. Any clay having a liquid limit above 50 should automatically be regarded as one of these "expansive" clays until it has been proven not to be.

CLASSIFICATION OF SOILS

Based on the grain size distribution and on the Atterberg limits, a standard classification for rather broad soil "groups" has been developed. The system has been adopted by ASTM under its designation D2487. The system is widely used throughout engineering and is probably the most widely accepted classification system in current use. The classification system is presented in Table 2-3. Note that the major subdivision is the division between fine-grained soils (silts and clays) and coarse-grained soils (gravels and sands). Note also that the classification of gravels and sands is based on the *coarse fraction* (the amount retained on the No. 200 sieve) rather than on the total weight of the sample.

The soil classification itself consists of two letters, shown in the column headed "Group Symbols." The first letter designates the dominant type of soil particles:

G indicates gravel.

S indicates sand.

M indicates silt.

C indicates clay.

The second letter of the group symbol is largely descriptive. For clean gravels and sands (less than 5% fines), the second letter indicates whether the gravel or sand is well graded (GW or SW) or poorly graded (GP or SP). If the gravel or sand contains large amounts of fines (more than 12% fines), the second letter indicates whether the fines are silts (GM or SM) or clays (GC or SC). If the amount of fines

TABLE 2-3 ASTM SOIL CLASSIFICATION SYSTEM (UNIFIED)

Major Divisions			Group Symbols	Typical Names	Classification Criteria		
Coarse-Grained Soils More than 50% retained on No. 200 sieve*	Gravels 50% or more of coarse fraction retained on No. 4 sieve	Clean Gravels	GW	Well-graded gravels and gravel-sand mixtures, little or no fines	Classification on basis of percentage of fines: Less than 5% pass No. 200 sieve GW, GP, SW, SP; More than 12% pass No. 200 sieve GM, GC, SM, SC; 5% to 12% pass No. 200 sieve Borderline classification requiring use of dual symbols	$C_u = D_{60}/D_{10}$ Greater than 4 $C_z = \frac{(D_{30})^2}{D_{10} \times D_{60}}$ Between 1 and 3	
			GP	Poorly graded gravels and gravel-sand mixtures, little or no fines		Not meeting both criteria for GW	
		Gravels with Fines	GM	Silty gravels, gravel-sand-silt mixtures		Atterberg limits plot below "A" line or plasticity index less than 4	Atterberg limits plotting in hatched area are borderline classifications requiring use of dual symbols
			GC	Clayey gravels, gravel-sand-clay mixtures		Atterberg limits plot above "A" line and plasticity index greater than 7	
	Sands More than 50% of coarse fraction passes No. 4 sieve	Clean Sands	SW	Well-graded sands and gravelly sands, little or no fines		$C_u = D_{60}/D_{10}$ Greater than 6 $C_z = \frac{(D_{30})^2}{D_{10} \times D_{60}}$ Between 1 and 3	
			SP	Poorly graded sands and gravelly sands, little or no fines		Not meeting both criteria for SW	
		Sands with Fines	SM	Silty sands, sand-silt mixtures		Atterberg limits plot below "A" line or plasticity index less than 4	Atterberg limits plotting in hatched area are borderline classifications requiring use of dual symbols
			SC	Clayey sands, sand-clay mixtures		Atterberg limits plot above "A" line and plasticity index greater than 7	
Fine-Grained Soils 50% or more passes No. 200 sieve*	Silts and Clays Liquid limit 50% or less		ML	Inorganic silts, very fine sands, rock flour, silty or clayey fine sands	Plasticity Chart (see below)		
			CL	Inorganic clays of low to medium plasticity, gravelly clays, sandy clays, silty clays, lean clays			
			OL	Organic silts and organic silty clays of low plasticity			
	Silts and Clays Liquid limit greater than 50%		MH	Inorganic silts, micaceous or diatomaceous fine sands or silts, elastic silts			
			CH	Inorganic clays of high plasticity, fat clays			
			OH	Organic clays of medium to high plasticity			
Highly Organic Soils			PT	Peat, muck, and other highly organic soils	Visual-Manual Identification, see ASTM Designation D 2488.		

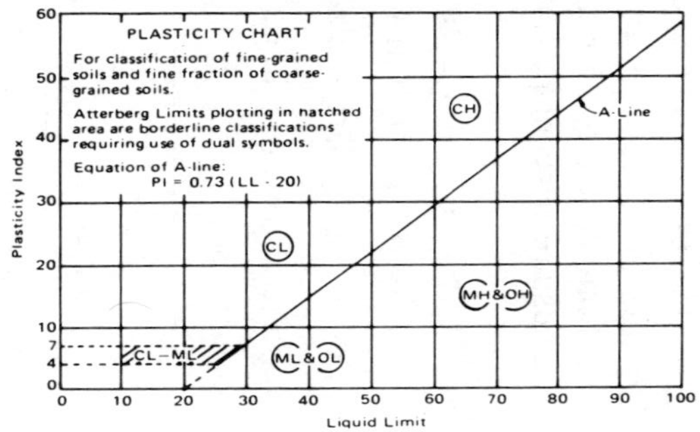

*Based on the material passing the 3-in. (75-mm.) sieve.

Source: Courtesy of Portland Cement Association; reprinted by permission.

is between 5 and 12%, dual symbols are used (e.g., GP-GM). For fine-grained soils, if the liquid limit is above 50, the second letter of the group symbol indicates high plasticity (MH or CH). If the liquid limit is below 50, the second letter of the group symbol indicates low plasticity (ML or CL). For soils of low plasticity with a PI between 4 and 7, a dual symbol is used (e.g., CL-ML).

A few examples will illustrate the classification system. Throughout the classification, close attention must be paid to the sieve sizes and whether the given percentages are percent passing or percent retained.

EXAMPLE 2-1: Soil Classifications
Unified Classification System ASTM D2487.
Grain sizes and Atterberg limits given.

Given the following criteria, classify the soils.

	Percent passing indicated sieve			
Sieve size	Sample 1	Sample 2	Sample 3	Sample 4
3 in.	100	200	90	100
1 in.	100	100	60	97
No. 4	95	100	44	62
No. 10	88	88	30	41
No. 40	81	76	16	33
No. 100	74	65	10	20
No. 200	65	54	6	13
Liquid limit	53	16	Nonplastic	21
Plastic limit	36	10		12

SOLUTION Sample 1 has 65% of its weight passing the No. 200 sieve; it is therefore classified as a fine-grained soil, either silt or clay. Its liquid limit is 53 and its PI is 17; from the plasticity chart it is seen to be a silt of high compressibility, designated MH. Note, however, that this "silt" contains 5% gravel (retained on No. 4 sieve) and 30% sand (between No. 4 and No. 200 sieves). Nonetheless, the soil still falls in a group characterized by the large amount of silt in the mixture and is classified overall as a silt.

Sample 2 has 54% of its weight passing the No. 200 sieve; it is classified as being fine-grained. Liquid limit is 16, PI is 6; the soil falls in the special group of soils given a dual symbol, CL-ML.

Sample 3 has only 6% fines; it is coarse-grained. The coarse fraction is seen to be 94 lb of a 100-lb sample. The gravel in a 100-lb sample is that part retained on the No. 4 sieve or 100 − 44 = 56 lb. The gravel portion of the coarse fraction is therefore 56/94 or 60% gravel. The coarse fraction is thus more than 50% gravel, so the soil is a gravel. The 60% diameter is seen to be 25 mm, the 30% diameter is 2 mm, and the 10% diameter is 0.15 mm. The uniformity coefficient is therefore $C_u = D_{60}/D_{10} = 0.167$; the coefficient of curvature is $C_z = D_{30}^2/D_{60}D_{10} = 1.1$. The gravel is seen to be well graded, with fines between 5% and 12% of its weight; from Table 2-3 it is classified GW-GM.

Sample 4 has only 13% fines; it is coarse-grained. Its coarse fraction is 87 lb of a 100-lb sample, with 38 lb of that being gravel. The gravel portion of the coarse fraction is thus 38/87 = 44%, or less than 50%; the soil is therefore a sand. It has more than 12% fines with the fines having a liquid limit of 21 and a PI of 9; the fines are classified as clay particles. The soil falls in the broad grouping of clayey sands, SC.

There is a second kind of soil classification that has been in use for many years and is still widely used, particularly when describing a soil verbally. This classification is based on the textural properties or consistency properties of the soil. The classification is shown in Table 2-4, largely self-descriptive; sands and

TABLE 2-4 CONSISTENCY AND TEXTURAL CLASSIFICATION

Sands: textural classification	Clays: consistency classification	Estimate of bearing capacity (ksf)
—	Very soft	Less than 0.5
Very loose	Soft	0.5 to 1.0
Loose	Medium	1.0 to 2.0
Medium	Stiff	2.0 to 4.0
Dense	Very stiff	4.0 to 8.0
Very dense	Hard	Over 8.00

CLAYS

1. A very soft clay can easily be penetrated several inches by fist.
2. A soft clay can be penetrated several inches by thumb.
3. A medium clay can be penetrated several inches by thumb with moderate effort.
4. A stiff clay can readily be indented by thumb, but penetrated only with great effort.
5. A very stiff clay can readily be indented by thumbnail.
6. A hard clay can be indented with difficulty by thumbnail.

SANDS

1. A very loose sand can be penetrated easily by a $\frac{1}{2}$-in. reinforcing bar pushed by hand.
2. A loose sand can be penetrated with difficulty by a $\frac{1}{2}$-in. reinforcing bar pushed by hand.
3. A medium sand can be penetrated readily by a $\frac{1}{2}$-in. reinforcing bar driven by a 5-lb hammer.
4. A dense sand can be penetrated about 1 ft by a $\frac{1}{2}$-in. reinforcing bar driven by a 5-lb hammer.
5. A very dense sand can be penetrated about 3 in. by a $\frac{1}{2}$-in. reinforcing bar driven by a 5-lb hammer.

clays in the same general strength group are entered on the same line. The textural and consistency classifications are still widely referenced in building codes and in older legal documents. Obviously, the classification is heavily dependent on the experience and bias of the person performing the test. Nonetheless, it is a useful tool and is still used widely in practice.

INDEX PROPERTIES

There are other properties of a soil, called *index properties,* that are useful indicators. One such index property has already been introduced, the water content $w = (W_W/W_S) \times 100$. There are others, of which the most important is probably the void ratio e, defined as

$$e = \frac{V_V}{V_S} \tag{2-5}$$

where V_V is the volume of voids in a sample and V_S is the volume of solids.

Another useful indicator is the porosity, n, defined as

$$n = \frac{V_V}{V} \tag{2-6}$$

where V_V is the volume of voids in a sample and V is the total volume. Another is the degree of saturation, $S\%$,

$$S\% = 100\left(\frac{V_W}{V_V}\right) \tag{2-7}$$

where V_W is the volume of water in a sample and V_V is the volume of voids.

The density of soil, γ, is the same as it is for other materials,

$$\gamma = \frac{W}{V} \tag{2-8}$$

where W is the weight of a sample and V is its volume. At times it is necessary to know the dry density of a soil, which is its density with all the water driven off. Dry density is denoted γ_d and is computed by

$$\gamma_d = \frac{W_S}{V} \tag{2-9}$$

where W_S is the weight of solids in a sample and V is the total original volume before the water was driven off.

The specific gravity of solids, G_S, is the specific gravity of the soil minerals, excluding air, gas, and water. It ranges from about 2.3 to 3.0, but usually falls between 2.5 and 2.8 for most soils. The specific gravity of a silica sand, for example, is typically about 2.63 to 2.68. The unit weight of the solids, γ_S (having no voids at all) is computed as $62.4 \times G_S$.

Computation of index properties can become confusing. It is simplified somewhat by using an ordered approach. A typical soil sample is shown in Fig. 2-8 following, where all the soil is shown lumped together as a separate mass. The water is similarly shown as a separate mass and the air is shown as a volume. In Fig. 2-8, volumes are shown on the left, weights on the right. Air is considered to weigh nothing in computing weights. To go from one side to the other, it is necessary to multiply or divide either by the unit weight of water, 62.4 pcf, or by the unit weight of solids, $62.4 \times G_S$.

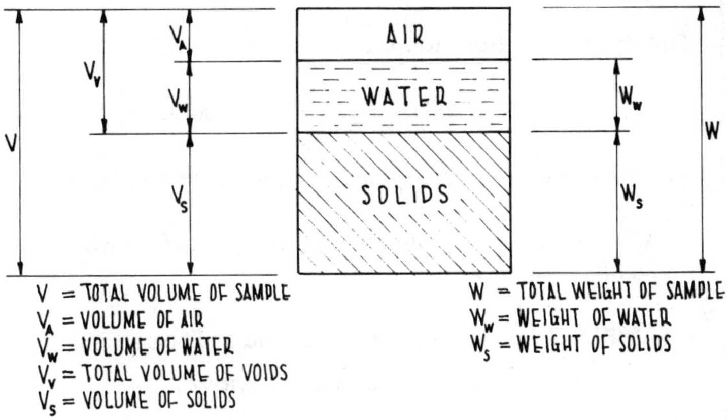

Figure 2-8 Index properties.

Some examples will illustrate the computation of index properties.

EXAMPLE 2-2: Computation of Index Properties
In-place density, G_S and dry density known.

The in-place density of a soil sample is measured and found to be 128 pcf. Its dry density has also been determined and found to be 119 pcf. The specific gravity of solids is 2.61. Compute the void ratio e and degree of saturation $S\%$.

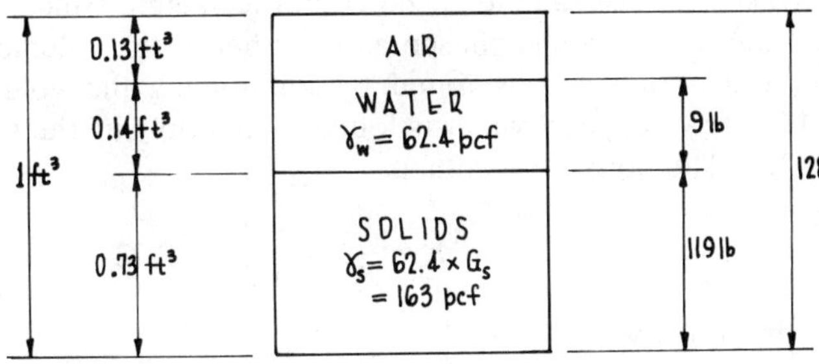

SOLUTION The sample is shown in the sketch. Since density is given per cubic foot, a volume of 1 ft³ is arbitrarily chosen; the density and dry density then become weights, shown on the right side of the sample. The weight of water is simply the total weight of the sample less the dry weight. The weight of solids, 119 lb, is divided by $62.4 \times G_S$ to obtain its volume, 0.73 ft³, which is entered on the left. Similarly, the weight of water is divided by 62.4 to obtain its volume, 0.14 ft³, also entered on the left. The remaining volume in 1 ft³ is air, $1.00 - 0.73 - 0.14 = 0.13$ ft³.

The index properties can now be computed. The void ratio e is found by definition,

$$e = \frac{V_V}{V_S} = \frac{0.13 + 0.14}{0.73} = 0.37$$

The degree of saturation is

$$S\% = 100\left(\frac{V_W}{V_V}\right) = 100 \times \frac{0.14}{0.27} = 52\%$$

All other index properties can be calculated similarly.

A second example will illustrate the procedure when the sample is saturated (all voids filled with water).

EXAMPLE 2-3: Computation of Index Properties
Void ratio and G_S known.

The void ratio of a saturated sample has been found to be 0.29. Specific gravity of solids is 2.71. Determine porosity n and dry density γ_d.

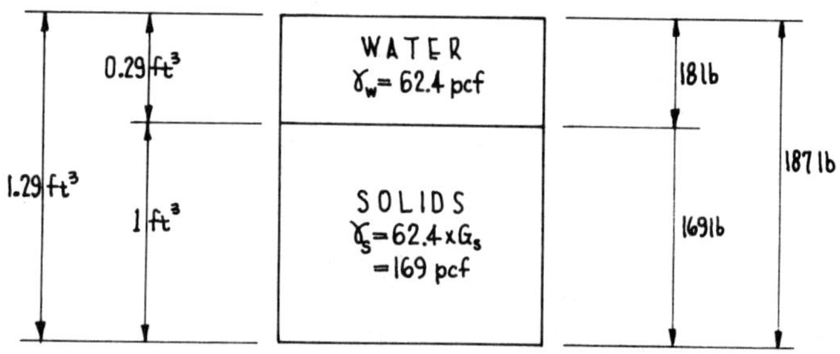

SOLUTION The sample is shown in the sketch. Since its void ratio is 0.29, the volume of solids is chosen as 1 ft³; hence the volume of voids is 0.29 ft³. The weight of solids is found by multiplying the volume of solids V_S by $62.4G_S$; the weight of water is found by multiplying the volume of water, V_W, by 62.4. The porosity n is then

$$n = \frac{V_V}{V} = \frac{0.29}{1.29} = 0.22$$

The dry density is

$$\gamma_d = \frac{W_S}{V} = \frac{169}{1.29} = 131 \text{ pcf}$$

Another example illustrates one of the minor mathematical complications that can occur in such calculations.

EXAMPLE 2-4: Computation of Index Properties
Water content, G_S and unit weight known.

The water content w of a sample has been found to be 12 and the unit weight to be 118 pcf; $G_S = 2.59$. Compute the void ratio e.

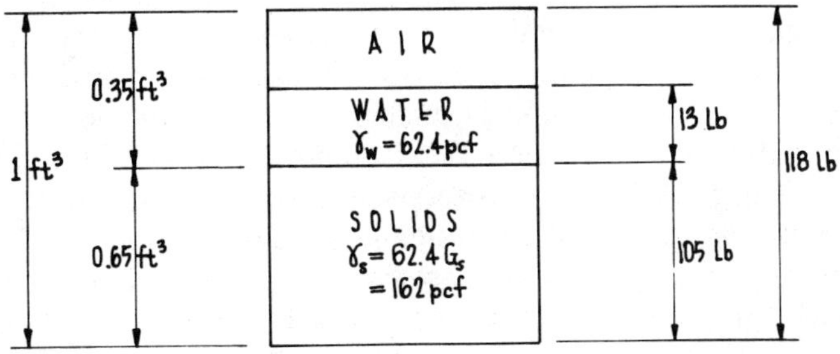

SOLUTION The weights are shown in the sketch for 1 ft³ of the sample. A direct calculation of volumes V_S and V_W in this case is not possible, since the weights W_S and W_W are not known. There are two equations, however, that contain the two unknowns, W_S and W_W:

$$w = \frac{W_W}{W_S} \times 100 = 12 \quad \text{or} \quad W_W = 0.12\,W_S$$

and

$$W = W_W + W_S = 118 \text{ lb}$$

These equations are solved simultaneously, yielding

$$1.12\,W_S = 118$$

Hence

$$W_S = 105 \text{ lb} \quad \text{and} \quad W_W = 13 \text{ lb}$$

The volumes are now readily found:

$$V_S = \frac{W_S}{\gamma_S} = \frac{105}{162} = 0.65 \text{ ft}^3$$

$$V_V = V - V_S = 1.00 - 0.65 = 0.35 \text{ ft}^3$$

The void ratio is then

$$e = \frac{V_V}{V_S} = \frac{0.35}{0.65} = 0.54$$

In general, the computation of index properties is readily accomplished using the format of the foregoing examples. Such computations seem popular in textbooks on soil mechanics, possibly because it is one of the few topics in soils where an exact numerical answer can actually be found.

PERMEABILITY, EFFECTIVE STRESS, AND SUBMERGENCE

The voids in a soil mass form an interconnected labyrinth of passages, a maze of "pores" through which water may flow or migrate whenever a pressure gradient is imposed. As the grain sizes of the soil become smaller and smaller, these passages become more numerous but also smaller and more restrictive. As the size of the passage becomes smaller, the flow of the porewater becomes slower as the frictional "drag" force between the water and the soil particles increases.

The *permeability* of a soil is its capacity to permit the flow of water through its pores when a pressure gradient is imposed. It is this flow of water through a soil mass that permits water wells to be recharged. It is this same flow that can create "quicksand" in some silts. It is the presence of this water under pressure that creates heave in basement floors and leaking in basement walls. For coarse-grained soils, the existence of this water can at times cause a significant loss of strength, a point that is explained more fully in subsequent discussions.

The apparatus traditionally used to demonstrate permeability in soils is shown in Fig. 2-9. In Fig. 2-9a the soil sample is saturated but not submerged. The overall average intergranular pressure $\bar{p}$, also called *effective pressure* or *effective stress,* is computed as

$$\bar{p} = H\gamma_{sat} \tag{2-10}$$

where γ_{sat} is the saturated unit weight and H is the height of the sample. It should be noted that the actual grain-to-grain contact pressure will be considerably higher than this average effective pressure and could approach infinity as the contact points become smaller and smaller.

Water is introduced into the sample as shown in Fig. 2-9b, but no flow of the water is yet induced. The soil particles in this case lose weight due to buoyancy. The loss in weight in turn causes a reduction in the average intergranular pressure $\bar{p}$ against the bottom of the container,

$$\bar{p} = H(\gamma_{sat} - \gamma_W) \tag{2-11}$$

where γ_W is the unit weight of water. Throughout the sample, the porewater pressure u_W is simply the hydrostatic pressure in the water; hence at the bottom of the sample,

$$u_W = (H_0 + H)\gamma_W \tag{2-12}$$

where H_0 is the height of water standing above the top of the sample.

Typically, the unit weight of soil is about 110 to 130 pcf, roughly twice the unit weight of water. The soil particles in Fig. 2-9b will therefore lose roughly half their original weight when submerged, with a corresponding reduction in inter-

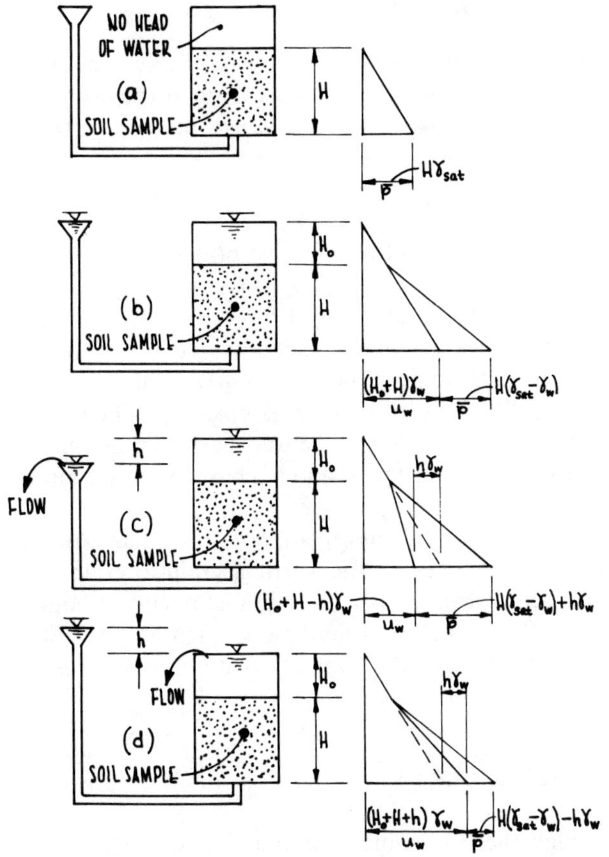

Figure 2-9 Effective pressure and porewater pressure.

granular pressure, as indicated by Eq. (2-11). This 50% reduction in intergranular pressure due to submergence will be shown later in this chapter to produce a corresponding 50% loss in strength in coarse-grained soils.

A downward flow of water through the sample is now induced as indicated in Fig. 2-9c. The hydrostatic head in the porewater at the bottom of the sample is readily computed:

$$u_W = (H_0 + H - h)\gamma_W \tag{2-13}$$

where the height h represents the loss in pressure head across the system due to frictional "drag" in the soil sample. Since the total pressure at the base of the container must remain unchanged, the effective pressure in the soil must increase by the amount of this drag; hence

$$\bar{p} = H(\gamma_{sat} - \gamma_W) + h\gamma_W \tag{2-14}$$

The downward flow of water thus creates an increase in the average pressure between the soil grains due to the frictional drag on the grains.

An upward flow of water through the sample is now induced as indicated in Fig. 2-9d. The hydrostatic head at the bottom of the sample is again computed,

$$u_W = (H_0 + H + h)\gamma_W \tag{2-15}$$

where the height h again represents the loss in head across the system due to the frictional drag on the soil particles. Since the total pressure at the bottom of the sample must remain unchanged, the effective pressure in the soil at that point must decrease by the amount of this drag; hence

$$\bar{p} = H(\gamma_{sat} - \gamma_W) - h\gamma_W \tag{2-16}$$

The head loss in this case further reduces the effective pressure $\bar{p}$.

For particular combinations of grain size, head loss, velocity of flow, and unit weight, the intergranular pressure $\bar{p}$ given by Eq. (2-16) can go to zero. In such circumstances, the soil particles become essentially weightless and the soil goes into a "quick" condition commonly called *quicksand,* although the soils usually associated with this condition are silts rather than sands. The occurrence of the bottom of an excavation turning "quick" due to excessive pumping rates is not a rare phenomenon; it can also happen when flooded basements are pumped at high rates.

It might be suspected as well that the lateral soil pressure against the wall of the container is also undergoing changes as the vertical grain-to-grain pressure changes. These changes in lateral earth pressures do in fact occur, although they are beyond the scope of a book on shallow foundations such as this. Standard references in general soil mechanics treat the subject of lateral earth pressures in detail, to put it mildly [3].

CONSOLIDATION AND SETTLEMENT IN CLAYS

Soils are not elastic. Although there is some rebound in a soil when a load is removed, the rebound does not restore the soil to its exact original size; there is always some residual deformation. To avoid any inferences that deformations are elastic or recoverable, all such deformations in soil due to loads are termed herein as "settlements."

A typical clay sample is shown in Fig. 2-10, magnified many times. The clay particles are shown as thin flat particles (like leaves) touching each other in random order, although the actual appearance of a particle is closer to that of an old battered book, with pages sticking out raggedly in all directions. The interstices between particles are filled with water and occasional pockets of trapped air.

Clay particles themselves are somewhat tough and flexible. Unlike sand grains, they can be deformed or bent before they break. As they bend, however, they develop more and more resistance to further deformation.

A clay that has never been subjected to any pressure greater than it now sustains, whether from overburden or from a foundation, is said to be *normally consolidated.* Such lightly condensed soils, compressed only by their own weight, occur in delta regions of large rivers when clay deposits are laid down by periodic flooding. The response of these soft compressible clays to a foundation load is reasonably predictable in comparison to other types of clay.

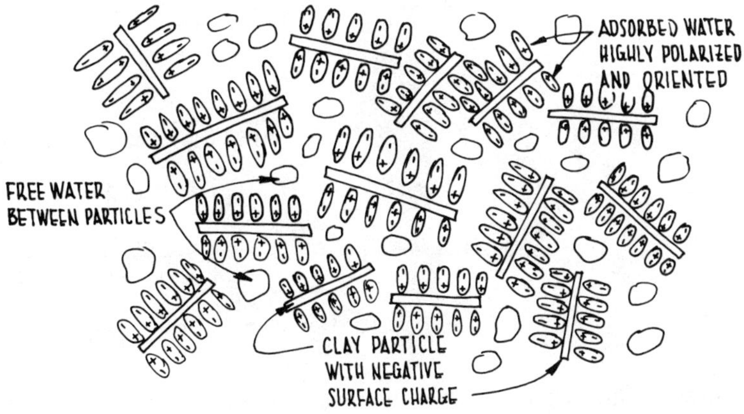

Figure 2-10 Typical clay sample.

The interstices, or pores, surrounding the particles in any clay soil are extremely small. The porewater is in thin layers and strips and is partially bonded to the clay. The porewater closest to the clay particles is quite viscous, acting like a very "thick" fluid, becoming thinner as distance from the particle increases. The water is trapped, of course, but can be forced out of the clay by the introduction of pressure.

A load placed on any clay soil, including normally consolidated clays, will at first be carried entirely by an increase in pressure in the trapped porewater. Over the months or years, as the "thinner" water is extruded outward away from the increased pressure, the clay particles gradually pick up more and more of the load. The clay particles will deform slightly as they pick up the load, developing more and more intergranular pressure as the excess water is sluggishly extruded from the soil mass under the load. Eventually, all the excess water will be extruded, the porewater pressure will return to its regular levels and the clay particles will carry all of the increase in load. At this point, the process is complete; the clay is now said to be *100% consolidated* at this increased pressure.

The process of making the very slow transfer of load from porewater to clay particles with the resultant decrease in volume is given the generic name *consolidation*. It is the mechanism whereby clay deforms under load. It explains why deformations occur so slowly in clays, sometimes taking several years before settlements finally stabilize.

The consolidation of a clay soil under a long building can also produce a particular kind of wide-area *distortion* settlement under the building. A typical example is shown in Fig. 2-11. Since the pressure under a building can only be dispersed outward, away from the building, the pressure under the center of the building will be considerably higher than at the edges. The consolidation of the clay, or reduction in volume, will therefore be higher at the middle of the building than at the edges. A long building founded on clay will therefore typically cup downward at its center as shown in Fig. 2-11, due to such distortion settlements.

The magnitude of distortion settlements can be severe for large long struc-

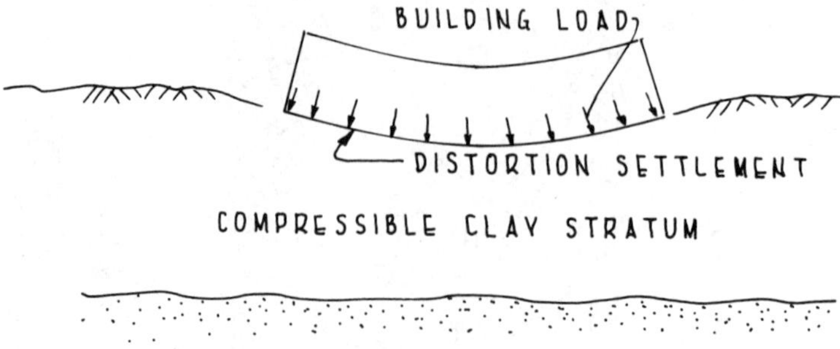

Figure 2-11 Distortion settlement in clay.

tures such as dams, embankments, and other such works. For the usual run of small buildings, however, distortion settlements are rarely large enough to be of concern. Notable exceptions can be found in some of the older "row" buildings on the normally consolidated soils of New Orleans, where distortion settlements are readily visible.

It should be noted that there is a distinct reduction in the volume of the clay as it consolidates. The loss in volume is of course the volume of porewater that is extruded and lost. The clay particles themselves are in a state of increased stress under this reduced volume.

If the load is removed at a later time, the restoring force offered by the deformed clay particles is not strong enough to pull air back into the clay mass and restore its size; the thick viscous water adjacent to the grains effectively prevents such an ingress of air. The clay mass is thus prevented from expanding until a source of water becomes available which can reenter the soil mass, restoring some part of the volume of porewater.

In this precompressed state, the clay is said to be *overconsolidated*. An overconsolidated clay is any clay that has been subjected to a greater level of consolidation pressure than it currently experiences. Of the two possible cases of consolidation that can be found in clay soils, either normal consolidation or overconsolidation, the more common case by far is that of overconsolidation. An overconsolidated clay can usually be recognized by its water content; if the water content of the in-place clay is less than the liquid limit, the clay is almost certainly overconsolidated.

The phenomenon of overconsolidation can also occur over the thousands of years of the geologic history of a clay stratum. When a clay stratum has been loaded by thousands of feet of overburden and the overburden is eventually eroded away, the release of the load creates restoring forces in the clay that can be quite large. The clay remains compressed, however, like a compressed spring, unable to expand until a source of water permits an ingress and restoration of some part of the porewater that had been extruded eons ago.

When a source of water is made available to an overconsolidated clay, the slow ingress of water allows the particles to unbend and to return closer to their original configuration. The expansion typically happens in small pockets at random locations, producing unpredictable heaving which can devastate buildings that have been built on the clay. In some geologically overconsolidated clays the expansion and heaving is a continuous and serious long-term problem; a typical example of such a clay is the red Permian clay of Texas, Oklahoma, and Kansas.

A clay can also become overconsolidated due to severe drying, or desiccation; restoring forces due to desiccation can also be quite large. A desiccated clay will usually be severely shrunken and cracked. Its water content may be even less than its plastic limit. Cracks commonly may be several inches wide and up to several feet deep and several yards long, forming a permanent plane of weakness in the clay mass. Desiccated clays are known for their rapid expansion when a water source finally becomes available; the effects on structures can be crippling.

To this point, the discussion of consolidation in clays has been limited to descriptions and observations. Based on these observations, a method for calculating settlements in a clay stratum can now be derived. The following method is not heavily theoretical; its basis is the results of tests performed on small samples of the clay.

Settlements in clay can only occur due to a reduction in the voids, shown as V_V in Fig. 2-12. The reduction occurs due to a change in pressure due to the footing load, where the average pressure within the stratum increases from p_0 to p_1.

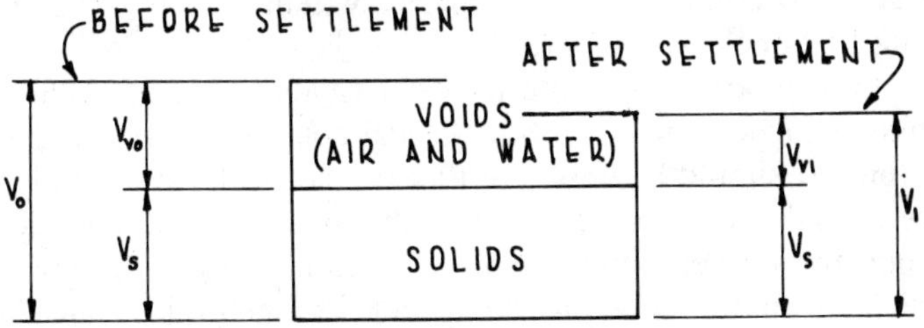

Figure 2-12 Change in void ratio.

The vertical strain induced in the clay due to the footing load is the deformation divided by the original height (width and length are unchanged):

$$\varepsilon = \frac{V_{V0} - V_{V1}}{V_0} = \frac{V_S e_0 - V_S e_1}{V_S + V_{V0}} = \frac{V_S e_0 - V_S e_1}{V_S(1 + e_0)} \tag{2-17}$$

Hence

$$\varepsilon = \frac{\Delta e}{1 + e_0} \tag{2-18}$$

where Δe = change in void ratio $e_0 - e_1$ in going from p_0 to p_1
e_0 = initial void ratio
e_1 = final void ratio

The total settlement S is then the strain ε times the total thickness of the stratum, H:

$$S = H \frac{\Delta e}{1 + e_0} \tag{2-19}$$

To apply Eq. (2-19) it is necessary only to know the settlement S_{test} in a test sample having a thickness H_{test} as it goes from an initial pressure p_0 to a final pressure p_1. Then, given the actual thicknesses H_{field} of the stratum itself, the settlement in the actual stratum in going from p_0 to p_1 in the field is found by simple ratios:

$$S_{field} = S_{test} \frac{H_{field}}{H_{test}} \tag{2-20}$$

where S_{field} is the actual settlement in the field stratum.

Unfortunately, the test procedure for obtaining the value of S_{test} in Eq. (2-20) is time consuming and relatively expensive; the details of the test procedure are discussed further in Chapter 3. To reduce the time required to obtain 100% consolidation in the test, the test sample is made quite thin, usually $\frac{3}{4}$ in. thick. The factor H_{field}/H_{test} in Eq. (2-20) is thus a very large number, typically falling between 200 and 400; it magnifies the test results (and errors) by that amount.

Due to such magnification of results, the sample must be very carefully taken and carefully prepared for the test. It must also be absolutely representative, without irregularities. An irregularity such as a root hole of only $\frac{1}{8}$ in. in diameter would be negligible in the actual soil mass, but if the selected sample were inadvertently to contain that irregularity within its $\frac{3}{4}$-in. thickness, the test results would be hopelessly distorted.

It is important to note in the derivation of Eq. (2-20) that the pressure is assumed to be constant both throughout the sample and throughout the field stratum. It should be apparent that the vertical pressure in the field stratum will vary considerably with depth, thereby limiting the application of Eq. (2-20) only to thin strata, over which the pressure can be assumed to be constant. However, a thick stratum of clay, say 20 ft thick, can be divided into layers about 2 to 4 ft thick and Eq. (2-19) or (2-20) then applied to each layer; the sum of these settlements is then the total settlement. The lab tests must of course be run for corresponding small increments of pressure. Such an approach using strips or layers of the stratum is demonstrated in Chapter 3 for computing settlements in thick deposits of clay.

Since the test procedure for settlements in clay are time consuming and expensive and, as noted, the results are only approximations at best, settlements in a clay stratum are frequently a vague point in the foundation design. Whenever such a set of conditions like these occur in soils or in any other area of engineering design, it is common practice to place more reliance on tried and proven empirical methods than to place a great deal of faith in lab tests and theoretical calculations. One such empirical method has evolved over the years around this question of settlements in clays.

Settlements in clays have long been linked to allowable bearing pressure. Long before the advent of theoretical soil mechanics, there was a belief in the industry that if some limiting factor of safety could only be found, an allowable bearing pressure for clay soils could be established such that the corresponding settlements simply would not be a problem. There was no attempt to define the relationship analytically between pressure and settlement; it was only presumed that if soil pressures were limited to some unknown but moderate value, the whole problem of settlements in clay would go away. The initial multifaceted problem in settlements was thereby reduced to a single problem in pressures, that of finding a factor of safety that would produce the desired value for the allowable bearing pressure.

Over the years, it gradually became accepted that this long-sought factor of safety could be taken as 3, that is, if clay soils were designed with a factor of safety of 3 to bearing capacity failure, the corresponding settlements would be within acceptable levels. This presumption has withstood the test of time and continues to be a useful guide; it is in fact widely accepted in today's engineering practice [4]. Although the exact magnitude of settlement has never been agreed upon openly, it can be assumed that the differential settlements between two footings will not be more than about $\frac{3}{4}$ in.

DEGRADATION AND SETTLEMENT IN SANDS

Sands are more elastic than clays, but it is not the elastic deformations of sands that are of primary interest. Like clays, it is the inelastic settlements of sands that are of primary concern to the foundation designer. These inelastic settlements are due to further degradation of the sand grains as they are chipped, broken, or crushed under an increase in load.

A typical sample of sand is presented in Fig. 2-13, shown at a greatly enlarged scale. The grains of sand are shown as irregular but generally rounded particles, contacting each other at a single point. The void spaces may or may not be filled with water. In very fine sands and nonplastic silts, any water in the voids may slow the settlement rate somewhat, similar to consolidation in clays.

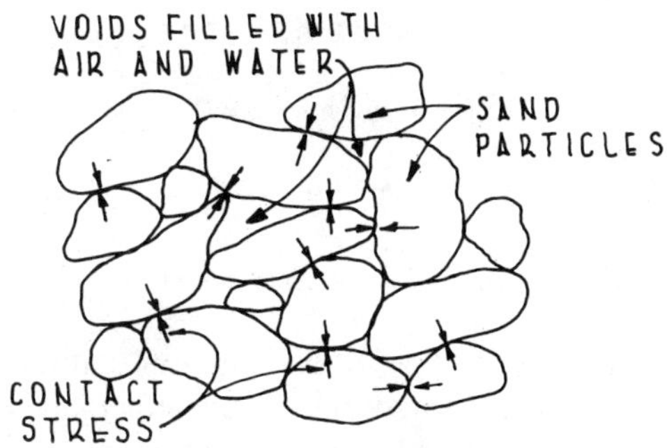

Figure 2-13 Typical sand sample.

Sand and silt particles are obviously friable since they have been undergoing degradation for thousands of years to get to their present size. The theoretical point-to-point contact stress between grains is infinitely large, so further chipping, crushing, cracking, and rounding of grains will occur as the additional load is applied. As the grain sizes are thus further reduced, the entire mass of sand undergoes a reduction in volume, thereby inducing settlements in the structure above.

The *degradation* of the particles is a slow process, propagated from particle to particle. As the first few particles crumble or chip, the load increases at other particles. As these in turn crumble or chip, the load shifts to other points. The resulting degradation can thus be quite slow, occurring over several weeks or up to several months following the addition of a load. Once a sand has undergone such degradation, there is no rebound when the load is removed. The sand has undergone a permanent change both in grain size and in volume. The only recovery is due to the elastic rebound of the particles.

For clays, there was a means presented in earlier discussions to compute settlements which, even with its disadvantages, was theoretically sound. For sands, the theory concerning settlements has not yet been accepted by the practice. A semiempirical method developed by Schmertmann [5] for computing settlements is presented in Chapter 3; it does, however, require the use of empirical constants that cannot be verified by conventional theory.

Distortion settlements, or large-area settlements, occur in sands as well as in clays but the distortion settlements in sands are due more to actual movements of the soil particles than to degradation. As the building is loaded, the particles at the sides of the building are less subject to confining pressures and can be pushed aside, allowing a much higher degree of settlement at the sides of the building than at the center. This effect more than offsets the effects of degradation; the building cups downward at the sides as shown in Fig. 2-14. As in clays, distortion settlement of small buildings founded on sands is not a major concern. Such settlements become a consideration only where the building is quite long, several hundred feet or more.

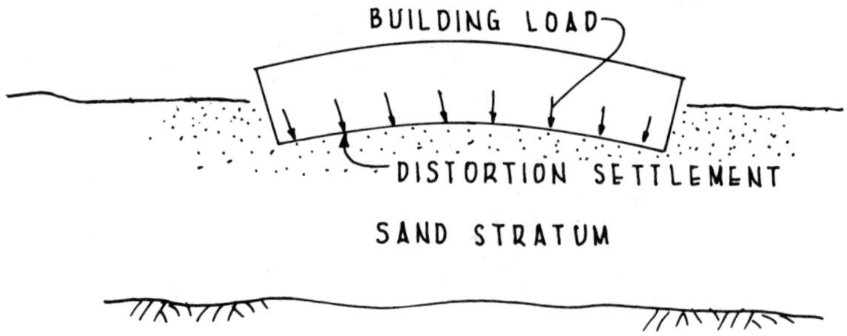

Figure 2-14 Distortion settlements in sand.

Elastic deformations in sand are a type of deformation that has no counterpart (or very little) in clays. Elastic deformations occur immediately upon application and release of the load. Like distortion settlements, they affect the entire building site and are not a major source of unequal settlements between footings; they are usually of little consequence in foundation design.

SHEAR STRENGTH OF CLAYS

The ability of a clay to carry a foundation is limited by the strength of the clay in shear. Shear strength of clay increases, however, as the clay becomes consolidated and the intergranular pressure increases. Theoretically at least, the shear strength of a clay carrying a foundation load and undergoing consolidation should increase slowly over the years.

For small buildings, however, the construction time is so short that no appreciable gain in strength can be realized before the building is completed and the loads are applied. Essentially, the clay must support the full design load of the building in its original natural state. Tests to determine the strength of the clay should not, therefore, include any benefits due to subsequent consolidation.

The strength of a soil is determined much like the strength of any other structural material: a sample is taken of the material and then tested to failure; the stress at failure is said to be its *ultimate strength*. For foundation design, a factor of safety is applied to this failure stress to obtain a "safe" allowable working stress (bearing pressure). The final foundation design is then based on this allowable working stress (pressure).

In making tests on soils, an additional influence occurs that is not present in other materials. The actual soil stratum is subject to a confining pressure laterally due to the overburden above; a typical soil sample taken from such a stratum is shown in Fig. 2-15, with the confining pressure p_3 indicated. The test procedure must, of course, account for this confining pressure.

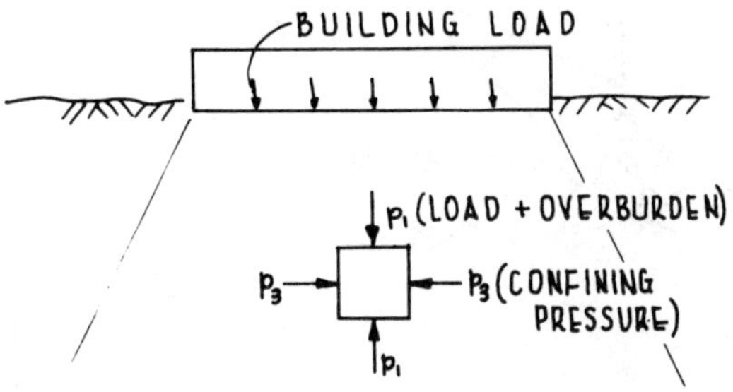

Figure 2-15 Typical confining pressure.

Clay soils have enough cohesion that a cored cylinder of the soil may be extracted intact and tested in a compression test machine. Such a test is shown in Fig. 2-16. The sample is encased in a membrane as shown, with the confining pressure p_3 reproduced in the surrounding airtight bell. When the load is applied to the top of the cored sample, the increase in vertical pressure is the applied load, given as q_u in Fig. 2-16. The total pressure p_1 is this value of q_u plus the confining pressure, $p_1 = q_u + p_3$. As q_u is increased to failure, the confining pressure p_3 remains unchanged.

In tests, it has been observed that clay will fail along a 45° plane as shown in Fig. 2-17; this point is verified analytically in Chapter 3. A square prism of the sample has been removed and shown at larger scale in Fig. 2-17. The area of the small sample is simply indicated as A. The vertical forces acting on the prism are summed, yielding

$$\Sigma F_V = 0 = p_1 A - \frac{(p_3 + \Delta p)A\sqrt{2}}{\sqrt{2}} - \frac{sA\sqrt{2}}{\sqrt{2}} \tag{2-21}$$

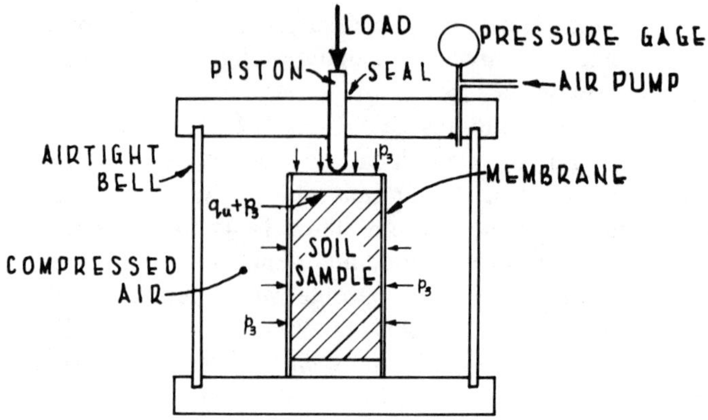

Figure 2-16 Triaxial compression test.

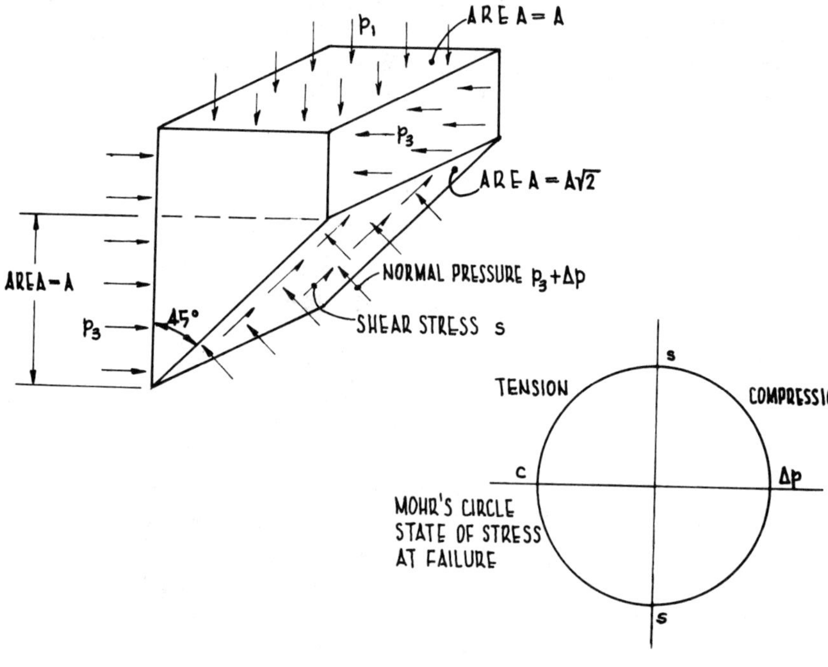

Figure 2-17 Stresses on the sample.

The result is simplified, using $p_1 = q_u + p_3$, yielding

$$0 = q_u - \Delta p - s \tag{2-22}$$

The horizontal forces are summed:

$$\Sigma F = 0 = p_3 A - \frac{(p_3 + \Delta p)A\sqrt{2}}{\sqrt{2}} + \frac{sA\sqrt{2}}{\sqrt{2}} \tag{2-23}$$

When simplified, the result is

$$\Delta p = s \tag{2-24}$$

Equation (2-24) is substituted into Eq. (2-12), yielding

$$s = \tfrac{1}{2} q_u \tag{2-25}$$

This very simple result states that the shear strength of the test sample at failure is one-half the applied vertical pressure q_u. It also reveals that the magnitude of the confining pressure p_3 has no effect on the shear strength of a clay soil. There is thus no need to apply any confining pressure in making the test, since it makes no difference in the outcome.

Since there is no need for confining pressure, there is no corresponding need for membranes or pressure bells. The cylindrical sample of soil can simply be placed in a compression test machine and loaded to failure. The test device then becomes quite simple, as shown in Fig. 2-18, requiring only a means to determine accurately the size of the sample and the load at failure. The test shown in Fig. 2-18 is called the *unconfined compression test*. It is one of the most common lab tests in use today to determine the strength of a clay soil. The shear strength s is simply half the unconfined compressive strength q_u, $s = \frac{1}{2} q_u$.

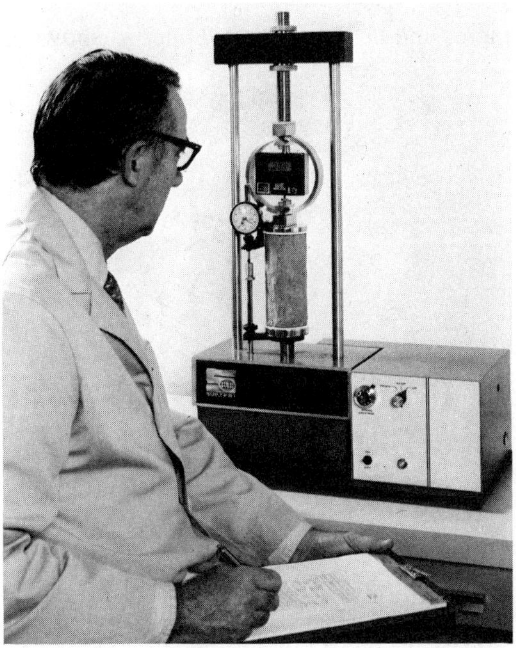

Figure 2-18 Unconfined compression test. (Courtesy of Soiltest, Inc.)

The stress denoted c in Fig. 2-17 is equal in magnitude to the stress Δp. It is called the *cohesion* of the clay and is usually denoted as c. It is shown in standard textbooks on strength of materials [26] that this value is equal in magnitude to the tensile strength of the material. As indicated in Eq. (2-24), the shear strength of a clay is thus numerically equal to the cohesive strength, or tensile strength, of the clay; hence

$$s = c \tag{2-26}$$

The unconfined compression test shown in Fig. 2-18 is widely used. The simple apparatus for running the test is available almost everywhere and is very easy to set up and use; the tests are commonly run on the tailgate of a Jeep. Although test results are sensitive to irregularities in the sample, the test is cheap enough and simple enough that several tests can be run on the same soil and any erratic results due to such things as pebbles or desiccation cracks can simply be discarded.

In recent years, another test has been developed that is even more simple and which is gaining wide acceptance by engineers. The test is called the *vane shear test* and is run in-place in the actual stratum; no samples are necessary. A sketch of a typical vane apparatus is shown in Fig. 2-19. The vane is simply driven to the desired depth, or placed in a drilled hole and driven the last 2 ft or so. The vane is then turned until the soil shears; the torque T required to cause shear failure in the soil is read and recorded. The solution of a simple calculation gives the shear strength at failure:

$$T = s\pi dh + \text{end effects} \tag{2-27}$$

where d and h are the diameter and height of the cylinder as shown in Fig. 2-19.

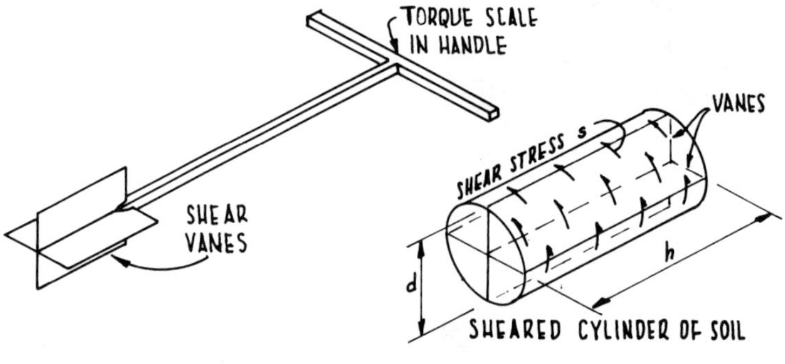

Figure 2-19 Vane shear test.

The vane shear test has the distinct advantage that there is no chance of having a sampling error; the test is run in place. Erratic results can occur when a pebble or a desiccation crack is encountered, so in practice, several tests are run at a test site and any erratic results are discarded. As the required depth of the test increases, the test obviously becomes more difficult and expensive.

It should be noted that the vane shear test measures the total shear strength of the soil. For pure clays having no sand content, the cohesion c is of course equal to this shear strength. For mixtures of soils having both a clay and a sand component, the test does not provide a way to distinguish between the strength due to friction in the sand and the strength due to cohesion in the clay; for such soils, the angle of internal friction ϕ and the cohesion c are not found in this test.

SHEAR STRENGTH OF SANDS

It is almost axiomatic that the strength of a sandy soil is directly proportional to its relative density. Improvements in its gradation to fill more of the void spaces with correctly sized particles will improve the relative density of the sand and will thereby increase its strength. Conversely, any development that serves to reduce the density, such as "fluffing" the sample or removing fines from the gradation, will serve to reduce the strength of the sand.

When a sand is being sampled for a strength test, it is very important that the sand particles remain in their in-place orientation. Even small disturbances can change the distribution of the particles, thereby altering the density and subsequently distorting the test. In sands, a representative sample *must* be undisturbed.

It is very difficult, however, to obtain a sample of sand without disturbing it. The sand simply will not cling together well enough to allow a sample to be extracted and handled during sampling and testing. As more sophisticated methods are introduced to reduce such disturbances, the cost increases commensurately.

It should be apparent that the unconfined compression test that is so widely used for clays cannot be used for sands. The triaxial test with its confining pressure would work, but its cost, along with the likelihood that the sample will be disturbed, discourages its use. For sands, a different type of test is usually made, the direct shear test.

A typical apparatus for a direct shear test is shown in Fig. 2-20. The sample of sand is placed in the ring and the overburden pressure σ is applied to the sample. The upper half is sheared sideways across the lower half and the shearing force at failure is recorded; the shear stress at failure is then the shearing force divided by the sheared area.

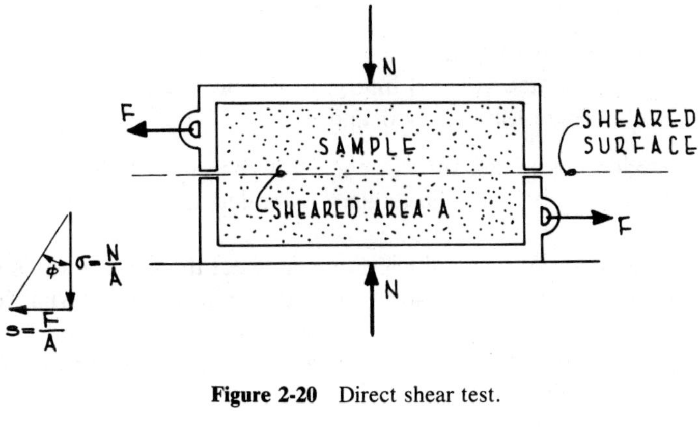

Figure 2-20 Direct shear test.

The shear stress at failure will increase as the overburden pressure σ is increased. Unlike clays, the strength of sand is thus directly proportional to the amount of overburden or overpressure. Consequently, the increase in pressure caused by a large footing on sand will actually cause an increase in the strength of the underlying sand.

The analogy between the direct shear test and the friction devices from high school physics is readily seen. A comparison is shown in Fig. 2-21, where the friction line between the block and the table (from high school physics) is directly analogous to the friction line between upper and lower halves of the direct shear device (from soil mechanics). The major difference is that the line of contact between sand grains is considerably rougher.

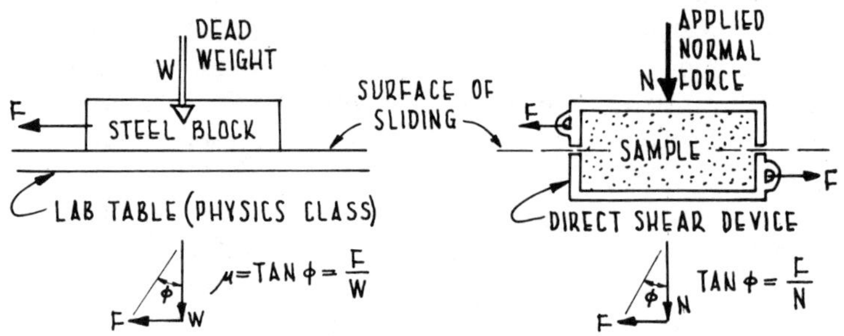

Figure 2-21 Friction analogy.

The friction force is the coefficient of friction μ times the normal force N:

$$F = \mu N \tag{2-28}$$

The coefficient μ is thus the tangent of the angle ϕ, as shown in Fig. 2-21,

$$\tan \phi = \mu = \frac{F}{N} \tag{2-29}$$

If the forces shown in Fig. 2-21 are divided by the sheared areas, the results are stresses. In soil mechanics, the equation is written in terms of stress, or pressure,

$$s = \sigma \tan \phi \tag{2-30}$$

where s = shear strength
σ = average vertical intergranular pressure
ϕ = angle formed by σ and s

The angle ϕ is a measure of the "roughness" of the sand particles and their tendency to interlock; the angle ϕ is called the *angle of internal friction* of the sand.

The angle of internal friction is a direct measure of the strength of the sand. A sand in its most dense state can have an angle of internal friction up to about 44°. A sand in its loosest state can have an angle of internal friction down to about 28°. The end result of the direct shear test is the determination of the magnitude of this angle of internal friction.

It should be evident from Eq. (2-29) that the magnitude of the shear stress s at failure is dependent on the magnitude of the intergranular pressure σ; if σ increases, s increases. At lower depths of a sand stratum where the overburden pressure is higher, the vertical intergranular pressure is higher; hence the sand gains strength with depth. Similarly, the presence of a large footing will actually cause an increase in the strength of the sand it rests on, since it causes an increase in the intergranular stress in the sand under the footing.

Conversely, however, any factor that causes a reduction in the intergranular pressure in a sand will produce a loss in shear strength s. It was noted earlier that submergence of a sand will reduce the intergranular pressure roughly by half; a reduction of the shear strength must occur in the same proportion. Consequently, a foundation on sand should be designed using the submerged strength of the sand if there is even a remote possibility that the site could become flooded during the life of the structure.

The direct shear test shown in Fig. 2-20 has been used widely for many years to determine the angle of internal friction of sands. Even so, it suffers from the same disadvantage as other tests used to determine the strength of sands, the difficulty in keeping a sample undisturbed throughout the procedure. Even a slight jolt of the sample can change its density, thereby distorting its angle of internal friction measurably.

In recent years, indirect test methods have been developed that have all but replaced the direct shear test. In the United States, the oldest and best-known example of such indirect tests is the *standard penetrometer test* (SPT). A more recent development is an analogous test called the *Dutch cone*. The two tests can be directly correlated, but there are several variations of the Dutch cone in use and care must be exercised in correlating such tests.

In this indirect test, a device on the end of a rod is driven into the sand stratum. In the SPT, the device is a standard sampling tool called a split spoon sampler, shown in Fig. 2-22. In the Dutch cone test, it is a specially designed cone, also shown in Fig. 2-22. For either case, the resistance of the sand to the penetration of the device is a measure of the strength of the sand.

To perform the standard penetration test, the sampler is driven into the sand using a hammer of 140-lb weight falling a distance of 30 in. The number of blows required to drive the point 1 ft into the stratum is the blow count N for the test; it can be correlated very accurately to the angle of internal friction of the soil. Such a correlation is shown in Fig. 2-23.

Unlike the SPT, the Dutch cone is not driven with a hammer. It is pressed into the soil by a hydraulic ram at the rate of 2 cm/sec. The results of the Dutch cone test are given as the resistance value stated as a pressure, kg/cm^2 or ksf. Average values of Dutch cone resistance are given in Fig. 2-23.

Since a penetration test is performed in place, any possibility of sampling errors and sample disturbances is eliminated. The tests are fast and inexpensive; multiple tests can be performed and erratic results discarded. Although the SPT is still the most widely used of these tests in the United States, the Dutch cone test is gaining in popularity and seems likely to replace the SPT at some time in the future.

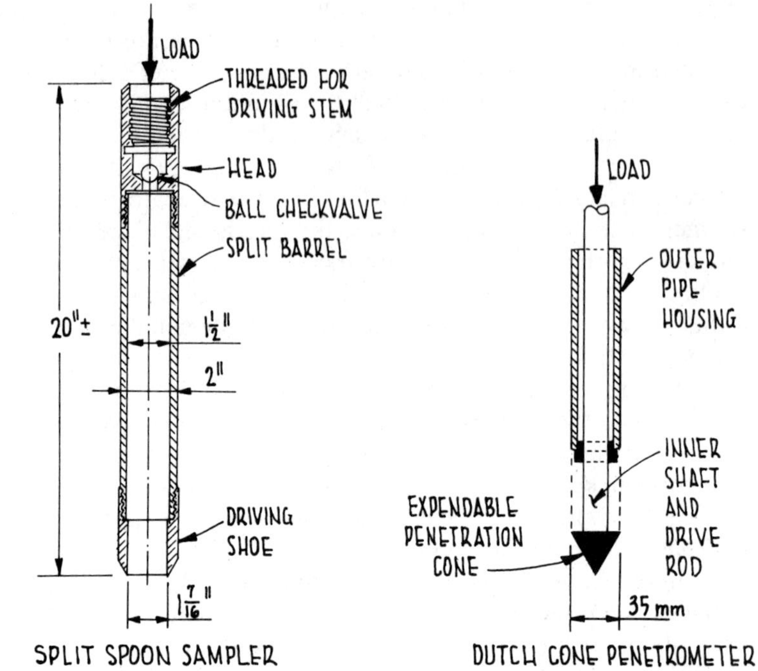

Figure 2-22 Standard penetration devices.

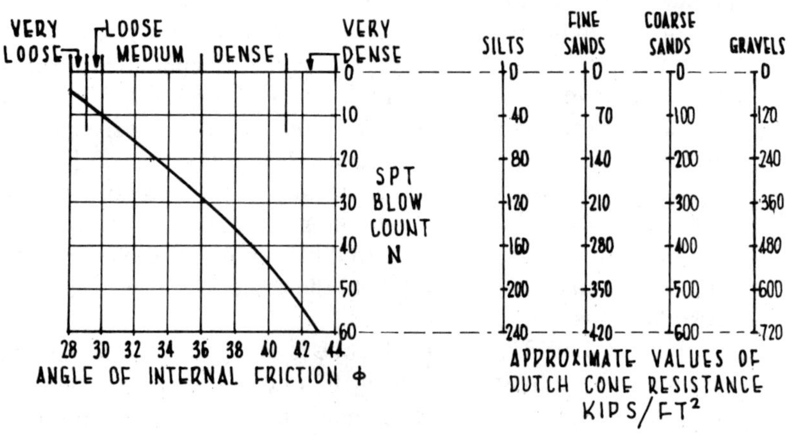

Figure 2-23 Standard penetration correlation. (From Ref. 5.)

It should be noted that the penetration tests do not provide the value of the angle of internal friction directly. The relationship between blow count N and the angle ϕ is usually given by an empirical graph or chart such as that shown in Fig. 2-23. It should also be noted, however, that Fig. 2-23 is valid only for clean sands, or soils having no clay content.

For soils that contain both sands and clays, the penetration tests (like the vane shear test) do not provide a way to distinguish between the strength due to friction and the strength due to cohesion. Only the overall strength is measured; the individual values for the angle of internal friction ϕ and the cohesion c are never known.

THE COULOMB EQUATION

As a point of review, the shear strength of a clay has been shown to be numerically equal to the cohesive strength of the clay,

$$s = c \tag{2-26}$$

Similarly, the shear strength of a sand has been shown to be its frictional resistance,

$$s = \sigma \tan \phi \tag{2-30}$$

For pure sands or pure clays, it is thus quite simple to find the shear strength of the soil from the foregoing relations.

Most soils, however, are neither pure sands nor pure clays. As indicated in the classification chart of Table 2-2, most soils are mixtures of sands and clays, being distinguished only by the amount of material being retained or being passed by the No. 200 sieve. The shear strength of such a mixture should obviously be some combination of the two.

The shear strength of the mixture is in fact the direct sum of its shear strength due to cohesion and its shear strength due to friction:

$$s = c + \sigma \tan \phi \tag{2-31}$$

While there may be some occasional inconsistencies in directly summing these shear strengths, the result is generally adequate for the design of shallow foundations. Equation (2-31) is the Coulomb equation; it is the general solution for the shear strength of mixtures of sands and clays.

SUMMARY

The following concepts are presented in the foregoing discussions.

- By the nature of their geologic origins, soils can be expected to be heterogeneous and inconsistent; their engineering properties can be defined only within broad brackets.
- Engineering properties of soils are distinctly different for coarse-grained soils than for fine-grained soils; the distinction between these two broad classifications of soils is based only on grain size.
- Coarse-grained soils are retained on the No. 200 sieve (0.075 mm clear); fine-grained soils pass the No. 200 sieve.
- Coarse-grained soils are distinguished by the gradation of their particles; fine-grained soils are distinguished by the amount of plasticity they exhibit.

- Not all fine-grained soils develop plasticity; nonplastic silts respond to load similar to a fine sand.
- Soils are classified in broad soil groups either in accordance with the grain size and distribution (if coarse-grained) or in accordance with the amount of plasticity exhibited (if fine-grained).
- Permeability of a soil is its ability to permit passage of water through its pores; permeability can have a distinct effect on other engineering properties.
- Under an increase in load, clays undergo a reduction in volume by slowly extruding porewater; the clay mass thereby deforms enough that the clay particles eventually receive all the increase in load.
- Under an increase in load, sands undergo a reduction in volume by slowly chipping, crushing, or breaking at the points of contact of the individual grains; the sand mass thereby deforms until contact areas between grains are large enough (or numerous enough) to carry the increase in load.
- The reduction of volume in fine-grained soils due to an increase in load is called consolidation; the reduction of volume is coarse-grained soils due to an increase in load is called degradation.
- The shear strength of clays is due entirely to the cohesion, or tensile strength, of the clay.
- The shear strength of sands is due entirely to confining pressure on the sand particles and the resultant interlocking, or friction, between particles.
- The shear strength of mixtures of clays and sands is the sum of the two strengths, stated mathematically as Coulomb's equation, $s = c + \sigma \tan \phi$.

REVIEW QUESTIONS

2-1. Name the four broad soil groups.

2-2. Organic soils such as peat and lignite are not included in standard engineering classifications. Why?

2-3. What is degradation?

2-4. What are some of the stronger influences that are continually breaking rock into grains of soil?

2-5. What influences can produce chemical changes in rock rather than mechanical breakage?

2-6. Why is soil so likely to be stratified?

2-7. Describe the soils in the A, B, and C horizons.

2-8. What feature makes a soil well graded?

2-9. When is a soil said to be uniformly graded?

2-10. Why are clays and silts not described as being well graded or uniformly graded?

2-11. Give the sieve sizes that fall at natural breakpoints between cobbles and gravels, between gravels and sands, between coarse sands and fine sands, and between coarse-grained soils and fine-grained soils.

2-12. What is meant by the D_{30} size of a soil?

2-13. What are the uniformity coefficient and the coefficient of curvature? How are they derived?

2-14. What values of C_u and C_z would indicate a well-graded gravel? A well-graded sand?

2-15. Describe the physical meaning of the liquid limit in a soil.

2-16. Describe the physical meaning of the plastic limit in a soil.

2-17. Verbally describe the plasticity index in terms of water content.

2-18. What is the A-line in classifying soils?

2-19. What index property is used in defining liquid limit and plastic limit?

2-20. Soils having a liquid limit above a certain value are said to be highly compressible. What is that value?

2-21. What is the classification of a soil whose PI falls below the A-line and whose liquid limit is above 50?

2-22. What screen size separates gravels and sands from silts and clays?

2-23. What is meant by a textural classification?

2-24. To what kinds of soils does a consistency classification apply?

2-25. What is water content? Void ratio? Porosity?

2-26. Define permeability in a soil.

2-27. Define effective pressure.

2-28. How does submersion affect effective pressure?

2-29. How does downward flow of water in a soil stratum affect the effective pressure? Upward flow?

2-30. How does a soil become "quick"?

2-31. What is consolidation?

2-32. How is consolidation different from degradation?

2-33. What is a normally consolidated clay?

2-34. What is an overconsolidated clay?

2-35. In clay soils, does a building undergoing distortion settlement cup upward or downward? Why?

2-36. What is the hazard inherent in placing a foundation on an overconsolidated clay?

2-37. What is meant by the term "desiccation"? How is desiccation related to overconsolidation?

2-38. How can the void ratio be reduced in a clay soil?

2-39. What is the magnification factor between lab and field results when running lab tests for settlements in clay soils?

2-40. At what factor of safety in clay soils will settlements generally be limited to 1 in.? How was this value arrived at?

2-41. How does long-term settlement occur in sands?

2-42. From what physical property does clay derive its strength?

2-43. What is the relationship between the unconfined compression strength of a clay and its tensile strength?

2-44. What is the relationship between the cohesive strength of a clay and its unconfined compression strength?

2-45. Along what angle to the direction of the axial load will a sample of clay fail in the unconfined compression test?

2-46. What are the advantages of the vane shear test over the unconfined compression test?

2-47. Why is the direct shear test not a very satisfactory means to determine foundation conditions in sands?

2-48. What is the angle of internal friction in soils?

2-49. What happens to the shear strength of the underlying sand as footing pressure is increased?

2-50. Describe the SPT.

2-51. What is the Dutch cone test?

2-52. In what types of soil is a penetration test most accurate? A vane shear test?

2-53. Write the Coulomb equation for shear strength in soils and define each variable.

2-54. When submerged, which soil undergoes the higher loss in strength, sand or clay?

2-55. Under load, which soil settles more slowly, sand or clay?

2-56. In a sand, does a building undergoing distortion settlement cup upward or downward? Why?

2-57. How is the allowable bearing strength on a clay soil affected when the footing is placed at a lower depth?

2-58. How best can clay deposits be compacted?

2-59. How does the shape of a footing affect the bearing strength of a sand? Of a clay?

2-60. Without referring to the classification chart, determine the classification of a soil whose PI falls above the A-line and whose liquid limit is 41.

PROBLEMS

Problems 2-1 through 2-6. Plot the gradation curves on semilog scale for the given soils. Compute C_u and C_z for the coarse-grained soils.

Sieve size	Percent passing the indicated sieve size					
	Soil 1	Soil 2	Soil 3	Soil 4	Soil 5	Soil 6
3 in.	100	80	100	100	100	100
0.75 in.	100	58	100	100	88	100
No. 4	100	39	95	100	72	100
No. 10	100	28	90	100	54	100
No. 40	91	13	80	100	26	99
No. 60	87	9	75	92	19	97
No. 200	73	0	62	6	10	91
w_L	67	—	39	19	31	81
w_P	25	—	28	19	31	35

Problems 2-7 through 2-12. Classify the soils of Problems 2-1 through 2-6 using the unified soil classification system.

Problems 2-13 through 2-18. Classify the given soils in the unified system. Interpolate as necessary to obtain D_{60}, D_{30}, and D_{10}. If D_{10} is less than 0.75, sketch the gradation curve and extrapolate.

Sieve size	Percent passing the indicated sieve size					
	Soil 13	Soil 14	Soil 15	Soil 16	Soil 17	Soil 18
3 in.	100	100	100	100	100	86
0.75 in.	99	87	100	100	84	59
No. 4	80	57	95	100	62	31
No. 10	53	39	85	100	49	11
No. 40	9	23	77	79	45	2
No. 60	5	20	74	75	42	0
No. 200	2	14	72	59	31	0
w_L	—	54	84	73	36	—
w_P	—	39	62	15	33	—

2-19. A soil is described in a telephone conversation as a soft clay. What range of bearing values should be expected for a foundation on this soil?

2-20. A soil is described as a dense sand. What would be the description of a clay having the same general level of bearing strength?

2-21. A soil is classified as a medium sand. What range of values for the angle of internal friction can be expected?

2-22. A stratum of sand is vibrated until it becomes more dense, going from a loose sand to a medium sand. What happens to its strength because of this vibration?

Problems 2-23 through 2-28. Compute the remaining index properties for each of the given soils.

	Soil 23	Soil 24	Soil 25	Soil 26	Soil 27	Soil 28
G_S	2.68	2.61	2.63		2.60	
w			18.00		21.00	
γ			119 pcf	124 pcf		118
γ_d						106
e	0.61	0.41		0.71		0.56
n						
$S\%$	Sat	46.00		Sat	Sat	

2-29. A fine sand has a Dutch cone penetration resistance of 149 ksf. Approximately, what is its angle of internal friction?

2-30. A Dutch cone test on a silt indicates a penetration resistance of 62 ksf. What should be the SPT blow count for this soil?

3

Mechanics of Soils

The mechanics of soil failure and foundation settlement is the subject of this chapter. As a general rule, the discussions are limited to those topics that are directly related to foundation design. One exception to this rule is the subject of recompaction. Recompaction is included since it is often necessary to disturb and recompact the foundation material when a building site is being graded. The effect of such recompaction on the foundation design must be evaluated during design stages and is therefore included here.

Placing a load on a footing creates an increase in the bearing pressure which will be dispersed downward and outward into the soil mass. At some distance downward, the load will be so widely dispersed that any further increases in soil pressure can safely be ignored. In keeping with the overall accuracy of calculations involving soils, these increases are usually ignored when they have diminished to about 10 to 15% of the contact pressure under the footing. It is this dispersion of pressure into a soil mass that is properly the first item of study in soil mechanics.

BOUSSINESQ PRESSURE DISTRIBUTION IN A SOIL MASS

The dispersion of a foundation load into a soil mass is treated in the mathematical theory of elasticity as a concentrated load at the surface of a semi-infinite elastic solid. The original solution to this problem was given by Boussinesq in 1885 [6]. In his solution, the solid is presumed to be homogeneous, isotropic, and elastic, with these ideal properties extending indefinitely in all directions.

A concentrated load P is assumed to be applied at the surface of the solid as shown in Fig. 3-1. The increase in vertical stress, $\Delta\sigma_z$, at a depth z below the surface and at a lateral distance r from the load P is given by

$$\Delta\sigma_z = \frac{3P}{2\pi}\left[\frac{z^3}{(r^2 + z^2)^{5/2}}\right] \tag{3-1}$$

A solution to Eq. (3-1) is readily possible on a small programmable calculator or a personal computer. A manual solution is quite tedious since the concentrated load P must be taken as the load at each square foot of a foundation, then the results of these solutions summed to find the total increase at a single point. Then the process must be repeated for the next point and the next, and so on.

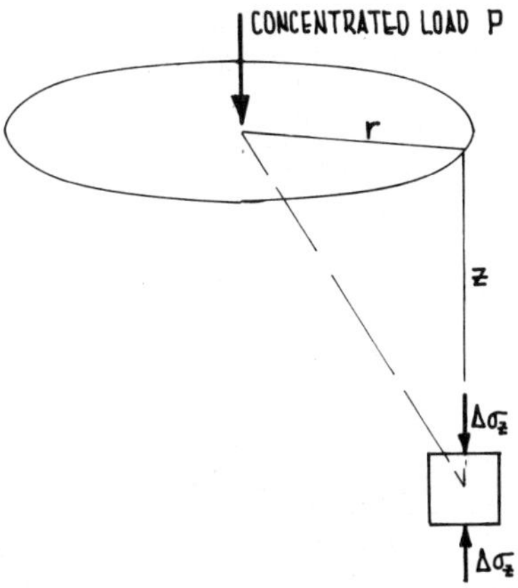

Figure 3-1 Stresses in an elastic solid.

The solution to Eq. (3-1) is shown graphically in Fig. 3-2 for a footing of width B. At one side is the solution for a strip footing of infinite length and at the other is the solution for a square spread footing. The "contour" lines show the pressure at various distances from the footing in terms of the contact pressure p and the footing width B.

Alternatively, the variation in pressure at various horizontal levels in the underlying soil can be plotted as shown in Fig. 3-3. The variation in pressure along the vertical axis at the centerline of the footing is also shown. Note that all dimensions are shown as multiples of the footing width B and that pressures are shown as increments of the average contact pressure p.

It should immediately be apparent that few soils could meet the requirements of the Boussinesq solution. Soils are not homogeneous, isotropic, or elastic, nor do they usually extend very far in any direction without an interruption of

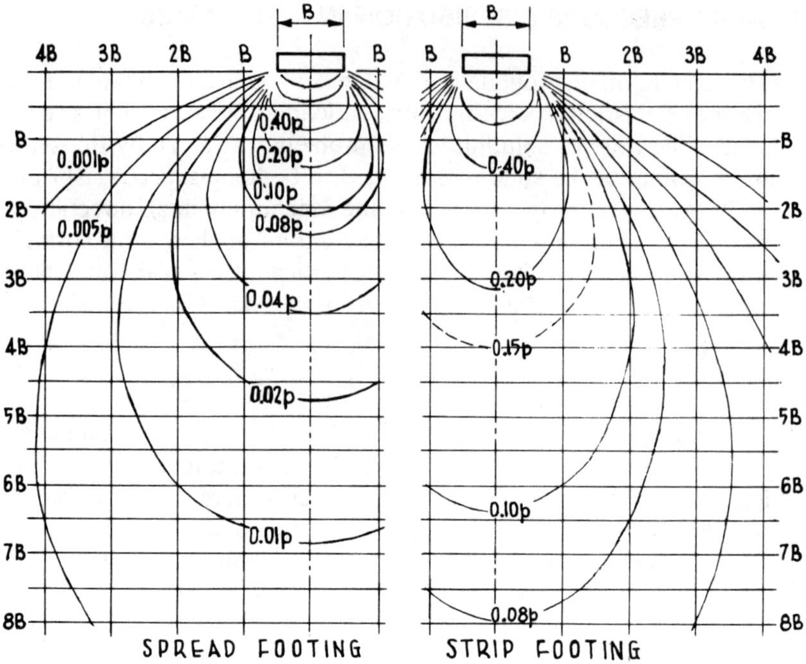

Figure 3-2 Boussinesq pressure distribution.

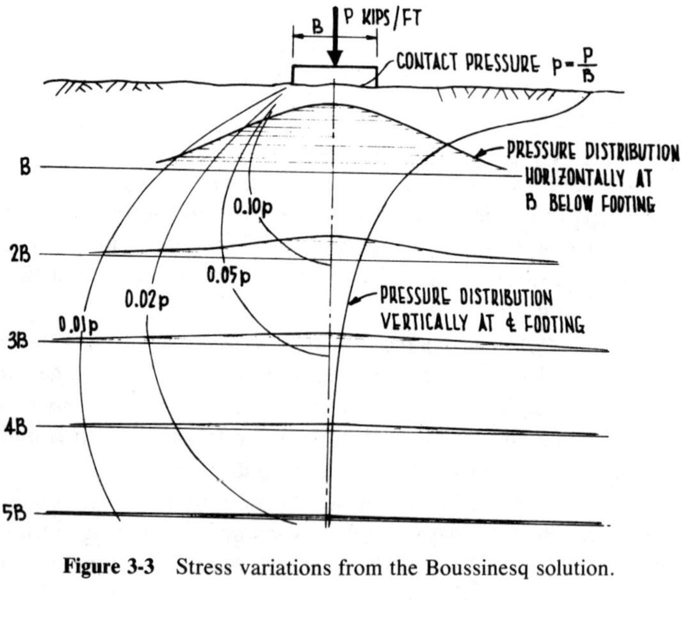

Figure 3-3 Stress variations from the Boussinesq solution.

some kind. Nonetheless, the idealized pressure distribution represented by the Boussinesq solution affords a standard against which the pressure distribution in any soil can be compared.

It is common practice in soil mechanics to represent the multiplicity of three-dimensional Boussinesq contour bulbs of Fig. 3-2 by only one or two contour bulbs, as indicated in Fig. 3-4. In such a representation, the outermost contour bulb is usually taken to be the $0.10p$ contour bulb for spread footings and the $0.15p$ contour bulb for strip footings. The overall accuracy of the analysis and of soil properties in general does not justify consideration of pressures smaller than about 10 or 15% of the contact pressure.

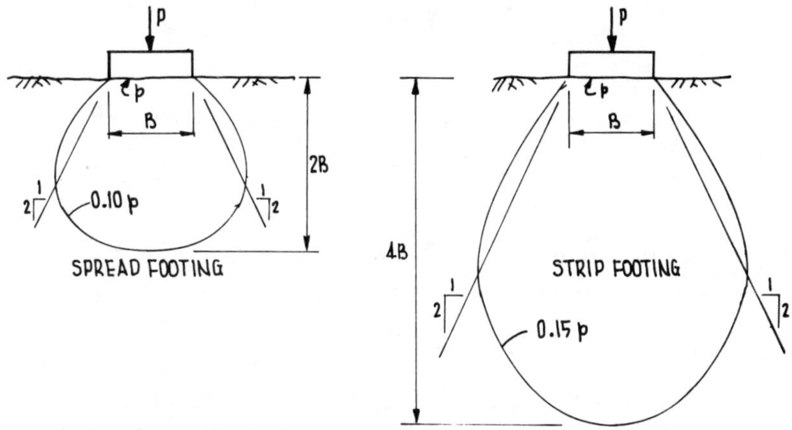

Figure 3-4 Abbreviated Boussinesq pressure bulbs.

It should be noted from Fig. 3-4 that for spread footings, the pressure contour of $0.10p$ extends downward to about $2B$ below the bottom of the footing. For a strip footing, the $0.15p$ pressure contour extends downward to a depth of about $4B$, or roughly twice as deep. The difference is attributed to the fact that pressure under a spread footing can be dispersed in all four directions while under a long strip footing it can be dispersed in only two directions; it therefore takes roughly twice as much depth to dissipate the soil pressure under a strip footing as it does under a spread footing.

The three-dimensional pressure bulb that exists under an entire building may be visualized as shown in Fig. 3-5a, based on the overall average increase in pressure under the entire building. Or it may be visualized as shown in Fig. 3-5b, based on the increase in pressure at the individual footings (*much* larger than the overall average). The overlapping pressure bulbs of the individual footings cause a buildup and overall dispersion of pressure that eventually produces the same overall pressure bulb shown in Fig. 3-5a.

The actual Boussinesq pressures are rarely used in design practice. They would almost never be used directly in the design of foundations for small buildings. Even so, the concepts and the parameters of the Boussinesq solution form the basis for most of the approximate methods used in the design of foundations; one such approximate method is introduced in Chapter 7. Some examples will

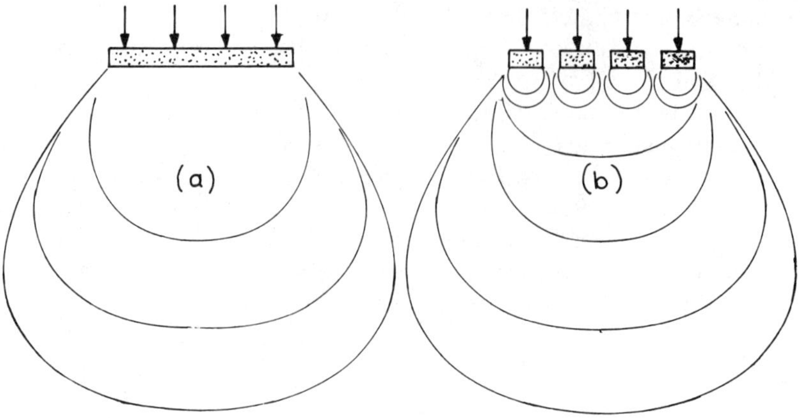

Figure 3-5 Overall pressure distribution.

illustrate the use of the Boussinesq pressure distribution as applied both to buildings and to footings.

EXAMPLE 3-1

A building 60 ft square exerts an average overall pressure over the building footprint of 300 psf. At what depth below the building will the average increase in pressure be reduced to one-half the increase at the top? To one-fourth the increase at the top?

SOLUTION From Fig. 3-2, the contour at $0.50p$ is seen to extend downward to about $0.75B$ and the contour at $0.25p$ extends downward to about $1.3B$. At a depth of $0.75 \times 60 = 45$ ft, the increase in pressure will be about one-half that at the surface, or 150 psf increase. At a depth of $1.3 \times 60 = 78$ ft, the increase will be about one-fourth that at the surface or 75 psf increase.

EXAMPLE 3-2

A column load of 120,000 lb is carried by a spread footing 6 ft 3 in. square. Determine the depth at which an increase in soil pressure of 150 psf will occur below the footing.

SOLUTION The contact pressure under the footing is seen to be 120,000 lb/6.25^2 ft^2 = 3070 psf. The pressure increase at 150 psf is then 150/3070, roughly 5% of the contact pressure. From Fig. 3-2, it is seen that the 5% contour extends downward below the footing to a depth of about $3.2B$, or $3.2 \times 6.25 =$ 20 ft deep. An average increase in pressure of 150 psf will therefore occur at about 20 ft under the footing (ignoring effects of overlapping pressure bulbs from other footings).

AT-REST PRESSURES IN A SOIL MASS

When pressures are uniform, there is no shear in the underlying soil. Until a disturbance occurs, the soil is in a state of principal stress, termed its *at-rest* condition. The vertical pressure at any point in the at-rest condition is simply the pressure due to the uniform load above that point.

The existence of a uniform vertical pressure in a soil stratum does, however, produce a lateral soil pressure within the stratum. The lateral soil pressures at any depth in the stratum can be likened to the pressures generated by an equivalent fluid as shown in Fig. 3-6. If the unit weight of this equivalent fluid is known, the lateral pressures at any depth is simply the hydrostatic pressure due to the equivalent fluid,

$$p_3 = \gamma_{\text{fl}} h \tag{3-2}$$

where p_3 = lateral pressure
γ_{fl} = unit weight of the equivalent fluid
h = standing head of the equivalent fluid

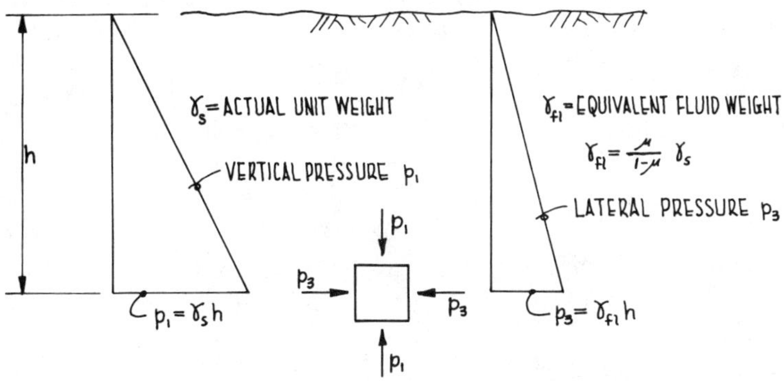

Figure 3-6 Pressures in the equivalent fluid.

For a uniform soil mass, the theoretical unit weight γ_{fl} of the equivalent fluid can be computed by the methods of the theory of elasticity:

$$p_3 = \frac{\mu}{1-\mu} \gamma_S h \tag{3-3}$$

where

$$\gamma_{\text{fl}} = \frac{\mu}{1-\mu} \gamma_S$$

and where γ_S is the unit weight of the soil and μ is Poisson's ratio for the soil. For many soils, a Poisson's ratio between 0.2 and 0.4 is common. Hence the theoretical equivalent fluid weight given by Eq. (3-3) ranges between 25 and 67% of its actual weight. This value is not supported very well by test results, however, which indicate that a closer range of lateral earth pressure occurs in most soils.

An experimentally derived range of values for the unit weight of the equivalent fluid is from 40 to 50% of the actual unit weight of the soil γ_S,

$$p_3 = k\gamma_S h \tag{3-4}$$

where $k\gamma_S$ is the unit weight of the equivalent fluid, γ_{fl}, with k between 0.4 and 0.5.

Like other materials of construction, soils are subject to creep. Hence if a soil mass is rigidly contained under a uniform pressure, then subjected to a transient disturbance, it will eventually return to its at-rest state following the release of the disturbance. The mechanism that permits such "flow" is, of course, creep.

Some examples will illustrate the concepts of at-rest pressures in a soil mass, to include the effects of long-term creep if appropriate.

EXAMPLE 3-3

Soil weighing 116 pcf is backfilled against a rigid basement wall. After a long period of time, what will be the lateral pressure against the wall at a depth of 6 ft?

SOLUTION When all movement has stopped and the soil finally returns to its at-rest pressure, the lateral pressure at a depth of 6 ft will be

$$\begin{aligned} p_3 = k\gamma_S h &= (0.4 \text{ to } 0.5) \times 116 \times 6 \\ &= 280 \text{ to } 350 \text{ psf} \end{aligned}$$

EXAMPLE 3-4

Insofar as computing lateral at-rest pressure in a soil is concerned, what is the equivalent fluid weight of a soil weighing 108 pcf and having a Poisson's ratio of 0.28?

SOLUTION From Eq. (3-3),

$$\begin{aligned} \gamma_{fl} &= \frac{\mu}{1-\mu}\gamma_S = \frac{0.28}{1-0.28} \times 108 \\ &= 42 \text{ pcf} \quad \text{(theoretically)} \end{aligned}$$

From experimental data, the equivalent fluid weight will probably be between 40 and 50% of its actual weight, or

$$\begin{aligned} \gamma_{fl} &= (0.4 \text{ to } 0.5)\gamma_S \\ &= 43 \text{ to } 54 \text{ pcf} \quad \text{(empirically)} \end{aligned}$$

EXAMPLE 3-5

A basement wall is backfilled with a sandy soil weighing 124 pcf. What happens to the lateral pressures against the wall when the area becomes flooded?

SOLUTION After all movement ceases and the soil reaches at-rest pressure (without flooding), the equivalent fluid weight of the soil will be 40 to 50% of

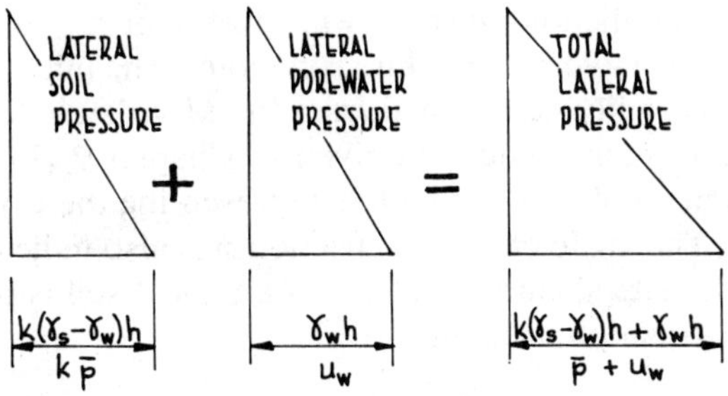

124 pcf or 50 to 62 psf, for an average of 56 psf. When flood stage occurs, the unit weight of the soil is reduced by its buoyancy, becoming $\gamma_S - \gamma_W = 124 - 62.4 = 61.6$ pcf. The equivalent fluid weight of the soil then becomes 40 to 50% of this reduced value for an average value of 28 pcf. However, at flood stage the porewater pressure must be added to the soil pressure, giving a total of $28 + 62.4 = 90$ psf. The increase in equivalent fluid weight at flood stage is therefore $90 - 56 = 34$ psf, which represents a total increase of 61% in lateral pressure against the basement wall.

A permanent load occurring over a small portion of a soil mass (such as a foundation load) will create a permanent disturbance to the at-rest state. A shear stress will then be added to the principal normal stresses of the at-rest state, creating an entirely new state of stress. An analysis of that state of stress follows.

GENERAL SHEAR FAILURE IN A SOIL MASS

It has been observed that failure in soils carrying a foundation load will be primarily a failure in shear rather than in compression or tension. A typical failure of this type is shown in Fig. 3-7, where the soil is displaced sideways along a "sheared"

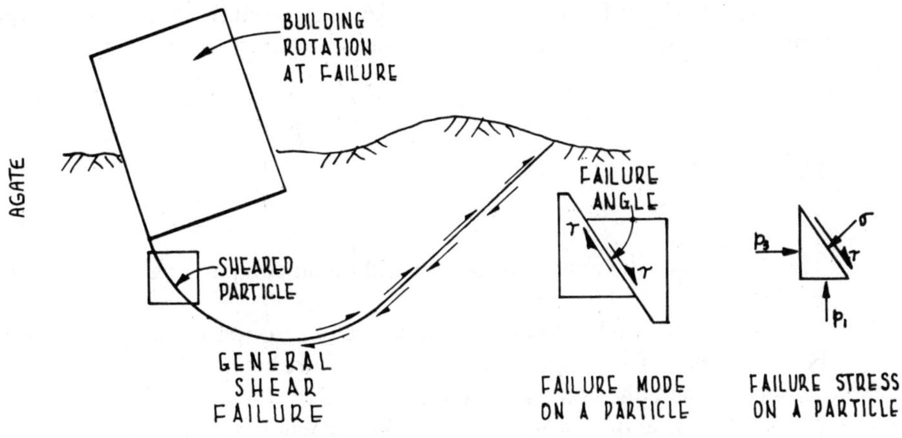

Figure 3-7 General shear failure.

surface. The sheared surface will always occur to one side or the other; the possibility of a two-sided failure is remote. The initial failure occurs at some angle θ which is readily determined from the Mohr's circle state of stress.

The Coulomb equation given in Chapter 2 [Eq. (2-30)] has the form of a straight line; a plot of such a line representing the Coulomb equation is shown in Fig. 3-8. The angle of internal friction ϕ is also indicated, as well as the cohesion c. The ultimate shear strength s' of a typical soil is plotted also, intersecting the Coulomb equation at point A.

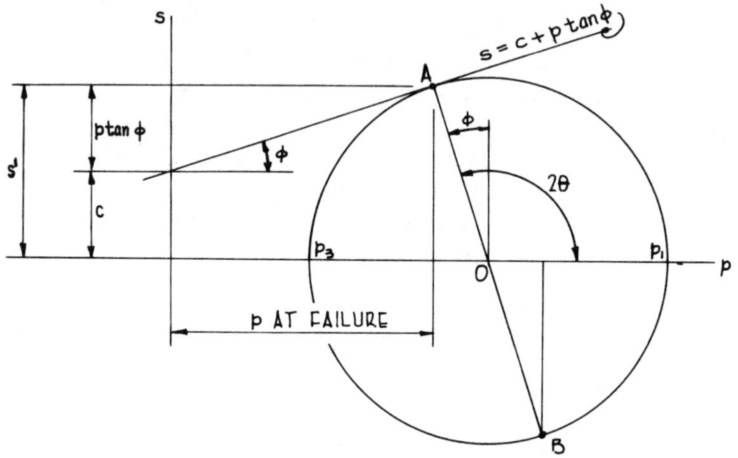

Figure 3-8 Mohr's circle of stress at failure.

A line is now constructed through point A, perpendicular to the line representing the Coulomb equation. The result is line AB, which forms the diameter of a Mohr's circle. The Mohr's circle is now drawn with its center at point O, tangent to the Coulomb equation at point A. This circle defines the state of stress at failure; any larger circle would somewhere exceed the ultimate shear stress s' and any smaller circle would not produce a value of s high enough to cause failure.

The angle at failure is shown as 2θ, measured from the state of the original principal stress. The value of the angle θ at failure is derived directly from the geometry,

$$2\theta = 90° + \phi$$

Hence

$$\theta = 45° + \tfrac{1}{2}\phi \qquad (3\text{-}5)$$

The location of the angle θ at failure and the accompanying stresses are shown in Fig. 3-9.

When the angle of internal friction ϕ is zero (pure clays), the state of stress becomes that of principal shear, with a failure angle of 45° and $\sigma_1 = \sigma_3$. Conversely, if the cohesion c is zero (clean sands), the angle of failure remains 45° + $\frac{1}{2}\phi$ but the value of ϕ would likely be larger where there is no clay component in the soil. These cases are shown for comparison in Fig. 3-10.

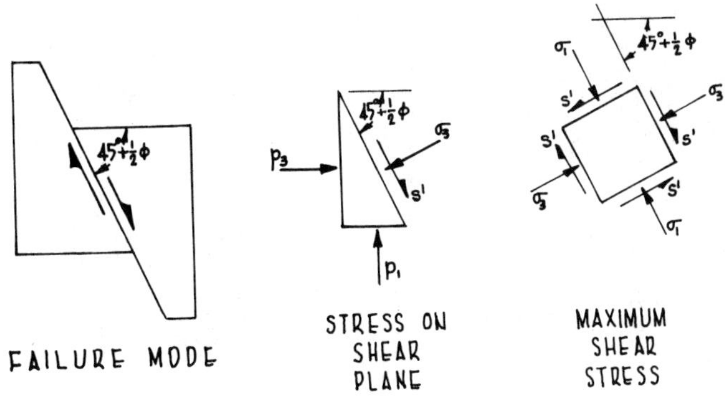

Figure 3-9 State of stress at failure.

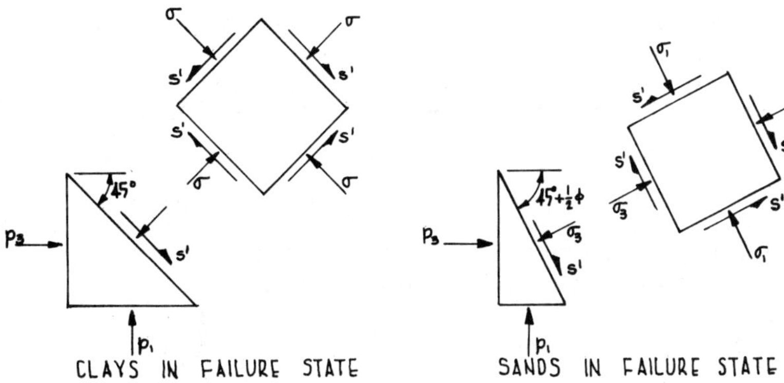

Figure 3-10 Failure states of clays and sands.

The relationship between major and minor principal stresses acting on a soil particle at failure is readily derived from the geometry of Mohr's circle. These principal stresses p_1 and p_3 are indicated on the Mohr's circle shown in Fig. 3-11. From the geometry, a relationship involving the principal stresses p_1 and p_3 is noted in the triangle DAO,

$$\frac{p_1 - p_3}{2} = \left(\frac{p_1 + p_3}{2} + c \cot \phi\right) \sin \phi \tag{3-6a}$$

Hence

$$p_1 = p_3 \frac{1 + \sin \phi}{1 - \sin \phi} + 2c \frac{\cos \phi}{1 - \sin \phi} \tag{3-6b}$$

The following identities allow transformation of the trigonometric terms of Eq. (3-6b) into the failure angle $45° + \frac{1}{2}\phi$:

$$\frac{\cos \phi}{1 - \sin \phi} = \tan\,(45° + \tfrac{1}{2}\phi) \qquad \frac{1 + \sin \phi}{1 - \sin \phi} = \tan^2(45° + \tfrac{1}{2}\phi) \tag{3-7}$$

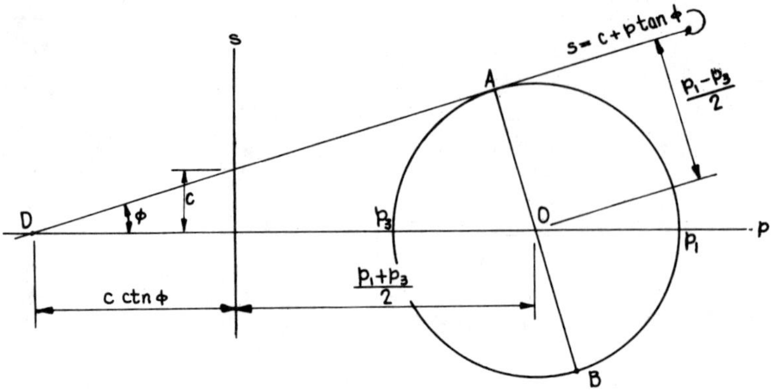

Figure 3-11 Typical failure state.

When these identities are substituted into Eq. (3-6b), the relationship between principal stresses becomes

$$p_1 = p_3 \tan^2(45^\circ + \tfrac{1}{2}\phi) + 2c \tan(45^\circ + \tfrac{1}{2}\phi) \tag{3-8a}$$

or

$$p_3 = p_1 \cot^2(45^\circ + \tfrac{1}{2}\phi) - 2c \cot(45^\circ + \tfrac{1}{2}\phi) \tag{3-8b}$$

Equations (3-8a) and (3-8b) can be used to compute lateral pressures in a soil in its failure state. Some examples will illustrate such computations.

EXAMPLE 3-6

Determine the equivalent fluid weight at failure for two sands having an angle of internal friction of 29° and of 36°, respectively, and unit weights of 121 pcf and 126 pcf, respectively.

SOLUTION The equivalent fluid weight can be given as a coefficient as indicated in Eq. (3-4),

$$p_3 = k\gamma_S H$$

It is noted that $\gamma_S H$ is the value of the vertical pressure p_1; hence

$$\frac{p_3}{p_1} = k$$

From Eq. (3-8b), therefore, with $\phi = 29^\circ$ and $c = 0$ for sands,

$$k = \cot^2(45^\circ + 14.5^\circ) = 0.35 \text{ at failure}$$

For $\phi = 36^\circ$, and again with $c = 0$ for sands,

$$k = \cot^2(45^\circ + 18^\circ) = 0.26 \text{ at failure}$$

The equivalent fluid weights for the two sands at failure are, therefore,

$$\gamma_1 = 0.35 \times 121 = 42 \text{ pcf}$$

$$\gamma_2 = 0.26 \times 126 = 33 \text{ pcf}$$

It is noted that the values of k at failure are considerably different from the values in the at-rest condition, given with Eq. (3-4).

EXAMPLE 3-7

A stratum of clay 24 ft thick carrying a spread footing failed when the contact pressure was 4950 psf. A sample of the clay was tested, showing an unconfined compression strength of 1920 psf. Determine the lateral pressure in the clay immediately below the footing when failure occurred.

SOLUTION Immediately below the footing, with $\phi = 0$ for clays, $c = q_u/2$ and $\cot 45° = 1$, from Eq. (3-8b),

$$p_3 = p_1 - \frac{2q_u}{2} = 4950 - 1920 = 3030 \text{ psf}$$

The lateral pressure in this clay was therefore 61% of the vertical pressure at the time when failure occurred, considerably different from the at-rest values given with Eq. (3-4).

ULTIMATE BEARING PRESSURE ON A SOIL MASS

The foundation bearing pressure that will cause failure in the underlying soil mass can now be derived. The foundation under consideration in the following derivation is an infinitely long footing, representing a strip footing or a grade beam. Loads are vertical; the case of an inclined load is deferred to a later section.

The actual failure mode is shown in Fig. 3-12a, with an approximate failure mode shown in Fig. 3-12b. A solution based on the approximate geometry has been found to be within an acceptable range of accuracy for soils. The average pressure along the failure line is taken at middepth of the soil mass shown acting on the particles at A and B in Fig. 3-12b. However, the vertical pressure in the soil directly under the footing must be diminishing in accordance with the Boussinesq pressure distribution (Fig. 3-4). With an approximate slope of 1 horizontal to 2 vertical for this portion of the Boussinesq pressure bulb, the pressure p'_{1A} at middepth of the failure block due only to footing loads is then

$$p'_{1A} = \frac{Q}{B + \frac{1}{2}B \tan (45° + \frac{1}{2}\phi)} = \frac{Q}{B}\left[\frac{1}{1 + \frac{1}{2}\tan (45° + \frac{1}{2}\phi)}\right] \tag{3-9}$$

It is noted that $Q/B = q_0$, which is the contact pressure under the footing. With this substitution, the final expression for p'_{1A} becomes

$$p'_{1A} = q_0 \frac{1}{1 + \frac{1}{2}\tan (45° + \frac{1}{2}\phi)} \tag{3-10}$$

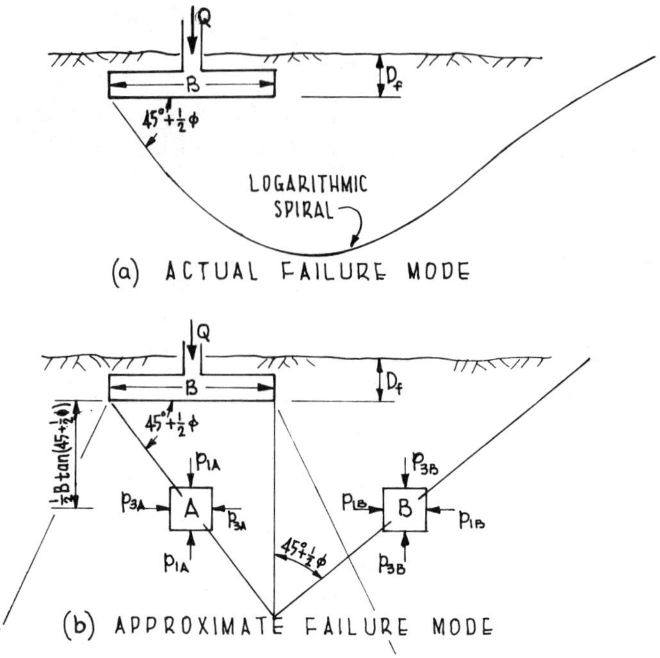

Figure 3-12 Actual and approximate failure modes.

The total vertical pressure p_{1A} due both to footing loads and to overburden is simply the sum of the two:

$$p_{1A} = q_0 \frac{1}{1 + \frac{1}{2} \tan (45° + \frac{1}{2}\phi)} + \gamma D_f + \tfrac{1}{2}\gamma B \tan (45° + \tfrac{1}{2}\phi) \qquad (3\text{-}11)$$

The confining pressure at A, p_{3A}, is given by Eq. (3-8b), which for a strip footing becomes

$$p_{3A} = q_0 \frac{1}{1 + \frac{1}{2} \tan (45° + \frac{1}{2}\phi)} \cot^2(45° + \tfrac{1}{2}\phi) + \tfrac{1}{2}\gamma B \cot (45° + \tfrac{1}{2}\phi)$$

$$+ \gamma D_f \cot^2(45° + \tfrac{1}{2}\phi) - 2c \cot (45° + \tfrac{1}{2}\phi) \qquad (3\text{-}12)$$

This confining pressure p_{3A} just as failure impends is assumed to be equal to the horizontal pressure at B. At B, however, the increase in applied pressure p_{1B} is now horizontal and the confining pressure p_{3B} is vertical. The confining pressure p_{3B} is seen to be the pressure due to the weight of soil above point B,

$$p_{3B} = \gamma D_f + \tfrac{1}{2}\gamma B \tan (45° + \tfrac{1}{2}\phi) \qquad (3\text{-}13)$$

where γ is the unit weight of the soil. (It has thus been tacitly assumed that the depth D_f is reasonably constant throughout the vicinity of the footing.)

The applied pressure p_{1B} just as failure impends may be stated in terms of the confining pressure p_{3B} through the use of Eq. (3-8a),

$$p_{1B} = p_{3B} \tan^2(45° + \tfrac{1}{2}\phi) + 2c \tan(45° + \tfrac{1}{2}\phi) \tag{3-14}$$

Substitution of Eq. (3-12) then yields

$$p_{1B} = \gamma D_f \tan^2(45° + \tfrac{1}{2}\phi) + \tfrac{1}{2}\gamma B \tan^3(45° + \tfrac{1}{2}\phi) + 2c \tan(45° + \tfrac{1}{2}\phi) \tag{3-15}$$

When p_{1B} as given by Eq. (3-15) is equated to p_{3A} as given by Eq. (3-12), the result yields the bearing stress under a strip footing just as failure impends,

$$\begin{aligned} q_0 = {} & \tfrac{1}{2}\gamma B[\tan^5(45° + \tfrac{1}{2}\phi) - \tan(45° + \tfrac{1}{2}\phi)][1 + \tfrac{1}{2}\tan(45° + \tfrac{1}{2}\phi)] \\ & + \gamma D_f[\tan^4(45° + \tfrac{1}{2}\phi) - 1][1 + \tfrac{1}{2}\tan(45° + \tfrac{1}{2}\phi)] \\ & + 2c[\tan^3(45° + \tfrac{1}{2}\phi) + \tan(45° + \tfrac{1}{2}\phi)][1 + \tfrac{1}{2}\tan(45° + \tfrac{1}{2}\phi)] \end{aligned} \tag{3-16}$$

Equation (3-16) is usually written in alternative form,

$$q_0 = \tfrac{1}{2}\gamma B N_\gamma + \gamma D_f N_q + c N_c \tag{3-17}$$

where, for a strip footing,

q_0 = contact pressure due only to loads

$N_\gamma = [1 + \tfrac{1}{2}\tan(45° + \tfrac{1}{2}\phi)][\tan^5(45° + \tfrac{1}{2}\phi) - \tan(45° + \tfrac{1}{2}\phi)]$

$N_q = [1 + \tfrac{1}{2}\tan(45° + \tfrac{1}{2}\phi)][\tan^4(45° + \tfrac{1}{2}\phi) - 1]$

$N_c = [1 + \tfrac{1}{2}\tan(45° + \tfrac{1}{2}\phi)][2[\tan^3(45° + \tfrac{1}{2}\phi) + \tan(45° + \tfrac{1}{2}\phi)]]$

γ = effective unit weight of the soil

B = width of a strip footing (or grade beam)

D_f = depth of overburden above the founding line

The factors N_γ, N_q, and N_c are called the *bearing capacity factors*. They have been under intensive study for many years. To date, no approximate solution, including that presented here, provides an entirely satisfactory mathematical solution for these factors.

Rather, a semiempirical graph for the bearing capacity factors is probably the most widely accepted solution. Such a graph is presented in Fig. 3-13 for a long footing. Generally accepted values proposed by Meyerhof [30] and Bell (reported in Ref. 31) are shown as a solid line, with the values of Eq. (3-17) plotted as the dashed line.

Other approximations and assumptions concerning the failure mode are sometimes used rather than the approximate failure mode shown in Fig. 3-12 and the assumptions made herein. For such analyses, the values for the factors N_γ, N_q, and N_c could be either higher or lower than those of Eq. (3-17). The factors N_γ, N_q, and N_c are under continuing study and verification [7].

The value q_0 given by Eq. (3-17) is the contact pressure that will cause failure under a long footing on a soil having both a sand and a clay component. The derivation does not include lateral loads on the footing, such as those due to wind or earthquake. Also, buoyancy effects on the effective unit weight must be included where appropriate, reducing the value of γ to its submerged value. A few examples will illustrate the use of Eq. (3-17).

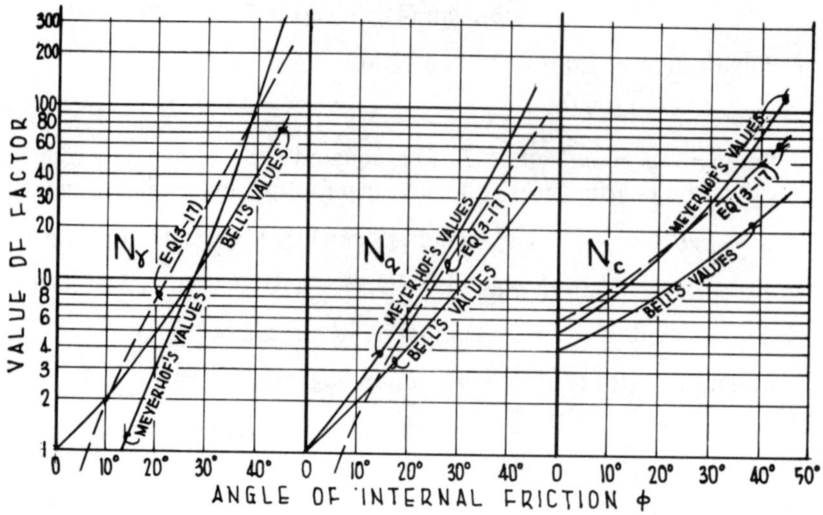

Figure 3-13 Bearing capacity factors.

EXAMPLE 3-8

A footing 8 ft square is located on a stratum of sand, with its founding line 2 ft below surface. The sand weighs 116 pcf and its angle of internal friction is 31°. How much load can the footing carry before the soil fails?

SOLUTION From Eq. (3-17) with $c = 0$ for sands and $45° + \frac{1}{2}\phi = 60.5°$, the contact pressure q_0 at failure is

$$q_0 = \tfrac{1}{2}\gamma B N_\gamma + \gamma D_f N_q$$

Now

$$\begin{aligned} N_\gamma &= (1 + \tfrac{1}{2}\tan 60.5)(\tan^5 60.5 - \tan 60.5) \\ &= 29.2 \end{aligned}$$

and

$$\begin{aligned} N_q &= (1 + \tfrac{1}{2}\tan 60.5)(\tan^4 60.5 - 1) \\ &= 16.5 \end{aligned}$$

Therefore,

$$\begin{aligned} q_0 &= \tfrac{1}{2} \times 116 \times 8 \times 29.2 + 116 \times 2 \times 16.5 \\ &= 17{,}400 \text{ psf} \end{aligned}$$

The bearing strength of the soil is therefore 17,400 psf; failure will occur at a load of $17.4 \times 8^2 = 1114$ kips.

EXAMPLE 3-9

What would be the strength of the soil in Example 3-8 if the footing size were reduced to 3 ft square?

SOLUTION For $B = 3.0$ ft and all other factors unchanged,

$$q_0 = \tfrac{1}{2} \times 116 \times 3.0 \times 29.2 + 116 \times 2 \times 16.5$$
$$= 8910 \text{ psf}$$

The bearing strength of the soil is therefore 8910 psf; failure will occur at a load of $8.91 \times 3^2 = 80.2$ kips. The strength of the soil is thus seen to be reduced sharply when the size of the footing is reduced. In this case, the soil loses almost half its bearing strength when the footing size is reduced from 8 ft to 3 ft.

EXAMPLE 3-10

What would be the reduction in the strength of a clay soil under square footing when the size of the footing is reduced from 8 ft to 3 ft?

SOLUTION For a clay soil, with $\phi = 0$, tan 45° = 1, from Eq. (3-17),

$$q_0 = cN_c = 6c$$

The bearing strength of the soil is seen to be unaffected by the size of the footing; it is a constant for all values of c.

EXAMPLE 3-11

A stratum of soil has an angle of internal friction of 16° and an unconfined compressive strength of 500 psf. The unit weight is 111 pcf. Determine the allowable bearing strength of this soil under a footing 7 ft 6 in. square, founded 2 ft 6 in. below grade, with a factor of safety of 2.5.

SOLUTION From Eq. (3-17), $45° + \tfrac{1}{2}\phi = 53°$, tan 53° = 1.33, $c = \tfrac{1}{2}q_u =$ 250 psf,

$$q_0 = \tfrac{1}{2}\gamma B N_\gamma + \gamma D_f N_q + cN_c$$

Now

$$N_\gamma = (1 + \tfrac{1}{2} \times 1.33)(1.33^5 - 1.33) = 4.7$$

$$N_q = (1 + \tfrac{1}{2} \times 1.33)(1.33^4 - 1) = 3.5$$

$$N_c = (1 + \tfrac{1}{2} \times 1.33)(2)(1.33^3 + 1.33) = 12.3$$

At failure,

$$q_0 = \tfrac{1}{2} \times 111 \times 7.5 \times 4.7 + 111 \times 2.5 \times 3.5 + 250 \times 12.3$$
$$= 6000 \text{ psf}$$

With a factor of safety of 2.5,

$$\text{allowable } p_a = \frac{q_0}{2.5} = 2400 \text{ psf}$$

EFFECTS OF FOOTING SHAPE ON ULTIMATE BEARING PRESSURE

When the footing is not long and continuous as assumed in the derivation of Eq. (3-17), corrections must be applied to the bearing capacity factors to compensate for end effects. Correction factors for various lengths and shapes are given in Table 3-1, where L is the length of the footing and B is its width. Interpolation can be used for values of L/B between 2 and 5. Some examples will illustrate the applications of Table 3-1.

TABLE 3-1 CORRECTION FACTORS FOR FOOTING SHAPES

Shape of foundation	Correction for q_0 for clays	Correction for q_0 for sands
Square, $L/B = 1$	1.20	0.85
Rectangular		
$L/B = 2$	1.10	0.90
$L/B = 5$	1.04	0.95
$L/B = \infty$	1.00	1.00
Round (diameter = B)	1.20	0.70

EXAMPLE 3–12

A grade beam 84 ft long and 18 in. wide is used to support a line of columns in a ML soil. What correction factor should be applied to the value of q_0 computed by Eq. (3-17)?

SOLUTION The value of L/B is $84/1.5 = 56$. From Table 3-1, it is seen that the correction factor for noncohesive soils with large values of L/B is 1, meaning that no correction is required.

EXAMPLE 3-13

A rectangular footing 3 ft by 6 ft is founded on an CL soil, having 53% of its grains passing the No. 200 sieve. What correction factor should be applied to the value of q_0 computed by Eq. (3-17)?

SOLUTION The value of L/B is $6/3 = 2.0$. From Table 3-1, it is seen that the correction factor for a cohesive soil with $L/B = 2$ is 1.1; hence the soil can take 10% more load than that indicated by Eq. (3-17). It should be noted, however, that if only a few less fines were present, the soil would be classified as a sand and the capacity would then be 10% less (rather than more) than that indicated by Eq. (3-17). For a soil mixture such as this, essentially halfway between a sand and a clay, an average of the two correction factors might be appropriate. The judgment of the designer is, of course, the final authority in such cases.

The ultimate bearing capacity given by Eq. (3-17) is that due only to considerations of the strength of the soil mass. Settlements or deformations of the foundation were never a consideration in the derivation, although limitations on such deformations must always be a constraint in the design of the building above. Subsequent discussions include these limitations due to settlements.

EFFECTS OF LATERAL LOADS ON ULTIMATE BEARING PRESSURE

Lateral loads on a building must eventually be transmitted to the soil by one or more of the footings. Such a load case introduces a shearing stress into the soil that was not considered in the failure analysis presented earlier. The overall effects of lateral loading can sometimes be among the more serious concerns in foundation design; a rather close look at these effects is warranted.

The state of stress in the soil under a footing is illustrated in Fig. 3-14, where a comparison is drawn between a footing carrying no lateral loads and a footing carrying a lateral load V. Since the soil in both cases will eventually fail in shear, it should be immediately apparent that the addition of the lateral load V could seriously aggravate the failure state of stress. The stressed particles in Fig. 3-14 are shown roughly at midheight of the failure circle or about at the point of average increase in stress.

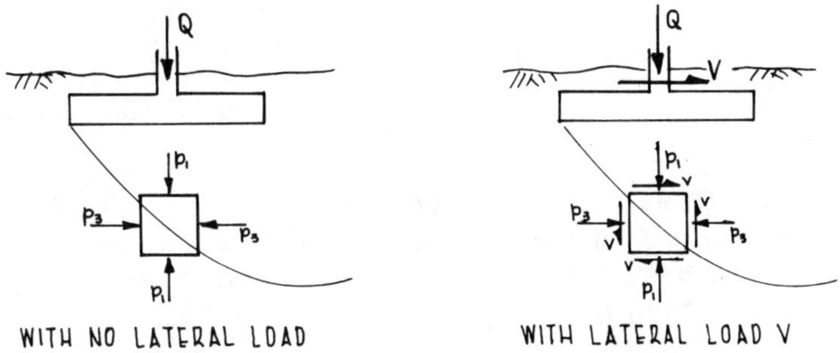

Figure 3-14 Addition of lateral force V to a footing.

The lateral force V is transmitted to the soil by friction. Consequently, the maximum lateral force V that can occur is the coefficient of friction times the vertical load Q,

$$\text{maximum } V = \mu Q \tag{3-18}$$

where μ is the coefficient of friction as sliding impends. For soils, the maximum value of μ is commonly taken at 0.3. The forces V and Q can be represented by an inclined force R as shown in Fig. 3-15, where the force R has a maximum angle of inclination of $\arctan(V/Q)$. Since the maximum value of V/Q is 0.3, the maximum angle of inclination is arctan 0.3 or 16.7°; an angle of inclination greater than this cannot occur.

The soil under the footing undergoes a sharp reduction in capacity due to the introduction of the lateral load. A great deal of study has been devoted to this problem, but there is not yet an accurate theoretical solution to predict the amount of loss. In practice, tables or graphs derived from experimental data are generally

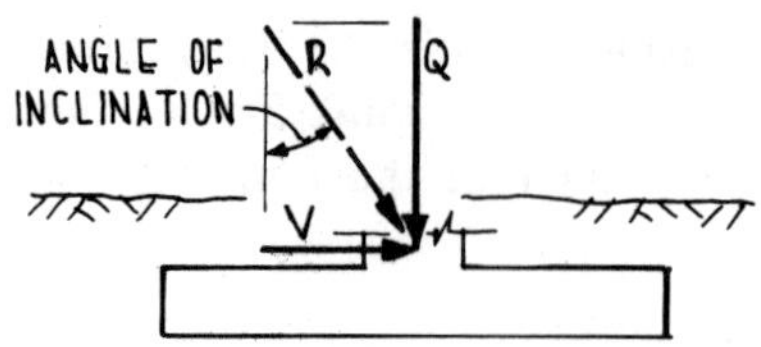

Figure 3-15 Inclined resultant force.

TABLE 3-2 PERCENT LOSS IN STRENGTH DUE TO A LATERAL LOAD V IN ADDITION TO A VERTICAL LOAD Q

			Sands angle of internal friction, ϕ			
V/Q	D_f/B	*Clays,* $\phi < 15°$	30°	35°	40°	45°
0.1	0	13	25	25	25	25
	0.5	11	20	21	22	23
	1.0	10	15	17	20	22
0.2	0	27	50	50	50	50
	0.5	23	40	42	45	47
	1.0	20	30	35	40	44
0.3	0	39	75	75	75	75
	0.5	34	60	63	67	70
	1.0	30	45	52	59	66

Source: Derived from the experimental work of Meyerhof [8].

used for design. Table 3-2 is one such table. The table gives the percent loss in bearing capacity due to the existence of a lateral load. The table has been extracted from graphs; its accuracy is somewhat limited.

When computing V/Q for use in Table 3-2, the "most probable" value of Q should be used; it represents the actual vertical load on the footing at the time the maximum wind load occurs or when the maximum earthquake load occurs. The load Q on a typical footing will not likely be at its absolute maximum value when these lateral loads occur, nor will it likely be at its absolute minimum. Means to estimate this "most probable" value are given in the discussions of Chapter 6.

It is noted from Table 3-2 that as much as 75% of the strength of the soil can be lost due to the presence of a lateral load. The consequences can be particularly severe for footings carrying shear walls, where the lateral load for an entire building can be concentrated onto only a few footings. This particular aspect of foundation design is also discussed in Chapter 6. An example will illustrate the use of Table 3-2.

EXAMPLE 3-14

A soil mixture classified as ML has 54% fines with an angle of internal friction of 19°. A footing 7 ft square is to be founded 2 ft below grade on this soil and is to be subjected to a lateral load of 23 kips concurrently with a vertical load of 90 kips. What would be the percent loss of strength in the soil under this load?

SOLUTION The value of V/Q is 23/90 = 0.26, D_f/B = 2/7 = 0.3. When Table 3-2 is entered with these values of V/Q and D_f/B, it is seen that the table does not fit these conditions very well. Technically, the soil is classified as noncohesive; hence the column for sands with $\phi = 30°$ would seem appropriate, even though ϕ is much too high. An interpolation for $D_f/B = 0.3$ in this column indicates that a reduction of 72% should be applied, which is a severe reduction. However, if a double interpolation were used with the column on clays, assuming $\phi = 0$ in that column, a reduction of 58% is indicated, still severe but probably more realistic. Whether the 72% reduction should be used or the 58% reduction should be used is a matter of judgment, to be resolved by the individual designer.

ALLOWABLE BEARING PRESSURE ON SANDS

The allowable bearing pressure on a sand may be limited by either of two considerations, strength or settlements. These two limitations are independent of each other; both must be evaluated before the allowable bearing pressure can be chosen. Although the numerically smaller value can usually be taken as the limiting value, there are occasions when the two approaches produce results that are significantly different. In such cases, the foundation may have to be designed for both cases and the more conservative of the two designs adopted as the final design.

The ultimate bearing capacity on a sandy soil is taken from Eq. (3-17) with cohesion c equal to zero:

$$q_0 = \tfrac{1}{2}\gamma B N_\gamma + \gamma D_f N_q \qquad (3\text{-}19)$$

It is well to note from Eq. (3-18) that the strength of a sand increases as the size of the footing increases (as B increases), and it also increases as the depth of burial increases (as D_f increases). These two observations will be recalled repeatedly in later discussions.

For design, the ultimate bearing capacity given by Eq. (3-18) is divided by an appropriate factor of safety (FS) to obtain a safe allowable bearing pressure p_a at working levels. Hence

$$p_a = \frac{q_0}{\text{FS}} = \frac{\gamma B}{\text{FS}}\left(\tfrac{1}{2}N_\gamma + \frac{D_f}{B}N_q\right) \qquad (3\text{-}20)$$

For sands, the factor of safety in Eq. (3-19) is normally taken as 2. A simple calculation using Eq. (3-20) will therefore yield the limiting value of bearing strength on a sand, but it should be noted that the only consideration so far has been strength; settlements have yet to be considered.

When settlements must be evaluated, the limiting value for allowable bearing pressure is not so easily computed. Settlements in sand are not only dependent on the angle of internal friction and elastic modulus, but are also strongly affected by grain size, grain shape, and grain hardness as well as the overall size of the footing and its depth of founding. There is presently no simple computation that will yield an allowable bearing pressure that corresponds to an acceptable settlement of, say, 1 in.

For design, the limiting value of soil pressure that will limit settlements to about 1 in. is largely an empirical value, derived from a combination of experience and judgment. A chart showing typical values of such soil pressures is given in Fig. 3-16, developed for a sand having a unit weight γ of 100 pcf. The chart is adapted largely from Peck et al. [4]; the approach has been used in the practice for a number of years with acceptable results. The horizontal lines in Fig. 3-16 represent the pressures at which the settlements are estimated to be in the order of 1

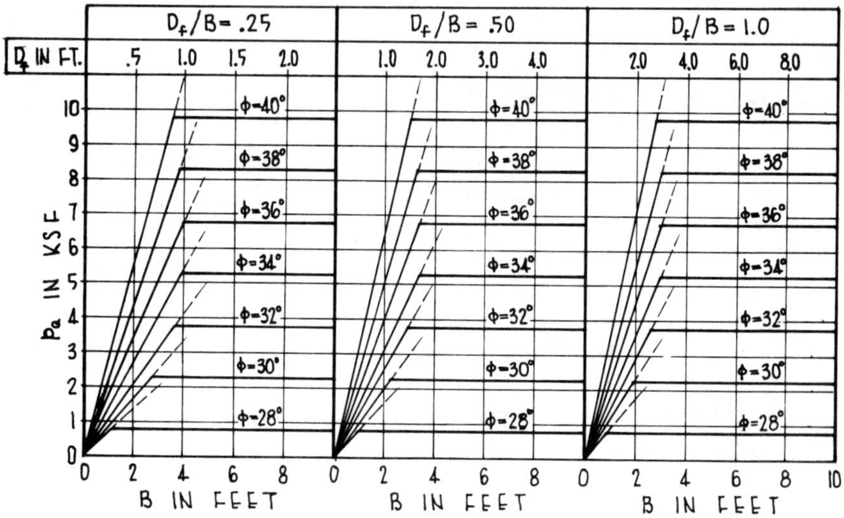

Figure 3-16 Pressure–settlement properties of sand. (Adapted from Ref. 4.)

Notes:

1. The value of B is the width of a square footing, the least width of a rectangular footing, or the diameter of a round footing; no correction factors for other footing shapes are necessary when settlements govern the design, but corrections for shape should be applied when strength governs.
2. Factor of safety to strength failure is taken as 2.
3. Unit weight of sand is taken as 100 pcf.
4. Horizontal portion of curve is estimated to be the soil pressure that will produce 1 in. of settlement in an unsubmerged sand.
5. The water table is presumed to be at least a distance B below the founding line.
6. Where the sand is subject to submergence, the allowable bearing values should be reduced to half the values given by the graphs.
7. The value of ϕ is assumed to be determined at a depth of about B below the founding line.
8. The effects of lateral loads are not included; the loss in strength due to lateral load is presented in later discussions.
9. Where settlements are not of concern, the sloping portion of the curve may be extended upward as indicated by the dashed lines.

in. The sloping portions are the pressures given by Eq. (3-20) for a factor of safety of 2. The two portions together provide a complete solution for the allowable bearing pressure in sands, to include both strength and settlement limitations.

It has been noted that when a sand is submerged it loses roughly half its strength. In addition, it should be noted that sand also loses roughly half its rigidity; hence settlements in a particular sand when it becomes submerged could be as much as twice those when it was dry. Consequently, the allowable bearing values of Fig. 3-16 should be cut in half when the sand is subject to submergence, regardless of whether the allowable bearing strength is limited by strength or by settlements.

In the development of Fig. 3-16, it was presumed that the water table was at least a distance B below the founding line. If the water table intrudes a significant distance into this zone, the soil should be considered to be submerged throughout the zone. It does not seem warranted in such an approximate solution to introduce refined calculations regarding percent of submergence.

The value of ϕ used in preparation of Fig. 3-16 was assumed to be determined at a distance 8 to 10 ft below the surface. If the actual tests were taken at a significantly different depth or if a significant amount of overburden is to be removed during site grading, an adjustment to the value of ϕ (or a new determination of ϕ) may be necessary.

Values of ϕ may be determined by any reasonable means. For convenience, a conversion table is given in Fig. 3-17, showing rough relationships between the angle of internal friction ϕ, the textural classification and the standard penetrometer count N.

As noted, the semiempirical graphs of Fig. 3-16 are based on a unit weight of sand of 100 psf. If the actual unit weight is significantly different from this, the allowable bearing pressure should be corrected accordingly. Since the allowable

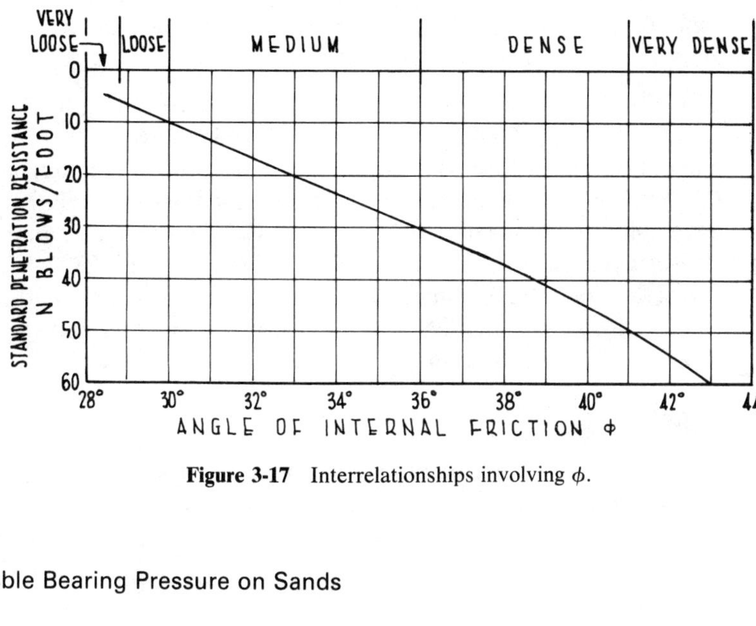

Figure 3-17 Interrelationships involving ϕ.

bearing pressure is a linear function of unit weight, the correction factor is quite simple:

$$\text{correction factor for weight} = \frac{\gamma_{\text{actual}}}{100} \tag{3-21}$$

The allowable bearing value should be multiplied by this factor.

EXAMPLE 3-15

A footing 8 ft wide is to be founded 2 ft deep in a sand having an SPT blow count of 16. The sand weighs 117 pcf and the building site is sometimes subject to spring flooding. Settlements should not exceed about 1 in. Determine the allowable bearing capacity of the sand under this footing.

SOLUTION For an SPT blow count of 16, ϕ is read from Fig. 3-17 as about 32°. For an angle of ϕ of 32°, B = 8 ft, and D_f/B = 0.25, read an allowable soil pressure of 3800 psf. The pressure is corrected to suit the actual weight of the soil of 117 pcf,

$$p_a = \frac{117}{100} \times 3800 = 4450 \text{ psf}$$

The pressure is again corrected for flood condition, where the soil loses half its strength,

$$p_a = \tfrac{1}{2} \times 4450 = 2225 \text{ psf}$$

Use a soil pressure of 2225 psf for this footing.

EXAMPLE 3-16

A very loose sand is used to support an isolated anemometer tower. Settlements are not a problem. The circular foundation is 5 ft across, buried 2.5 ft below grade. The sand weighs 112 pcf and the area is subject to flooding. Lateral loads can be as much as 10% of the vertical loads. Determine the allowable soil pressure.

SOLUTION For a very loose sand, ϕ is read from Fig. 3-17 as 28°. For ϕ = 28°, B = 5 ft, D_f/B = 0.5, p_a is read from Fig. 3-16 as 4000 psf. (Since settlement is of no concern, the sloping portion of the curve is projected on out to B = 5 ft.) For a circular shape, the correction factors of Fig. 3-1 are applicable, hence $p_a = 0.7 \times 4000 = 2800$ psf. For 10% lateral loads, a loss in strength of 20% is taken from Table 3-2; hence $p_a = 2800 - 0.20 \times 2800 = 2240$ psf. Apply correction for submergence, $p_a = \frac{1}{2} \times 2240 = 1120$. Apply correction for actual soil weight, $p_a = (112/100) \times 1120 = 1250$ psf. Use an allowable soil pressure of 1250 psf for this foundation.

EXAMPLE 3-17

What will be the approximate settlement of a footing 8 ft square located 4 ft below grade on a sand having an angle of internal friction of 31° and a unit weight of 121 pcf if the footing is subjected to a load of 100 kips? 200 kips? 300 kips? Effects of lateral loads and flooding are of no interest in this particular investigation.

SOLUTION For $\phi = 31°$, the pressure at roughly 1-in. settlement is found from Fig. 3-16 to be 3 ksf. The load that will produce 3 ksf is then $3 \times 8^2 = 192$ kips. Estimated settlement at 100 kips is found by ratio:

$$S_{100} = \frac{100}{192} \times 1 \text{ in.} = 0.52 \text{ in.}$$

Estimated settlement at 200 kips is found by ratio:

$$S_{200} = \frac{200}{192} \times 1 \text{ in.} = 1.04 \text{ in.}$$

Estimated settlement at 300 kips is found by ratio:

$$S_{300} = \frac{300}{192} \times 1 \text{ in.} = 1.56 \text{ in.}$$

EXAMPLE 3-18

For Example 3-15, what will be the allowable soil pressure if it is decided to change the factor of safety to 3?

SOLUTION From Example 3-15, $p_a = 2225$ for FS = 2. "Ultimate" $q_0 = \text{FS} \times p_a = 2 \times 2225 = 4450$ psf. At FS = 3, $p_a = q_0/\text{FS} = 4450/3 = 1480$ psf. At a factor of safety of 3, the allowable bearing pressure will be 1480 psf.

ALLOWABLE BEARING PRESSURE ON CLAYS

The allowable bearing pressure on a clay soil is derived in the same way as that for sandy soils. The ultimate bearing capacity for clays is again determined from Eq. (3-17), where, with ϕ approaching zero, the equation becomes

$$q_0 = 6c \tag{3-22}$$

Equation (3-22) indicates that for clays, the existence of a small angle of internal friction has no effect on strength. Similarly, the weight of the soil, γ, has no effect, and as a result the strength of clay is not substantially affected by submergence. Further, the term containing D_f has dropped out, indicating that the strength of a clay soil does not improve with depth. Similarly, the terms containing the footing size B have also dropped out, indicating that the strength of clay soils is unaffected by footing size. It must be remembered, however, that these conclusions are based solely on theoretical considerations for a heavy clay having little or no internal friction.

The correction factors for footing shape given in Table 3-1 remain valid for clays, however, yielding for a saturated clay

$$q_0 = 6c \qquad \text{(infinitely long)} \qquad (3\text{-}23a)$$

$$q_0 = 1.04 \times 6c = 6.24c \qquad (L/B = 5) \qquad (3\text{-}23b)$$

$$q_0 = 1.10 \times 6c = 6.60c \qquad (L/B = 2) \qquad (3\text{-}23c)$$

$$q_0 = 1.20 \times 6c = 7.20c \qquad \text{(square or round)} \qquad (3\text{-}23d)$$

Where settlements govern the design, the factor of safety to bearing failure for clays is normally taken as 3; at this factor of safety, settlements in clay can be expected to be less than about 1 in. If settlements are not of major concern, a factor of safety of 2 is generally considered to be adequate. A plot of Eq. (3-22) is shown in Fig. 3-18 for these two factors of safety, along with a summary of limitations concerning the use of the graph.

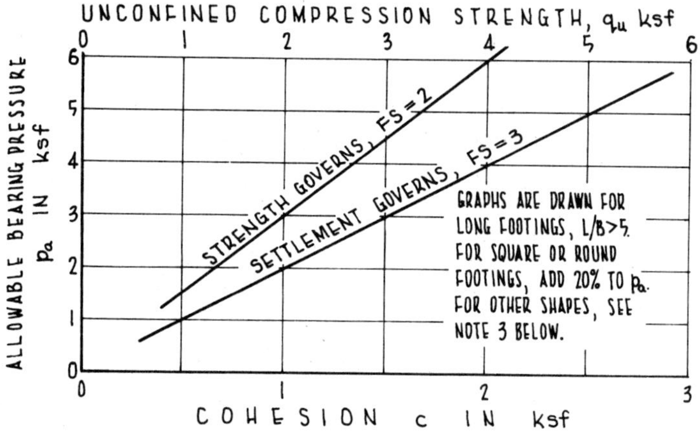

Figure 3-18 Bearing strength of clays.

Notes:

1. At a factor of safety of 3 in clays, settlements should not exceed 1 in.
2. Where a footing is subject to lateral loading, additional corrections may be necessary; see the discussions later in this chapter.
3. For $L/B = 5$, add 4% to P_a, increasing to 10% for $L/B = 2$, thereafter increasing to 20% for $L/B = 1$ (square or round footings).

Values of cohesion c for use in obtaining the allowable bearing pressure from Eq. (3-22) or from Fig. 3-18 may come from any acceptable source. A graph indicating the approximate relationship between cohesion, standard penetrometer values, consistency classification, and unconfined compression test values is given in Fig. 3-19.

It is common practice to take the allowable bearing pressure on clay soils to be twice the cohesive strength; this approximation is readily justified just by an examination of Fig. 3-18. Alternatively, the allowable bearing pressure on a clay soil can be taken to be equal to its unconfined compressive strength; this relationship is readily derived from Eqs. (2-25), (2-26), and (3-22). Such approximations are well within the usual range of accuracy in soils.

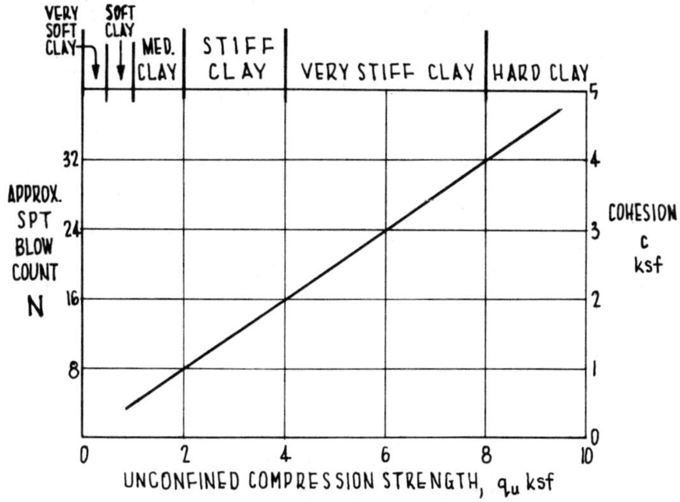

Figure 3-19 Interrelationships involving cohesion.

EXAMPLE 3-19

A footing 8 ft square is to be founded 2 ft below surface on a soft clay soil. Settlements should not exceed 1 in. Determine the allowable bearing pressure for this footing.

SOLUTION For soft clays, the usual range of cohesion c is seen from Fig. 3-19 to be 0.25 to 0.50. From Eq. (3-22),

$$q_0 = 6 \times (0.25 \text{ to } 0.50) = 1.5 \text{ to } 3 \text{ ksf}$$

Use a factor of safety of 3 to limit settlements to 1 in.,

$$p_a = \frac{q_0}{3} = 0.5 \text{ to } 1 \text{ ksf}$$

For square footings in clay, an increase in allowable pressure of 20% is allowed,

$$p_a = 0.6 \text{ to } 1.2 \text{ ksf}$$

Use an average pressure p_a of 0.9 ksf. Note that the size of the footing and the depth of burial have no effect on the bearing strength of the clay.

EXAMPLE 3-20

A medium clay is to be used to support an isolated water tank. Settlements are not a problem. The circular foundation is 10 ft in diameter, founded 16 in. below surface. The clay weighs 107 pcf. The area is not subject to flooding, but lateral loads can be as much as 12% of the vertical loads. Determine the allowable soil pressure.

SOLUTION From Fig. 3-18, for a factor of safety of 2, select an allowable soil pressure of 2.25 ksf, corresponding to a value of c between 0.5 and 1.0 (Fig. 3-19). Add 20% for a round footing,

$$p_a = 2.25 \times 1.2 = 2.70$$

From Table 3-2, deduct a loss in strength of 16% due to lateral loads,

$$p_a = 2.70 - 0.16 \times 2.70 = 2.27 \text{ ksf}$$

Use an allowable soil pressure of 2.27 ksf.

PRESUMPTIVE BEARING PRESSURES

Long before the advent of soil mechanics, foundations were designed successfully in many areas using soil pressures and foundation types proven by years of experience to be satisfactory within that particular locale; the theoretical analyses given in the preceding sections were neither known nor needed. The fact that soil mechanics now exists does not invalidate that earlier experience. In many of the older urban areas, the traditional types of foundations and allowable soil pressures for that area are quite valid today and their continued use is certainly justified.

In other areas, however, such experience (and records) simply do not exist. Bustling urban developments exist today where there was only a prairie or a swamp just a few years ago. In such a locale, there is no experience to draw on when choosing a foundation type or an allowable bearing pressure.

Often, however, there is a strong tendency to project the traditional proven methods into a new locale, bypassing the more staid practices of conventional foundation engineering. One such practice is the use of presumptive bearing pressures, or allowable soil pressures found to be acceptable in one geographic area and presumed to be acceptable for all other areas. A table of such presumptive pressures is presented in Table 3-3. It should be noted that the values given in

TABLE 3-3 PRESUMPTIVE BEARING PRESSURES

Material	Allowable bearing (ksf)
Sand	
Loose	1 to 3
Medium	3 to 6
Dense	6 to 12
Clay	
Soft	0.5 to 1.5
Firm	1.5 to 2.5
Stiff	2.5 to 6.0
Rock	
Fractured	10 to 20
Seamed	20 to 50
Massive	50 to 200

Table 3-3 are subject to the correction factors for footing shape given in Table 3-1. They are also subject to loss of strength due to inclined loads, as prescribed in Table 3-2.

Presumptive values are often prescribed by the building code for a city. One such building code used throughout much of the western United States is the Uniform Building Code [9], which contains presumptive values not only for bearing pressure but also for lateral pressures. Unfortunately, the code is often adopted by a city without being amended to suit local conditions; the same foundation code cannot possibly apply to the expansive clays of Wyoming in the same way it applies to the low-plasticity clays of California. In addition, it is never quite clear whether presumptive values are governed by strength or by settlement. The designer must therefore make a judgment in the matter. It will be seen in Chapter 6 that such a judgment could be expensive or dangerous or both.

Even though the use of presumptive soil pressures is frowned upon by soils engineers, they are in fact widely used, particularly for preliminary design. In most cases, presumptive pressures are overly conservative (but not always) and produce a foundation having a significantly higher cost than necessary. On small projects, however, the additional cost is usually far less than what it would cost to perform a soils investigation.

The designer who uses presumptive values in a design remains professionally responsible and personally liable for the adequacy of the design.

DISTRIBUTION OF CONTACT PRESSURES

It has been implied in earlier discussions that the contact pressure between a footing and the supporting soil is uniformly distributed. For short-term loading, this implication is not at all accurate. For long-term loading, it may come somewhat closer to being true.

The contact pressure under an absolutely rigid footing is shown graphically in Fig. 3-20. For sands, the pressure distribution is that shown in Fig. 3-20a and for clays, in Fig. 3-20b. As might be expected, the response of the two soils to this load condition is exactly opposite. It is, however, a localized effect, with the average pressure being the same for both cases.

The confining pressure at the edges of a rigid footing on sand is seen to diminish toward the edges of the footing, allowing the sand at the edges of the footing to move laterally, relieving somewhat the pressure at the edge of the footing. The end result is the pressure distribution shown in Fig. 3-20a, where the sand trapped near the center of the footing eventually must take more of the load. Placing the footing at a lower depth, thereby increasing D_f and the confining pressure, will help to produce a more uniform distribution.

No footing is entirely rigid, however. As the spread footing of Fig. 3-20a is loaded, the edges deflect upward, relieving further the pressure at the edge of the footing and thus aggravating the nonuniform pressure distribution. The effects of structural deflection of the footing are not relieved by placing the footing at a lower elevation; they occur whatever the founding depth D_f may be.

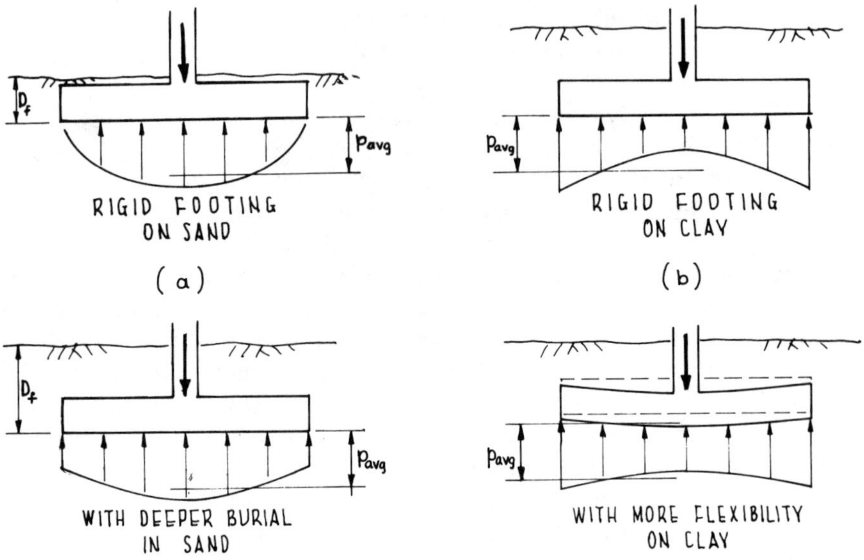

Figure 3-20 Contact pressure under a rigid footing.

The nonuniform distribution is relieved somewhat with time, as the sand at the higher pressures undergoes more degradation, more creep, and more settlement than the sand at the lower pressures. As the settlement increases at the center, the edges then pick up more load, thereby equalizing the pressures somewhat. It is doubtful, however, whether the pressure distribution ever approaches a uniform distribution.

In sands, this increase in vertical pressure at the center of a footing is not an insignificant amount. It can be as much as 40% above the average pressure [10] (at worst) and can increase the lateral earth pressures under the center of the footing by some 20%. Superficially, these increases would seem to justify including these effects in the design calculations.

It must be remembered, however, that the strength of sands is dependent on the confining pressure. As the vertical pressure increases, the confining pressure increases a corresponding amount. As a result, the strength of the sand toward the center of the footing actually becomes somewhat stronger than that at the edge of the footing.

When the foregoing effects of the increased degradation, increased creep and increased strength are considered together, the overall effect of nonuniform contact pressures in sands becomes much less serious. It is, in fact, rarely considered for ordinary sizes of footings in small buildings. The effects can become notable, however, for larger footings (B greater than about 10 ft) and for mat and raft foundations. Advanced textbooks in foundations treat the subject in detail [11].

In clay soils, the pressure variation is opposite to that of sands. In clays, the unequal deformations between the loaded clay particles and the adjacent unloaded clay particles along the edge of the rigid footing serve to increase sharply the pressures at the edge of the footing. The unloaded clay particles go into tension, pulling upward on the adjacent loaded particles, creating the pressure diagram shown in Fig. 3-20b. Placing the founding line at a higher or lower elevation will not relieve the condition; contact pressures in clay are relatively unaffected by confining pressures.

In contrast to sands, however, the nonuniform soil pressures in clays are relieved considerably by the upward deflections that must occur at the edges of the footing. Additionally, with the higher pressures at the edges of the footing, the clay must undergo a higher degree of consolidation and creep at the edges of the footing, thereby relieving the pressure concentrations even further. It seems likely that with time, the pressure distribution in a clay soil will become much closer to a uniform distribution.

As with sands, the effects of such nonuniform distribution of contact pressures for ordinary footings in clay soils is rarely a design consideration. It is not considered at all in this book. The problem does become more serious in larger foundations, however, to the extent that it becomes a routine design consideration in large-scale mat and raft foundations [11].

CONSOLIDATION SETTLEMENTS IN CLAYS

Consolidation has been shown to be the mechanism that produces settlement in clays. It has also been noted that properties of clay soils are distinctly variable and that the engineering properties of one clay may be significantly different from the engineering properties of another clay. Nonetheless, the response of a clay soil to an applied load does follow predictable patterns, and even though the degree of response may vary somewhat, it is readily possible to establish overall patterns of deformation and settlement. Such general patterns of deformation and settlement in clays are the topic of this section.

All deposits of clay begin their geologic lives as normally consolidated clays. Normally consolidated clays are those that have never been subject to more pressure than they now experience. It should not be inferred, however, that pressures in a normally consolidated clay are low; the pressures at the bottom of a deep deposit of clay can be quite large, and might be increased even more over its geologic life when hundreds of feet of other soils are deposited on top of it. The one distinctive characteristic about a normally consolidated clay is that during its geologic life, pressures have never been higher than they are now.

In any clay, as pressure p increases, the void ratio e decreases. A typical curve showing the relationship between e and p for a normally consolidated clay is shown in Fig. 3-21. Note that the scale for p is shown as a logarithmic scale.

Settlements even in a thick stratum of a normally consolidated clay are easily computed since the slope of the e–log p curve in Fig. 3-21 is seen to be reasonably linear (straight). Equation (2-19) applies:

$$S = H\frac{\Delta e}{1 + e_0} \tag{2-19}$$

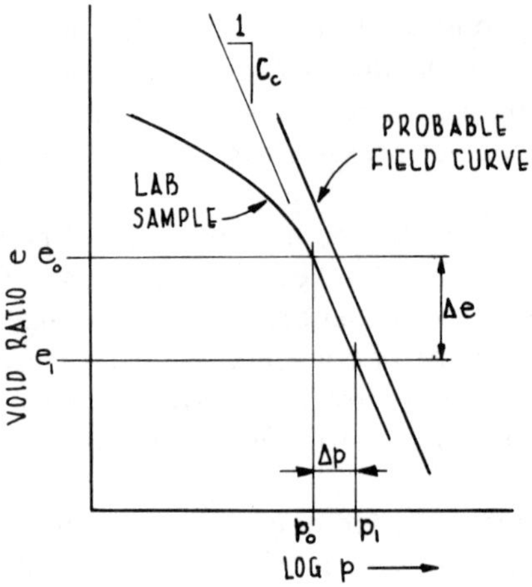

Figure 3-21 Normally consolidated clay.

where H is the thickness of the stratum in which the pressure has been increased and S is the settlement incurred in increasing the average pressure in the stratum from p_0 to p_1. The slope of the linear or straight portion of the curve, C_c, is readily expressed in terms of e and p,

$$C_c = \frac{e_0 - e_1}{\log_{10}p_1 - \log_{10}p_0} = \frac{\Delta e}{\log_{10}(p_1/p_0)} \tag{3-24}$$

C_c is called the *coefficient of consolidation* of the particular clay soil. Equations (2-19) and (3-24) are combined to eliminate Δe, yielding

$$S = \frac{C_c}{1 + e_0} H \log_{10} \frac{p_0 + \Delta p}{p_0} \tag{3-25}$$

Equation (3-25) yields directly the settlement in a stratum of clay of thickness H, having an average pressure p_0 and an average void ratio e_0, when an increase in pressure Δp is applied. The problem, of course, is in finding C_c, which would seem to entail running pressure–settlement tests to establish e_1. Such tests are not necessary, however. A reasonably accurate approximation of C_c has been developed by Skempton [12],

$$C_c = 0.009(w_L - 10) \tag{3-26}$$

where w_L is the liquid limit of the clay. An accurate approximation of settlements in a normally consolidated clay is therefore possible using Eqs. (3-25) and (3-26), in which only the liquid limit and the in-place void ratio need be known.

An example will illustrate the use of Eqs. (3-25) and (3-26). It should be noted that the pressure p_0 in Eq. (3-25) is taken to be the average pressure in the stratum, to include the effects of pressure dissipation due to the Boussinesq pressure distribution. Within the usual range of accuracy in soils, the average increase in pressure in this case is taken at midheight of the stratum.

EXAMPLE 3-21

A surface stratum of medium sand averaging 8 ft thick is underlain by a stratum of normally consolidated clay 22 ft thick as indicated. Unit weight of the sand is 127 pcf and of the clay is 112 pcf. The void ratio of the clay at midheight of the stratum is 0.81 and its liquid limit is 61. A building is to be located on the sand, having an average contact pressure of 600 psf, which diminishes to 120 psf at middepth of the clay stratum. Determine that part of the settlement due to consolidation of the clay.

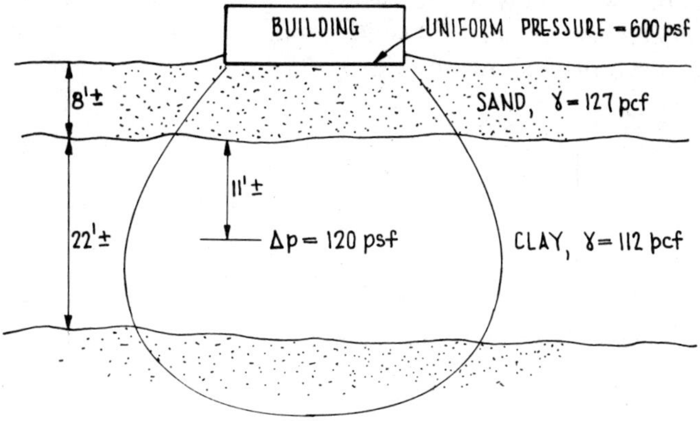

SOLUTION The settlement in a normally consolidated clay can be computed from Eq. (3-25) using Skempton's approximation for the coefficient of consolidation C_c.

$$S = \frac{C_c}{1 + e_0} H \log_{10} \frac{p_0 + \Delta p}{p_0}$$

Here

$$C_c = 0.0009(w_L - 10) = 0.459$$

$$e_0 = 0.81$$

$$H = 22 \times 12 = 264 \text{ in.}$$

$$p_0 = \gamma H_{\text{sand}} + \gamma H_{\text{clay}} = 127 \times 8 + 112 \times 11 = 2250 \text{ psf}$$

$$\Delta p = 120 \text{ psf}$$

So

$$S = \frac{0.459}{1 + 0.81} 264 \log_{10} \frac{2250 + 120}{2250} = 1.51 \text{ in.}$$

It should be noted that this deposit of clay is located at a significant depth below the building, where the pressure has diminished to only about 20% of the contact pressure. Even so, the settlement due to the clay stratum is quite high;

when the settlement due to the sand stratum is added, the overall settlements will probably be excessive. Such large settlements, even at low pressures, are common in normally consolidated clays; settlement can be expected to be a primary concern wherever these clays are encountered.

Refer again to the graph of the normally consolidated clay shown in Fig. 3-21. If the pressure in the deposit were to be decreased by removal of several hundred feet of overburden, the clay would then be relieved of much of its confining pressure and would then be classified as overconsolidated. The clay at a few feet below the new surface would rebound, but the rebound would not follow the original e–log p curve; it would rebound only partially, following a curve similar to the curve *BCD* shown in Fig. 3-22.

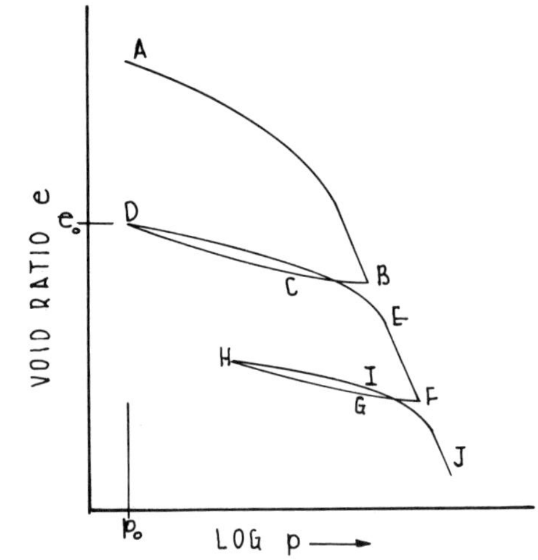

Figure 3-22 Rebound curves.

In its unloaded state, the overconsolidated clay would have a void ratio e_0 at its surface where pressure p_0 is zero. When the clay is reloaded, as by a footing load, the clay will recompress with the increase in pressure, following the curve *DEF*. If the footing load undergoes long-term fluctuations, the rebound–reload curves will follow an intermediate curve, such as *FGH* and *HIJ*, respectively.

To predict settlements in an overconsolidated clay, it is necessary for one to know the rebound–reload curve for the clay in the expected range of pressures. The usual procedure is to take a sample of the clay into the lab and subject it to pressures in the range of the anticipated footing pressures. Its deformation under lab conditions is recorded, providing a measure of the deformations to be expected in the field. Such load–deflection tests are called *consolidation tests* and are a very common part of a soils investigation in clay soils.

A typical soil sample in a consolidation test is about 3 in. in diameter and $\frac{3}{4}$ in. thick; a thicker sample would require much more time to consolidate. Even at $\frac{3}{4}$ in., the consolidation may be so slow that several days might be required to run the test. The extrapolation of results from a $\frac{3}{4}$-in. lab sample into an actual stratum 20 to 50 ft thick requires that the lab test be run with extreme care and precision.

To evaluate accurately the results of the consolidation test, it will be necessary to know the in-place void ratio corresponding to a particular pressure. Measurement of in-place densities is usually done at the time the samples are taken for the consolidation test; the in-place water content is determined in the lab and the in-place void ratios can then be computed.

In the consolidation test, the sample is fitted snugly into a circular confining ring and is loaded by a uniform pressure as indicated schematically in Fig. 3-23. The apparatus allows free flow of water out of the sample at top and bottom. The amount of deformation the sample undergoes when it is then loaded is observed and recorded at various points in time. When deformations cease at one level of pressure, another increment of load is added and the deformation is again observed and recorded under this increased pressure. The process is repeated for several increments of load until the anticipated range of pressures has been applied.

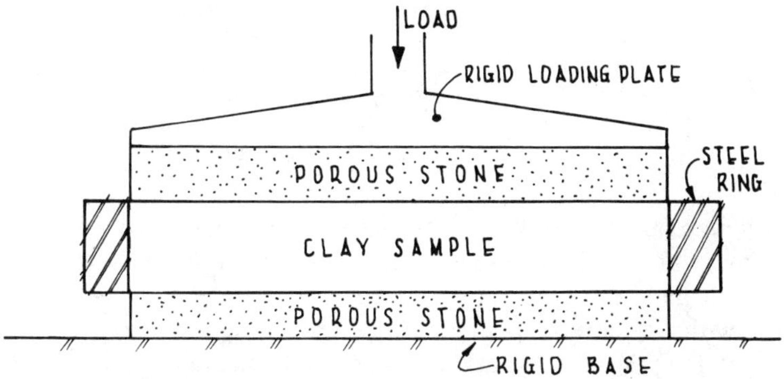

Figure 3-23 Consolidation test sample.

The results at each increment of load are plotted, providing a graph similar to that of Fig. 3-24a. Void ratios corresponding to the various pressures are readily computed once the in-place void ratio and pressure are known. The set of curves shown in Fig. 3-24 are commonly called the *e*–log time curves; a set of such curves is necessary, one curve for each increment of load.

There are two types of consolidation reflected in the *e*–log time curves of Fig. 3-24a. The first type is the primary consolidation, described by that portion of the curve that cups downward, *ABC'*. The second type is the secondary consolidation, described by the remainder of the curve, *C'DE*.

A graphical method is used to project the theoretical point *C*, where the two curves meet. A line is drawn through the inflection point *C'* tangent to both curves. Another line is drawn projecting the secondary consolidation curve as a straight line. The intersection of these two lines occurs at point *C*, which is taken to be the point where primary consolidation is 100% complete and secondary consolidation begins.

Secondary consolidation is of interest only rarely. The time involved can be upward of 50 to 100 years and is out of the usual range of concern in today's

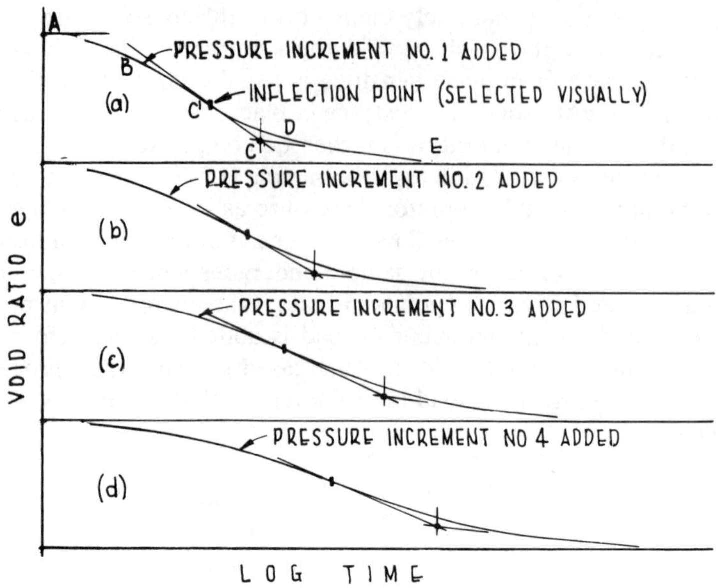

Figure 3-24 *e*–log time curves.

commercial construction. In general, references to consolidation will mean only primary consolidation, that part shown by curve *ABC* in Fig. 3-24a.

It must be remembered that Fig. 3-24a is drawn using a beginning pressure p_0, with an increment of pressure producing an increase to $p_0 + \Delta p$. At that beginning pressure and at that Δp, the clay reaches 100% consolidation when void ratio and pressure reach point *C*. At this point, essentially all the water has been forced out of the clay that is going to be forced out at this increase in pressure.

The initial portion of the primary consolidation curve (up to about 60 to 75% consolidation) is considered to be parabolic. The void ratio at the initial point *A* corresponding to 0% consolidation can therefore be projected for such a parabolic curve as indicated in Fig. 3-25. To project the curve to point *A*, find values of *e*

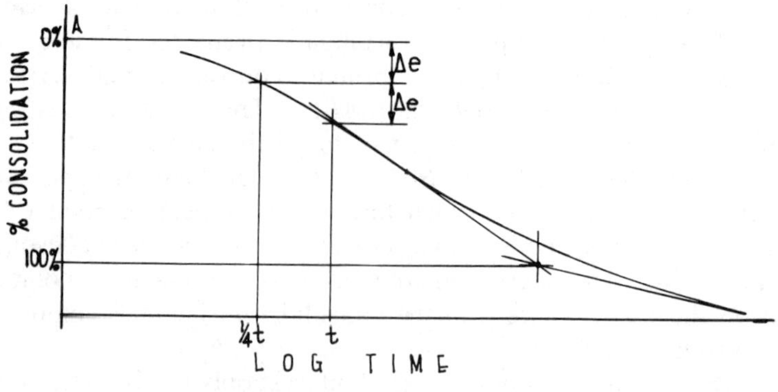

Figure 3-25 Point of zero consolidation.

and t for some arbitrary point in the first half of the consolidation curve, find the value of e corresponding to $\frac{1}{4}t$ as indicated, then lay off the distance Δe as shown. The intersection of the horizontal line with the ordinate defines the point A. With the points at 0% consolidation now known, the vertical scale of the graph is then laid off linearly on the vertical ordinate, providing the $U\%$ scale.

Once the set of e–log time curves are complete, the final consolidation curve can be drawn. The values of e and p corresponding to 100% consolidation are taken from the various e–log time curves and plotted as a separate graph. The result is the consolidation curve shown in Fig. 3-26.

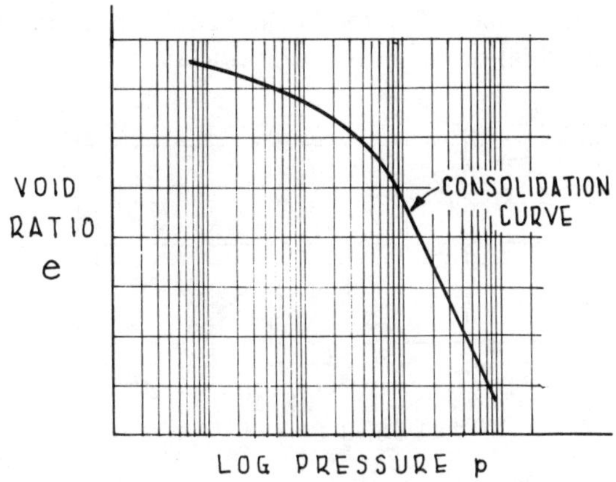

Figure 3-26 Consolidation curve.

The consolidation curve is commonly called the e–log p curve. It is roughly equivalent to the stress–strain curve used in studying structural steel or timber. It constitutes the load–deflection curve for a particular clay; each point of the curve defines the void ratio e at 100% consolidation under the total vertical pressure p. It has the distinct shortcoming, however, that the time required to reach 25%, 50%, or even 100% consolidation for various pressure increments is not usually indicated.

A consolidation curve run in the laboratory, such as that of Fig. 3-26, will be somewhat at variance with the actual field consolidation curve, since the lab sample starts recompression from zero ambient pressure and the in-place soil starts recompression from the in-place pressures. There are methods to correct the lab curve to better match the in-place field curve [11]. For the upper levels of a stratum where shallow foundations occur, the differences between the two curves are so small that they are ignored in the settlement calculations presented in this book.

Settlement in a clay soil is readily computed once the consolidation curve is available. Equation (2-19) applies,

$$S = H\frac{e_0 - e_1}{1 + e_0} \tag{2-19}$$

where e_0 is the void ratio at the beginning pressure p_0 and e_1 is the void ratio at the final pressure p_1. This settlement is of course the settlement at 100% consolidation; it may require several years for all of this settlement to occur. It was noted in the derivation of Eq. (2-19) that the equation must be applied to thin layers of the deposit of clay in order to minimize the variation in the average initial pressure p_0. An example of such a computation follows, where the stratum of clay is divided into 5 layers for the sake of the computations.

EXAMPLE 3-22

A footing is to be located 3 ft below surface level on a stratum of overconsolidated soft clay 18 ft thick, having a unit weight of 116 pcf. The consolidation curve for the clay is shown. The increase in contact pressure at the founding line is 3400 psf, diminishing to 400 psf at the bottom of the stratum. Estimate the settlement due to consolidation in the stratum of clay.

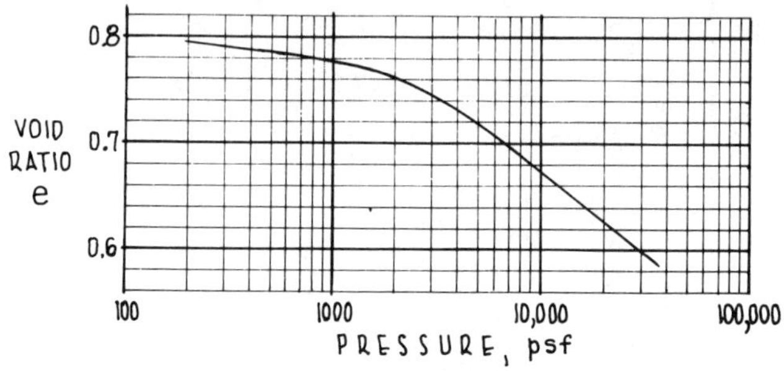

SOLUTION Equation (2-19) will be applied, using five layers of clay, each 3 ft thick. The pressure variation will be assumed to vary linearly from founding line to bottom of stratum, although a more accurate distribution might be made using the Boussinesq variation of Fig. 3-2. The increase in pressure at each level is shown in the following tabulation together with total pressure and corresponding void ratios.

Depth (ft)	Initial value (psf)		Average value					
	p_0	Δp	p_0	e_0	p_1	e_1	Δe	S (in.)
3	350	3400						
			525	0.785	3625	0.735	0.050	1.04
6	700	2800						
			870	0.780	3370	0.737	0.043	0.89
9	1040	2200						
			1215	0.775	3115	0.740	0.035	0.72
12	1390	1600						
			1565	0.765	2865	0.745	0.020	0.41
15	1740	1000						
			1914	0.760	2614	0.750	0.010	0.10
18	2088	400					Total S =	3.26

It is pointed out that unlike sands, the largest increments of settlement in clays occur adjacent to the founding line. In sands, the largest increments of settlement occur at some distance below the founding line. This comparison is discussed further when settlements in sands are computed.

The scale and accuracy of the consolidation curve is not always large enough to permit reading the void ratios to 0.001, but readings to 0.005 are usually reasonably possible, as was generally done in this example. Since the result is only an estimate at best, elaborate means to improve accuracy of readings do not seem justified.

Time was not a factor in Example 3-22. It was simply assumed that the consolidation was 100% complete. The settlements so computed were final settlements and were the maximum values that could occur under the given pressures.

It can often happen that settlements must be found for some intermediate time interval, or at a point in time long before 100% consolidation is complete. In such cases time does become a factor and the e–log time curves discussed previously must be resurrected. By itself, the consolidation curve of Fig. 3-26 can offer no help in judging the time required to achieve a specified level of consolidation.

The time required for water to be forced out of a layer of clay is proportional to the square of the distance it must travel,

$$t = K_c H^2 \tag{3-27}$$

where t = time required to achieve a given degree of consolidation U_i at a given pressure p_i

H = half-height of the layer, or the farthest distance water must travel to escape the clay layer

K_c = constant of proportionality, and is a constant for a particular clay

A typical sample is shown in Fig. 3-27, with typical field strata shown beside it. The farthest distance the water must travel to a free surface is the distance H, as shown. It is this value of H that is to be used in Eq. (3-27). Equation (3-27) affords a means to compare results between the lab sample and the field stratum,

$$\frac{t_{\text{lab}}}{t_{\text{field}}} = \frac{H^2_{\text{lab}}}{H^2_{\text{field}}} \tag{3-28}$$

where t_{lab} = time required to achieve a given degree of consolidation U_i at a given pressure p_i in the lab sample

t_{field} = time required to achieve the same degree of consolidation U_i at the same average pressure p_i in the field stratum

H_{lab} = farthest distance water must travel to a free surface in the lab sample

H_{field} = farthest distance water must travel to a free surface in the field stratum

Some examples will illustrate the use of Eq. (3-28).

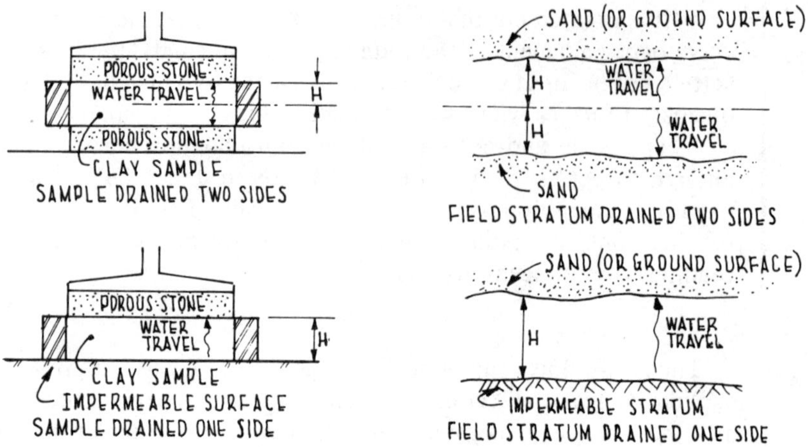

Figure 3-27 Travel distances for consolidation.

EXAMPLE 3-23

A stratum of clay 31 ft thick is underlain and overlain by deposits of sand. A sample of the clay 0.75 in. thick is tested in a consolidation test, drained both sides. In going from 3000 psf to 4000 psf, the lab sample reached 50% consolidation in 73 sec and 100% consolidation in 6.1 min. How long will it take the field stratum to reach 50% and 100% consolidation between these same pressures?

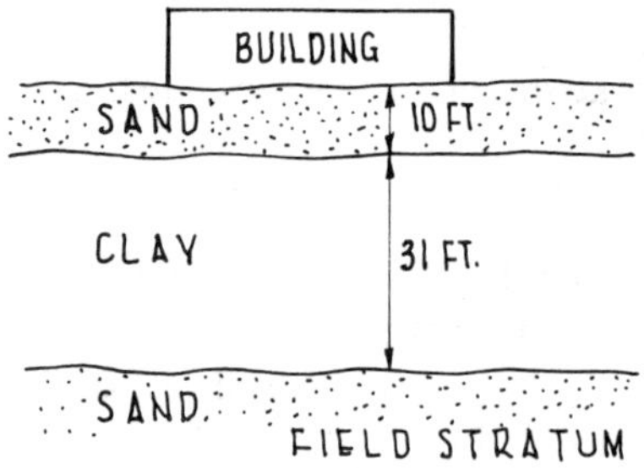

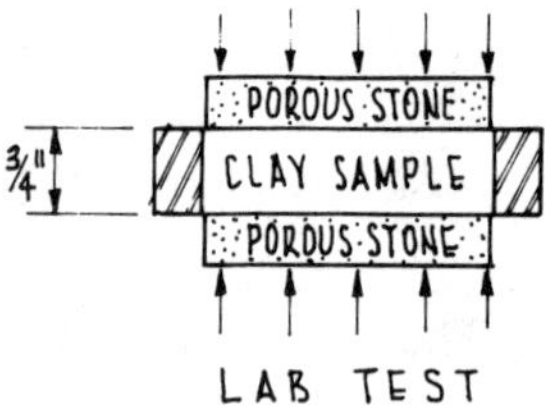

SOLUTION The solution is given by parabolic ratio:

$$\frac{t_{\text{lab}}}{t_{\text{field}}} = \frac{H^2_{\text{lab}}}{H^2_{\text{field}}}$$

At 50% consolidation,

$$\frac{73 \text{ sec}}{t_{\text{field}}} = \frac{(0.75/2)^2}{(31 \times 12/2)^2}$$

$$t_{\text{field}} = \frac{138{,}000}{0.5625} 73 = 17.9 \times 10^6 \text{ sec}$$

$$= 207 \text{ days or roughly 7 months to reach 50\% consolidation}$$

At 100% consolidation,

$$\frac{6.1 \text{ min}}{t_{\text{field}}} = \frac{(0.75/2)^2}{(31 \times 12/2)^2}$$

$$t_{\text{field}} = \frac{138{,}000}{0.5625} 6.1 = 1.5 \times 10^6 \text{ min}$$

$$= 2.85 \text{ years to reach } 100\% \text{ consolidation}$$

EXAMPLE 3-24

If the clay of Example 3-23 had been underlain by a stratum of impermeable shale, what would be the times required to reach 50% and 100% consolidation?

SOLUTION The only change in the solution is in the distance the water must travel in the field stratum. Since it must now travel in only one direction, H_{field} becomes 31 ft rather than 15.5 ft. At 50% consolidation,

$$\frac{73 \text{ sec}}{t_{\text{field}}} = \frac{(0.75/2)^2}{(31 \times 12)^2}$$

$$t_{\text{field}} = \begin{matrix} 27.6 \text{ months if drained one side compared to} \\ 7 \text{ months if drained two sides} \end{matrix}$$

At 100% consolidation,

$$\frac{61 \text{ min}}{t_{\text{field}}} = \frac{(0.75/2)^2}{(31 \times 12)^2}$$

$$t_{\text{field}} = \begin{matrix} 11.4 \text{ years if drained one side compared to} \\ 2.85 \text{ years if drained two sides} \end{matrix}$$

Obviously, improvements in field drainage may be well worth considering when one is trying to reduce consolidation times in clays.

EXAMPLE 3-25

A stratum of clay at the surface of a building site is 26 ft thick, underlain by sand, and has a unit weight of 112 pcf; in-place pressure at midheight of the clay stratum is 1460 psf. It is proposed to add fill in the general area of the building foundation, using 10 ft of imported sand weighing 115 pcf; when in place, this fill will produce a total pressure of 2610 psf at midheight of the clay stratum. A consolidation test, drained both sides, is run on a sample 0.75 in. thick between 1460 psf and 2610 psf; 50% consolidation was reached in 48 sec and 100% consolidation in 244 sec.

(a) If construction is delayed until the clay reaches 50% consolidation, how much delay is involved?

(b) If construction time for the building is scheduled for the 12 months following this initial delay, will the clay reach 100% consolidation by the time construction is finished?

SOLUTION

(a) At 50% consolidation, the time is given by parabolic ratio,

$$\frac{t_{\text{lab}}}{t_{\text{field}}} = \frac{H^2_{\text{lab}}}{H^2_{\text{field}}}$$

$$\frac{48 \text{ sec}}{t_{\text{field}}} = \frac{(0.75/2)^2}{(26 \times 12/2)^2}$$

$$t_{\text{field}} = 8.31 \times 10^6 \text{ sec}$$

$$= \begin{array}{l}\text{96 days delay between the time fill is}\\ \text{placed and construction is started}\end{array}$$

(b) At 100% consolidation,

$$\frac{t_{\text{lab}}}{t_{\text{field}}} = \frac{H^2_{\text{lab}}}{H^2_{\text{field}}}$$

$$\frac{244 \text{ sec}}{t_{\text{field}}} = \frac{(0.75/2)^2}{(26 \times 12/2)^2}$$

$$t_{\text{field}} = \begin{array}{l}\text{489 days from time fill was placed until}\\ \text{100\% consolidation is reached}\end{array}$$

Total elapsed construction time is 96 + 365 days = 431 days; therefore, consolidation will not quite be complete when construction is finished.

It has already been observed that the $U\%$–log time curves are parabolic for about the first 60 to 75% of the consolidation. At any two degrees of consolidation $U\%$ within the stratum, the variation of $U\%$ with time is then [13]

$$\frac{t_1}{t_2} = \frac{(U_1\%)^2}{(U_2\%)^2} \tag{3-29}$$

where t_1 is the time required to achieve the degree of consolidation $U_1\%$ at a given pressure p_i, over a given thickness H_i, and t_2 is the time required to achieve the degree of consolidation $U_2\%$ at the same pressure p_i over the same thickness H_i.

It is usually assumed that settlements are proportional to the degree of consolidation, that is,

$$\frac{U_1\%}{U_2\%} = \frac{S_1}{S_2} \tag{3-30}$$

where $U_1\%$ and S_1 are degree of consolidation and settlement at one point in time, and $U_2\%$ and S_2 are degree of consolidation and settlement at a second point in time.

The use of Eqs. (3-29) and (3-30) in conjunction with Eq. (3-28) will allow valuable predictions to be made concerning relative settlements. For example, when a building is to be built over a clay deposit, the reduction in void ratio and corresponding settlements due to the building loads can be estimated by some convenient method. The building site can then be preloaded with fill material (surcharged) over the several months that the project is in its design and bidding stages, the amount of surcharge being carefully selected such that it will produce the same reduction in void ratio that the building would eventually have produced. When the surcharge is later removed and the building erected in its place, the settlements will already be complete. This and other common construction problems in clays are presented in the following examples.

EXAMPLE 3-26

A fill material 14 ft thick is placed on a clay stratum 18 ft thick. The pressure at midheight of the stratum is 1035 psf before the fill is placed and 2885 psf after the fill is placed. The clay is underlain by a dense sand and is considered to be drained both sides. A consolidation test is run on a clay sample, 0.75 in. thick, indicating that between 1035 and 2885 psf the clay will reach 100% consolidation in 258 sec. A computation similar to that of Example 3-22 indicates that a total settlement of 2.99 in. will occur under this surcharge.

(a) Estimate the settlement that will have occurred at the end of 60 days.
(b) Estimate how long it will take to achieve a settlement of 2.00 in.
(c) Estimate how long it will take to obtain 50% consolidation.

SOLUTION Settlements are proportional to the degree of consolidation $U\%$; hence the settlements can also be found by parabolic ratio,

$$\frac{t_1}{t_2} = \frac{(U_1\%)^2}{(U_2\%)^2} = \frac{S_1{}^2}{S_2^2}$$

The time t_2 to achieve 100% consolidation in the field stratum is found by parabolic ratio,

$$\frac{t_{\text{lab}}}{t_{\text{field}}} = \frac{H_{\text{lab}}^2}{H_{\text{field}}^2}$$

$$\frac{258 \text{ sec}}{t_{\text{field}}} = \frac{(0.75/2)^2}{(18 \times 12/2)^2}$$

$$t_{\text{field}} = \begin{array}{l}\text{248 days for 100\% consolidation, at which}\\ \text{time settlement } S = 2.99 \text{ in.}\end{array}$$

(a) At the end of 60 days,

$$\frac{60 \text{ days}}{248 \text{ days}} = \frac{S_1^2}{2.99^2}$$

$$S_1 = 1.47 \text{ in. of settlement at 60 days}$$

(b) For a settlement of 2 in.,

$$\frac{t_1}{248 \text{ days}} = \frac{2^2}{2.99^2}$$

$$t = 111 \text{ days to achieve 2 in. of settlement}$$

(c) For 50% consolidation,

$$\frac{t_{50}}{248 \text{ days}} = \frac{50^2}{100^2}$$

$$t_{50} = 62 \text{ days to achieve 50\% consolidation}$$

EXAMPLE 3-27

A building is to be constructed over a soft clay deposit 44 ft thick having a unit weight of 107 pcf. The deposit is underlain by loose sand. The average sustained pressure exerted by the building at its founding line is about 475 psf. Settlement due to this increase was computed using the methods of Example 3-22, indicating a settlement of 3.6 in. can be expected under this surface loading of 475 psf.

It is proposed to surcharge the building site and surrounding area with sand weighing 117 pcf to produce at least 3 in. of initial settlement. The surcharge material is sand weighing 117 pcf; the depth of surcharge could be made any depth from 4 ft thick to 12 ft thick. Three *e*–log time tests are run, beginning from the initial pressure at midheight of the stratum of $107 \times 22 = 2350$ psf. The first test adds a pressure equivalent to 4 ft of sand ($4 \times 117 = 470$ psf), the second test adds a pressure equivalent to 8 ft of sand ($8 \times 117 = 940$ psf), and the third test adds a pressure equivalent to 12 ft of sand ($12 \times 117 = 1400$ psf). The consolidation time at 100% consolidation for the three pressures were 38 sec, 50 sec, and 66 sec, respectively. Consolidation samples were standard 0.75-in. samples, drained both sides.

Determine a surcharge thickness that will produce the desired 3 in. of settlement in 2 months, which is the estimated startup time for actual construction after the contract is awarded. (It will be shown in a subsequent section on comparative settlements that for equal sizes of footings, settlements are approximately proportional to the footing pressure; that is, if footing pressures are doubled, settlements are doubled. Consequently, if the settlement under a pressure of 470 psf is 3.6 in., the settlement under 940 psi will be 7.2 in., and under 1400 psf will be 10.8 in.)

SOLUTION The time to reach 100% consolidation with 4 ft of surcharge is given by a pressure increase of 470 psf, or

$$\frac{t_1}{t_2} = \frac{H_1^2}{H_2^2}$$

$$\frac{t_1}{38} = \frac{(44 \times 12)^2}{(0.075)^2}$$

$$t_1 = 218 \text{ days or } 7.26 \text{ months}$$

At the end of this time, the settlement will be 3.6 in. The time required to get 3.0 in. of settlement at this pressure is found by parabolic ratio,

$$\frac{t_1}{218} = \frac{3^2}{3.6^2}$$

$$t_1 = \begin{array}{l}151 \text{ days or 5 months to get 3 in. of settlement} \\ \text{using 4 ft of fill}\end{array}$$

From this result it is seen that a surcharge height greater than 4 ft will be required if a settlement of 3 in. is to be realized in 2 months.

For a surcharge height of 8 ft,

$$\frac{t_1}{50} = \frac{(44 \times 12)^2}{0.75^2}$$

$$t_1 = 287 \text{ days or } 9.56 \text{ months}$$

At the end of this time, settlement will be 7.2 in.; hence the time required to obtain 3 in. of settlement is

$$\frac{t_1}{287} = \frac{3^2}{7.2^2}$$

$$t_1 = 50 \text{ days for 3 in. of settlement using 8 ft of fill}$$

The results are interpolated between a pressure of 470 psf at 151 days to obtain 3 in. of settlement and a pressure of 940 psf at 50 days to obtain 3 in. of settlement, yielding for 60 days,

$$p = 893 \text{ psf, which corresponds to a fill height of } 893/117 = 7.6 \text{ ft}$$

It is therefore estimated that a fill height of 7.6 ft for 60 days will produce a settlement of 3 in. in this stratum of clay.

In Chapter 1 it was stated that this book deals with ordinary soils (whatever that means) having an ordinary response to loads. An ordinary response to load was taken to be a relatively smooth load–deflection curve, with no sharp breaks. Such ordinary soils will not have sharp and unexpected increases in deformation due to small increases in load.

Not all clays respond to load in such a predictable way. One example of such an unusual clay is shown in the consolidation curves of Fig. 3-28, where a lightly preconsolidated clay is compared both to a normally consolidated clay and a heavily preconsolidated clay. The curves are greatly exaggerated to better illustrate a point.

In going from a void ratio such as e_0 to a void ratio e_1, all three clays generally follow predictable patterns with no sharp breaks. In going from e_1 to e_2, however, the lightly preconsolidated clay suddenly deforms a much larger amount under a very small increase in pressure; a building would not normally be designed to accommodate such deformations. To avoid such surprises, care must be taken

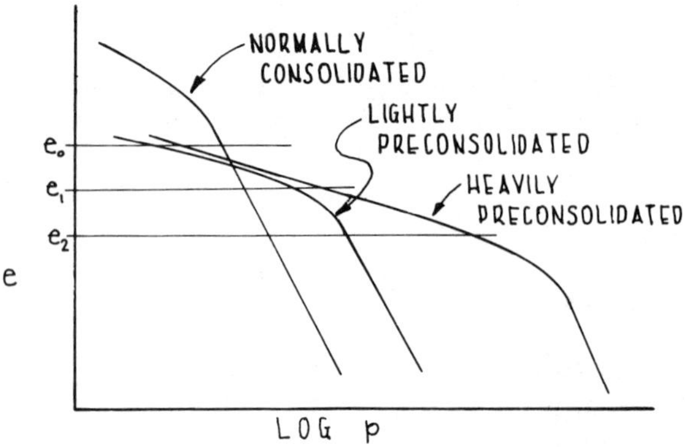

Figure 3-28 Levels of preconsolidation.

to recognize the "breakpoint" or "bend" in lightly preconsolidated clays. By comparison, building loads on heavily preconsolidated clays are not often large enough to reach the breakpoint (as indicated on the curve of heavily consolidated clay in Fig. 3-28), so such sharp increases in deformations are less likely in heavily preconsolidated clays than in lightly preconsolidated clays.

A second type of clay that does not qualify as an ordinary clay is a *sensitive* clay, a clay that has a very loose (flocculent) lay-up of particles, like a card castle. A typical but exaggerated consolidation curve for a sensitive clay is shown in Fig. 3-29, where it is compared to a more ordinary preconsolidated clay.

Deflections occurring in the steep portion of the consolidation curve for sensitive clays or silts can be disastrous in buildings. Such soils are considered to be among the more treacherous of all foundation materials. The design of foundations in sensitive clays and silts is best referred to experienced professionals.

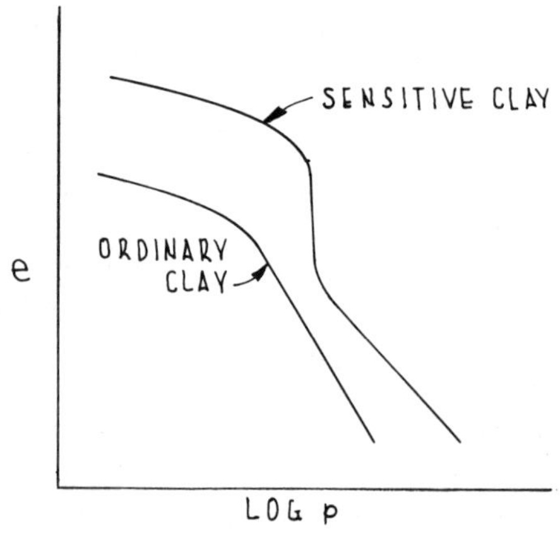

Figure 3-29 Consolidation curve for sensitive clay.

DEGRADATION SETTLEMENTS IN SANDS

The most widely used method in today's practice for computing settlements in sand is that developed by Schmertmann [5,14]. The method is a mixture of rational and empirical bases which provides results in reasonable agreement with observations. The original approach was made in 1970, but was amended in 1978 to include (among other refinements) the difference in settlement characteristics between spread footings and strip footings.

In the Schmertmann approach, settlement is modeled after the elastic theory. The settlement S is taken to be the integral of the strains ε_z over the depth of the pressure bulb,

$$S = \int_0^\infty \varepsilon_z dz = \int_0^\infty \frac{\sigma_z}{E_z} dz \tag{3-31}$$

where, by definition, $\varepsilon_z = \sigma_z/E_z$, in which σ_z is stress (pressure) and E_z is the elastic modulus. The pressure σ_z is a variable, but it can also be expressed as the increase in contact pressure p under the footing (a constant) times a variable influence factor I_z, $\sigma_z = pI_z$; hence

$$S = p \int_0^\infty \frac{I_z}{E_z} dz \tag{3-32}$$

The influence factor I_z represents the variation in strain along the depth of the pressure bulb. In this region, the pressure is rapidly decreasing and the strength of the sand is changing accordingly. The depth of the pressure bulb in Schmertmann's approach is, as usual, taken to be $2B$ for spread footings and $4B$ for strip footings; the upper limit of integration becomes $2B$ for spread footings and $4B$ for strip footings. It is of interest to note in Schmertmann's solution that pressures and settlements are directly proportional; that is, when contact pressures increase 50 or 75%, settlements increase 50 or 75% correspondingly.

In addition to the rational integral given by Eq. (3-32), Schmertmann adds two empirical correction factors. The first correction factor accounts for variations due to depth of overburden D_f,

$$C_1 = 1 - 0.5\left(\frac{\gamma D_f}{p}\right) \tag{3-33}$$

where γ is the unit weight of the overburden soil. The second correction factor accounts for effects of time, which affects settlements in sands as well as in clays,

$$C_2 = 1.2 + 0.2 \log_{10}(t \text{ years}) \tag{3-34}$$

With these factors, the final equations become:

$$\text{For spread footings:} \quad S = C_1 C_2 p \int_0^{2B} \frac{I_z}{E_z} dz \tag{3-35a}$$

For strip footings: $$S = C_1 C_2 p \int_0^{4B} \frac{I_z}{E_z}\, dz \tag{3-35b}$$

These integrals can be evaluated for thin layers over the depth of the pressure bulb (a finite integration) once the influence factor I_z and the modulus E_z are defined.

The influence factor I_z is developed by Schmertmann, based on theoretical considerations in combination with known test results. The theoretical variation of I_z for a spread footing in a purely elastic half-space is shown in Fig. 3-30a [32]. Test results [33] and computer solutions that support or modify the theoretical solution are shown in Fig. 3-30b. Graphs for a strip footing would be similar, but the curves would be expanded vertically over a depth of $4B$.

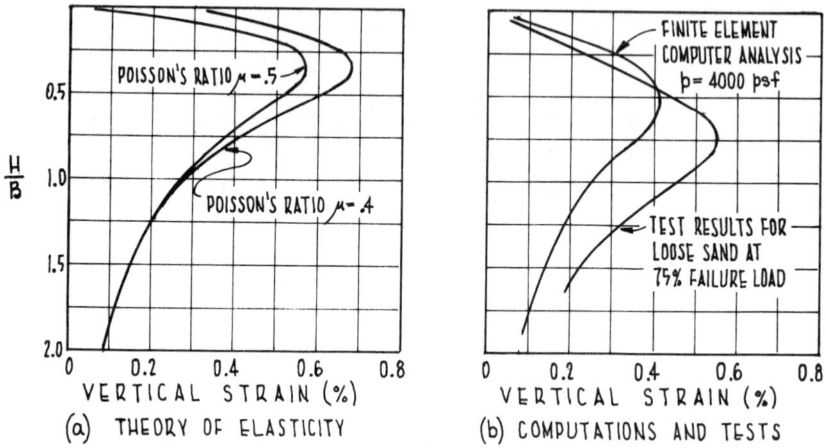

Figure 3-30 Vertical strain variation.

It is of particular interest to note that the greatest contribution to settlements in sand comes at some distance below the contact surface of the footing rather than at the contact surface itself, where pressure is highest. This depth where the maximum contribution occurs is seen in Fig. 3-30 to be about $0.5B$ below the contact surface. Also, the peak value of the influence factor is seen to be about 0.6 in the theoretical solution of Fig. 3 30a; the magnitude of this peak value is closer to about 0.5 in the experimental results indicated in Fig. 3-30b.

A simplified diagram for the variation in I_z as developed by Schmertmann is given in Fig. 3-31. The depth of influence is set at $2B$ for spread footings and $4B$ for strip footings. If the actual sand stratum is not deep enough to accommodate the entire curve, the lower part of the curve should simply be truncated at the bottom of the stratum.

With the factor I_z now defined, the only remaining factor in Eqs. (3-35) to be evaluated is the modulus E_z. Schmertmann in 1970 gave an empirical relationship for calculating E_z based on the Dutch cone penetration values; he made a further refinement to this relationship in 1978. The final result is

For square footings: $$E_z = 2.5 q_c \tag{3-36a}$$

For strip footings: $$E_z = 3.5 q_c \tag{3-36b}$$

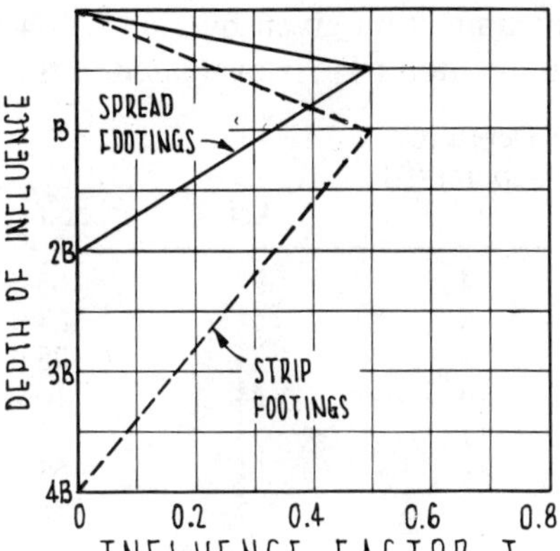

Figure 3-31 Simplified factor I_z.

where q_c is the Dutch cone penetration resistance in the same units as p. (Where SPT blow count is the only test result available, an approximate conversion between SPT blow count and Dutch cone resistance can be made using Fig. 2-23.)

An example will illustrate the evaluation of Eq. (3-35a) for computing settlement of a spread footing on sand. Evaluation of Eq. (3-35b) for a strip footing would be similar, but the evaluation would be made over a depth of $4B$ rather than over a depth of $2B$. For a finite evaluation of the integral, Eq. (3-35a) becomes

$$S = C_1 C_2 p \sum_{0}^{2B} \frac{I_z}{E_z} (\Delta z) \tag{3-37}$$

EXAMPLE 3-28

A footing 7 ft square is founded 3 ft deep in a thick stratum of medium sand, $\gamma =$ 124 pcf. Contact pressure under the footing is 2610 psf. The log of Dutch cone resistance values is shown. Compute the settlements to be expected at the end of one year, 5 years, and 10 years.

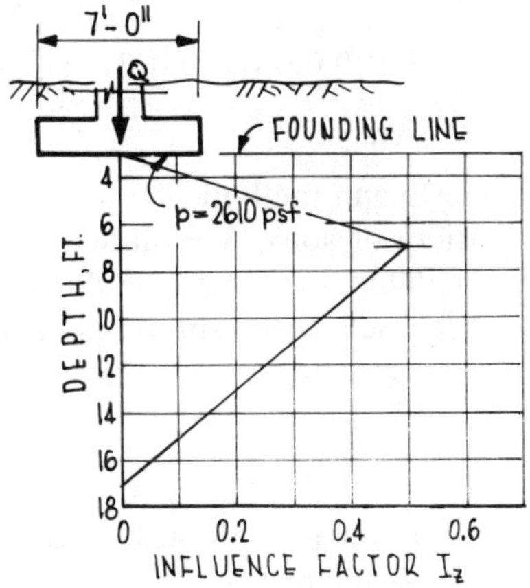

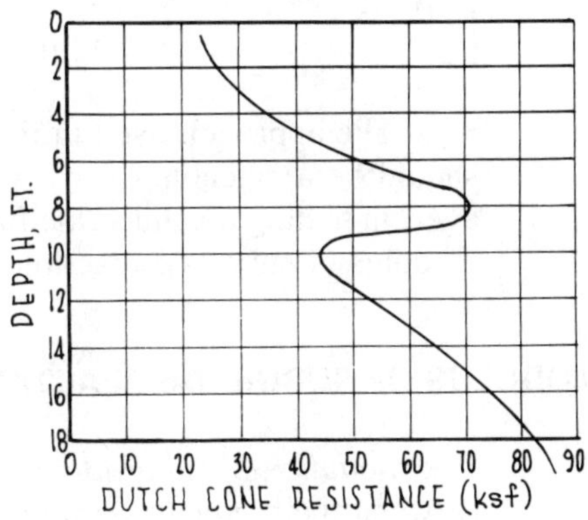

SOLUTION The finite summation given by Eq. (3-37) will be evaluated for layers 2 ft thick. The summation term is evaluated in the following tabulation.

Layer	Δz ft	Depth to midpoint (ft) ft	q_c ksf	E_z ksf	I_z	$\frac{I_z}{E_z}\Delta z$
1	2	4	34	85	0.13	0.0037
2	2	6	51	128	0.38	0.0059
3	2	8	71	178	0.45	0.0051
4	2	10	44	110	0.35	0.0064
5	2	12	53	133	0.25	0.0038
6	2	14	61	153	0.15	0.0020
7	2	16	75	188	0.15	0.0005
					$\sum_0^{2B}$ =	0.0274

The value of C_1 is given by Eq. (3-33),

$$C_1 = 1 - 0.5\left(\frac{\gamma D_f}{p}\right) = 1 - 0.5\left(\frac{124 \times 3}{2610}\right) = 0.93$$

The value of C_2 is given by Eq. (3-34),

$$C_2 = 1.2 + 0.2 \log_{10}(t \text{ years})$$

For 1 year: $C_2 = 1.20$
For 5 years: $C_2 = 1.34$
For 10 years: $C_2 = 1.40$

Final values of settlement are given by Eq. (3-37).

At 1 year: $S = 0.93 \times 1.2 \times 2.61 \times 0.0274 = 0.0798$ ft $= 0.96$ in.

At 5 years: $S = 0.93 \times 1.34 \times 2.61 \times 0.0274 = 0.0891$ ft $= 1.07$ in.

At 10 years: $S = 0.93 \times 1.40 \times 2.61 \times 0.0274 = 0.0931$ ft $= 1.12$ in.

The approach used in the Schmertmann method is sound theoretically and should be applicable over a wide variety of soils. The difficulty remains, however, in finding a valid value for the modulus, E_z. This modulus is also called the modulus of subgrade reaction; it is discussed further in the next section.

MODULUS OF SUBGRADE REACTION

The *modulus of subgrade reaction* of a soil is the constant of proportionality between the contact pressure p and the settlement S,

$$K = \frac{p}{S} \tag{3-38}$$

For a settlement S of 1 in., the modulus of subgrade reaction is therefore seen to be the bearing pressure on the soil that will produce a settlement of 1 in. Such a bearing pressure has been discussed at length in this chapter, but the term "modulus of subgrade reaction" was never applied to it.

The modulus of subgrade reaction is somewhat akin to the modulus of elasticity in conventional structural materials, in that it defines the relationship between stress (pressure) and deformation (settlement). Its units are usually given in kips per cubic foot, although units of kips per square foot per inch of settlement would be more meaningful. Unlike the modulus of elasticity, the modulus of subgrade reaction is only partially elastic (if it is elastic at all), with most of the settlement it describes being a permanent change in void ratio in the soil mass.

It has been seen throughout the earlier discussions that settlements in soil can be dependent on a variety of factors. Settlements are affected not only by the contact pressure, but by the size of the footing, by the depth of founding, by the water content in the soil, by the friability of sand grains, and sometimes even by the pattern of deposition of clay grains. Obviously, the modulus of subgrade reaction for a soil cannot be constant; it must vary throughout the soil mass due to all the foregoing field conditions. Any attempt to define the modulus of subgrade reaction as a constant must therefore include rather sweeping simplifications concerning actual field conditions and their effect on the settlements.

The modulus is also heavily time dependent; a soil will usually deform much less under a short-term load than it will under the same load applied over a long time. In clays especially, but also to a lesser extent in sands, it has been shown that loads must be sustained over weeks or even months to produce appreciable levels of settlement. Any effort to define the modulus of subgrade reaction must somehow allow for this dependency on time. Due largely to such variability, earlier efforts in the practice to define a constant modulus of subgrade reaction for a soil have generally met with failure. Consequently, the modulus has essentially fallen from use in contemporary practice, although it still appears in some of the older, more traditional settlement formulas. It is not used in this book except as a reference value in some of the derivations, and even there it is used as a variable rather than as a constant.

COMPARATIVE SETTLEMENTS IN ALL TYPES OF SOILS

It seems evident that every particle of soil that lies within the Boussinesq pressure bulb will undergo an increase in pressure when the footing load is applied and will make some contribution to the total settlement of the footing. Consequently, if the modulus of subgrade reaction could be determined accurately at each level within the pressure bulb, the contribution at each level could be determined and the sum of all these increments would then be the total settlement. The problem, of course, lies in trying to find an accurate value for the modulus of subgrade reaction at each level of the pressure bulb.

It has been noted that settlements in soil are not elastic. Nonetheless, it has been shown repeatedly [15–17] that the distribution of compressive stress within a soil mass corresponds reasonably well to that which will occur in an elastic mass under the same loads and boundary conditions. In fact, the more successful methods of predicting settlements in a soil mass (such as Schmertmann's) utilize an elastic solution, recognizing that it is true only as long as the soil is undergoing an increase in compression; when the load is released, the rebound curve will not be even close to an elastic rebound curve. The modulus of elasticity used in Schmertmann's method might therefore be better termed a modulus of compression deformation.

It has been shown as well that for both clays [18] and sands [5], the methods of the theory of elasticity may be applied to soil masses undergoing an increase in compression. Since the only settlements of concern are the maximum settlements, which can only occur under this increased compression, the methods of the theory of elasticity will therefore be generally applicable throughout our limited range of interest. The Boussinesq pressure bulb, for example, is itself an elastic solution; it is quite true that it is valid only under an increase in compression when it is applied to soils, but it is only in this range of increased compression that it is of interest.

From the theory of elasticity, the vertical settlement S of any point on the surface of an elastic half-space subjected to a uniformly distributed load over a circular or rectangular area can be computed as [19]

$$S = K_0 p B \tag{3-39}$$

where p = contact pressure
B = least width of the rectangle or diameter of the circle
K_0 = coefficient of proportionality

The value of K_0 at any level of the pressure bulb can be computed by

$$K_0 = \int_0^H \frac{C_0}{R_s}\, dz \tag{3-40}$$

where C_0 is a coefficient depending on footing shape, rigidity, depth of founding, length-to-width ratio, Poisson's ratio, pressure influence, and other variables. R_s is the modulus of subgrade reaction at the point z below the footing and z is the level within the Boussinesq pressure bulb being evaluated.

By itself, Eq. (3-39) has the same drawback as any of the other solutions for settlement—it is difficult or impossible to evaluate K_0 accurately. If, however, the equation is used to compare two settlements within the same soil mass rather than to compute absolute values of settlement, the unknown coefficient K_0 cancels out, leaving

$$\frac{S_1}{S_2} = \frac{p_1 B_1}{p_2 B_2} \tag{3-41}$$

where the subscript 1 denotes footing number 1 and the subscript 2 denotes footing number 2. Equation (3-41) is valid only for like values of K_0 for the two footings, that is, for two circular footings, for two square footings, for two strip footings, or for two grade beams; the two footings should also be founded at the same general depth and should be in the same general range of rigidities.

Theoretical values for K_0 have been computed for various footing shapes, length-to-width ratios, and rigidities [19]. Although these absolute values for K_0 may themselves be subject to question, their relative values are believable. For example, the values of K_0 for circular and square footings are known to be quite close, varying only a few percentage points under various conditions [20]. It can therefore be concluded that the general settlement characteristics between square and round footings are similar, and that their settlements will generally be within a few percentage points of each other; their settlements within a particular soil may therefore be considered to be roughly the same, at least within the usual range of accuracy in soils [20].

Relative values of K_0 for strip footings compared to square footings are not as close, however. There is, in fact, a distinct difference in settlement characteristics between square or round footings and long strip footings. The value of K_0 for strip footings has been found to be on the order of 1.6 to 2.2 times the value for spread footings under various conditions [19], indicating that settlements under strip footings will be roughly 1.6 to 2.2 times those of spread footings having the same width and contact pressure. (This difference was noted earlier in the Schmertmann analysis for degradation settlements in sand.)

For the sake of comparisons, the value of K_0 for strip footings is taken in this book to be twice its value for spread footings. Equation (3-39) then becomes for the two cases:

Square and round footings: $$S = K_0 pB \tag{3-42a}$$

Strip footings and grade beams: $$S = 2K_0 pB \tag{3-42b}$$

It should be apparent that settlement characteristics for strip footings would be similar to those for grade beams; hence Eq. (3-42b) will apply to grade beams as well.

Equations (3-42a) and (3-42b) are used in the examples of Chapter 7 when comparing settlements between any two footings within a group of footings. No attempt is made in those examples to compute the absolute value of settlements; hence the actual absolute values of K_0 are never needed. Some brief examples will illustrate the use of Eqs. (3-42a) and (3-42b).

EXAMPLE 3-29

A footing of 6 ft square is loaded to a pressure of 3100 psf. An adjacent footing is 8 ft square, loaded to a pressure of 2700 psf. The larger footing is estimated to settle about 1 in. under the given pressure. What will be the estimated settlement of the smaller footing?

SOLUTION Relative settlements are given by Eq. (3-42a).

$$\frac{S_1}{S_2} = \frac{p_1 B_1}{p_2 B_2}$$

where subscript 1 refers to the 6-ft footing and subscript 2 refers to the 8-ft footing.

$$\frac{S_1}{1\text{ in.}} = \frac{3100 \times 6}{2700 \times 8}$$

S_t = 0.86 in. of settlement under the 6-ft footing.

EXAMPLE 3-30

A column footing 8 ft square is located adjacent to a wall footing 5 ft wide. The column footing is loaded to 2800 psf. At what contact pressure will the wall footing settle the same amount as the column footing?

SOLUTION Relative settlements are given by Eqs. (3-42a) and (3-42b).

$$\frac{S_1}{S_2} = \frac{p_1 B_1}{2 p_2 B_2}$$

where subscript 1 refers to the column footing and subscript 2 refers to the wall footing. For $S_1 = S_2$,

$$1 = \frac{2800 \times 8}{2 \times p_2 \times 5}$$

$$p_2 = 2240 \text{ psf}$$

This pressure will produce settlements in the wall footing roughly equal to the settlement in the column footing.

COMPARISON OF RESPONSE OF SOILS TO BEARING LOADS

A summary of points is listed in Table 3-4, in which the response of clay to a foundation load is compared to the response of sand. The same table was presented at the beginning of Chapter 2 to illustrate the wide difference in engineering properties between sands and clays. It is shown again here as a means to summarize the points presented in this chapter.

The first six points of Table 3-4 are concerned with the strength of the soil, the last four with settlements and compaction. The points correspond; that is, point 2 in sands corresponds to point 2 in clays. The 10 points listed in Table 3-4 serve to illustrate some of the sharp contrasts between the two types of soils.

Table 3-4 deals with the two extremes in types of soil, a pure sand and a pure clay. Most soils are neither, but are an indistinct blend of the two. Such mixtures of sands and clays may have some of the properties of both. It thus becomes necessary to distinguish whether a mixture of the two soils will respond to load mostly as a sand or mostly as a clay. One way to draw such a distinction is simply to look at the unified soil classification for the soil: G, S, and ML soils will respond to load mostly as noncohesive soils, or sands; C and MH soils will respond to load mostly as cohesive soils, or clays. If there is any doubt, the foundation should be

TABLE 3-4 COMPARISON OF RESPONSE TO LOAD

Footings on sand or cohesionless silts	Footings on clay or cohesive silts
1. Sand strength increases with confining pressure, whether due to overburden or to surface loads.	1. Clay strength is relatively constant regardless of magnitude of confining pressure.
2. Sand strength is due entirely to friction, measured by the angle of internal friction ϕ.	2. Clay strength is due entirely to cohesion (tensile strength).
3. Sand strength is relatively insensitive to footing shape.	3. Clay strength is influenced considerably (up to 20%) by footing shape.
4. Sand strength increases markedly (doubled or tripled) with depth of burial of footing.	4. Clay strength is relatively insensitive to depth of burial of footing.
5. Loss of strength in sand is significant if overburden (confining pressure) is removed or eroded away.	5. Little loss of strength is caused by removal of overburden.
6. Strength of a dry sand is cut in half when the sand is submerged.	6. Strength of clay soils is relatively unaffected by submergence.
7. Settlements in sands occur soon after application of load, measured in weeks or months.	7. Settlements in clay occur *very* slowly following application of load, measured in months or years.
8. Settlements in sand can occur under relatively short-term loads.	8. Settlements in clays are relatively unaffected by short-term loads.
9. Settlements in a dry sand are essentially doubled if the sand is submerged (or if water table rises).	9. Settlements in clays are markedly affected (but not doubled) if the clay is submerged.
10. Deposits of sand are best compacted by vibration and submergence, with some pressure.	10. Deposits of clay are best compacted by long-term surcharge pressure.

designed for both types and the more conservative result chosen for the final design.

The points listed in Table 3-4 are qualitative rather than quantitative. They affect design concepts and design philosophy rather than design calculations. Anyone engaged in foundation design should be intimately familiar with each point in Table 3-4 as well as the ramifications implied by each point.

RECOMPACTION OF DISTURBED SOILS

It is not always possible to place foundations on undisturbed soil. Quite often, a building site is subject to extensive grading and earthmoving before the foundations are started. The soil directly below the footing might therefore be disturbed or even relocated; hence its engineering properties will have been changed from those assumed for the project design.

It was noted earlier that the structural properties of a soil are closely linked to its density. It follows that if a soil is disturbed, but its density is later restored to a value at least as high as its original in-place density, its strength and settlement properties may have changed, but they will be at least as good as those before the disturbance. Such an approach is adopted here as a very general design guide.

Packing a disturbed soil into a reduced volume, thereby increasing its density, is called compaction. There are two primary factors affecting the amount of increase in density that can be achieved, the compactive effort and the water content. At any given water content in the soil, simply increasing the compactive effort will increase the density. Conversely, however, if the compactive effort is fixed, there exists a particular water content at which the highest possible density can be achieved; this particular water content is the *optimum moisture content* corresponding to that particular level of compactive effort.

The optimum moisture content applies generally to clays, or to soils having some degree of plasticity. Moisture sometimes helps in compaction of sands also, but for an entirely different reason. The following discussions presume that the soil being compacted has some degree of plasticity, but that it is not necessarily a heavy clay; there will be both air and water in its void spaces.

The apparatus used in making a determination of optimum moisture content is shown in Fig. 3-32. The loose soil is mixed with a small amount of water and placed in the cylinder in five lifts (volume of the cylinder is $\frac{1}{30}$ ft^3). Each lift is compacted with 25 blows of a hammer weighing 10 lb falling 18 in. (compactive energy is therefore fixed). The weight, volume, and water content of the compacted sample is determined and its dry unit weight, W_s/V, is computed.

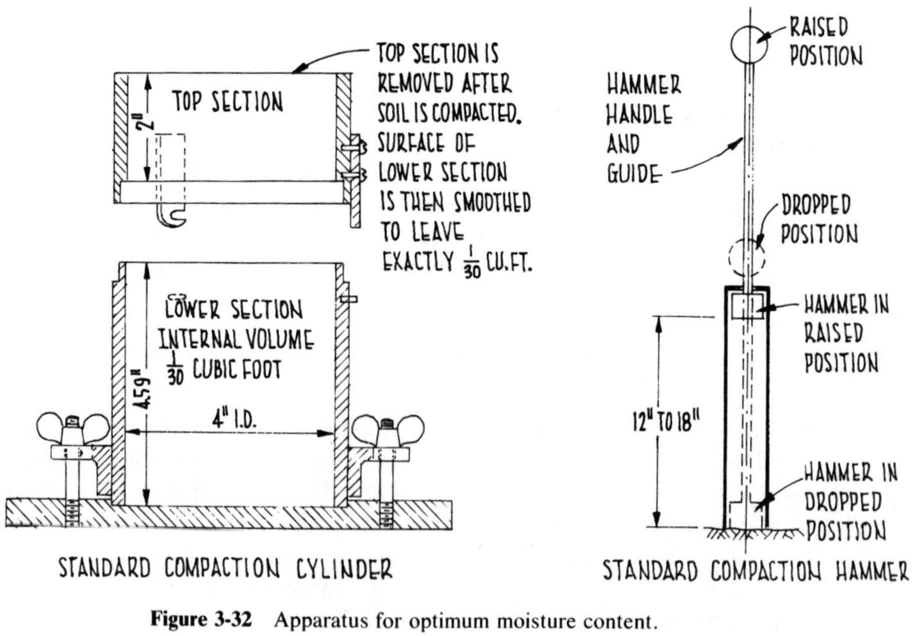

Figure 3-32 Apparatus for optimum moisture content.

The sample is then removed; it may be remixed with the unused portion of the sample, but it is preferable not to do so. More water is then mixed thoroughly into the remaining sample. The test is then run again for this higher water content, yielding a second dry unit weight, W_s/V, associated with this higher water content but with the same compactive effort. A third sample is run at a yet higher water content, thus providing the dry unit weight at three levels of moisture but with only one level of compactive effort.

The results of the three tests are plotted on a graph as indicated in Fig. 3-33 and connected by a parabolic curve as shown. It may sometimes be necessary to run more than three tests to get water contents both above and below optimum as shown. The skill of the technician who is running the test can be a distinct asset in running such tests, since a good technician can often judge by the feel of the soil mixture whether it is above or below optimum moisture content.

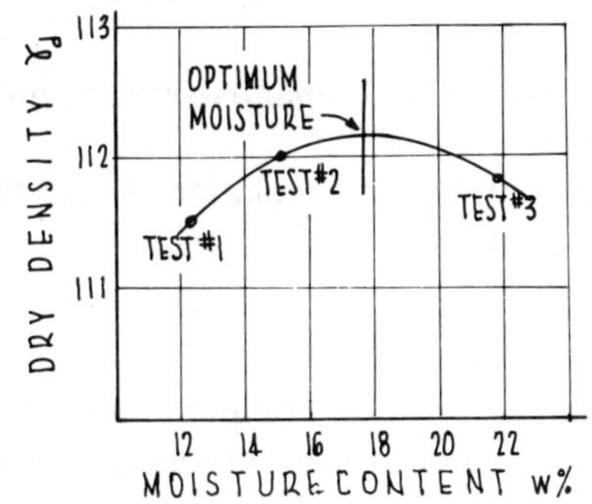

Figure 3-33 Optimum moisture curve.

The highest point on the curve of Fig. 3-33 is the optimum moisture content for the given compactive effort. The compactive effort (10-lb hammer falling 18 in. in five lifts) is that specified by ASTM test designation D1577-70. (This level of compactive effort in the lab corresponds to a compactive effort in the field that would be produced by modern heavy sheepsfoot rollers compacting the soil in lifts of about 10 in. thick.)

Other levels of compactive effort might be used, however, to correspond to other anticipated levels of compactive effort in the field. In other sets of specifications (AASHTO [21], Proctor [22]), the compactive effort might be either higher or lower than that specified by ASTM D1557-70. Typical optimum moisture curves for higher and lower compactive efforts are shown in Fig. 3-34 for comparison.

Another curve is shown in Fig. 3-34, that of zero air voids. If enough compactive effort were to be provided to drive all the air out of the voids in the sample, the result would be the zero air voids curve. The curve is obtained by computing the dry unit weight γ_{d0}, assuming zero air voids at the given water content,

$$\gamma_{d0} = \frac{W_S}{V_S + V_W} \tag{3-43}$$

When plotted against the water content, these unit weights give the zero air voids curve shown in Fig. 3-34. The zero air voids curve represents the highest unit

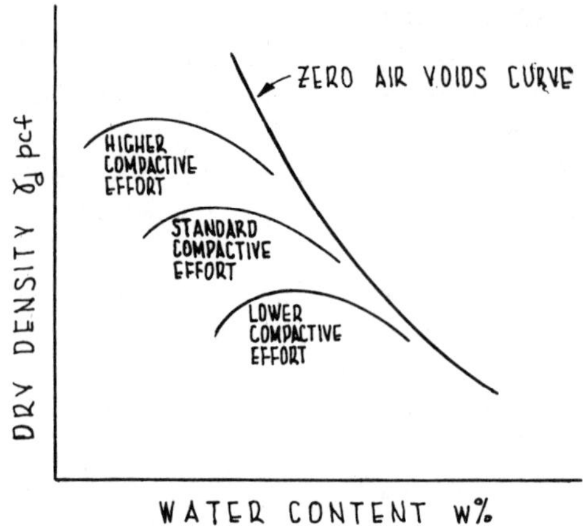

Figure 3-34 Comparison of optimum moisture contents.

weight that could be obtained at the indicated water content with an unlimited amount of compactive effort.

An example will demonstrate the calculation of both the dry density and zero air voids at various water contents.

EXAMPLE 3-31

A standard test for optimum moisture content produced the listed results. Plot the moisture–density curve, choose the optimum moisture content, and plot the zero voids curve. $G_s = 2.67$.

Sample 1: $\gamma = 114$ pcf, $w = 13\%$
Sample 2: $\gamma = 119$ pcf, $w = 16\%$
Sample 3: $\gamma = 124$ pcf, $w = 19\%$
Sample 4: $\gamma = 125$ pcf, $w = 22\%$

SOLUTION Compute the dry density for sample 1:

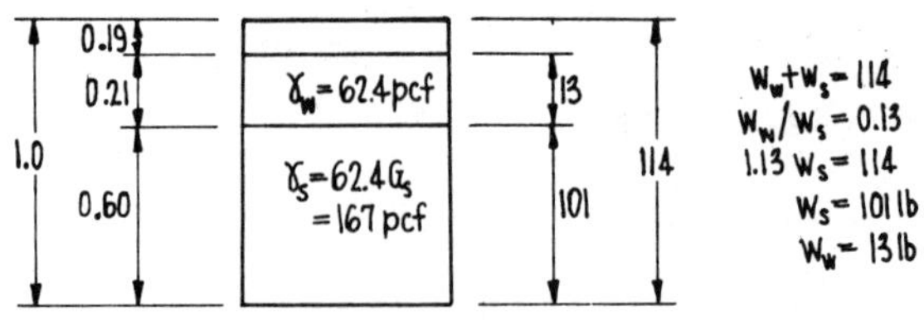

$$\gamma_d = \frac{W_S}{V} = \frac{101}{1} = 101 \text{ pcf}$$

At zero air voids,

$$\gamma_{d0} = \frac{W_S}{V_S + V_W} = \frac{101}{0.60 + 0.21} = 125 \text{ pcf}$$

Similar computations for the other samples yield, for

Sample 2: $\gamma_d = 103$ pcf, at zero air voids, $\gamma_{d0} = 117$ pcf
Sample 3: $\gamma_d = 104$ pcf, at zero air voids, $\gamma_{d0} = 111$ pcf
Sample 4: $\gamma_d = 102$ pcf, at zero air void, $\gamma_{d0} = 104$ pcf

Results are plotted as indicated:

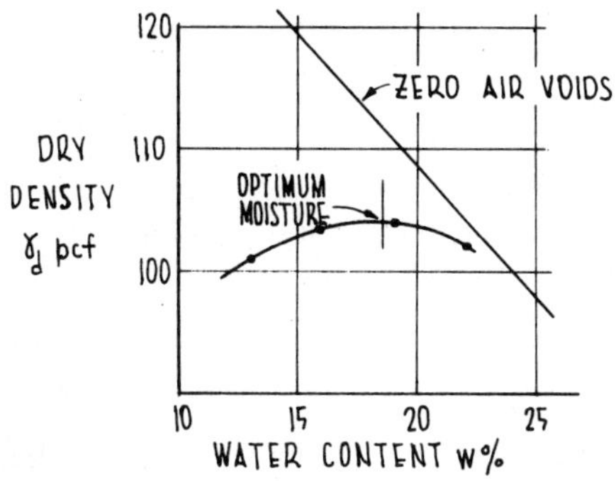

Optimum moisture is read from the graph at 18.5%.

It is worth remembering that in the more plastic soils, the optimum moisture content falls very close to the plastic limit. Working a soil in the field at a water content close to its plastic limit can sometimes result in surface slicks, where equipment can slide or get stuck. As a result, contractors will often compact a plastic fill by keeping it slightly on the "dry" side of optimum, adding a bit more compactive effort to get the required unit weight at the less-than-optimum moisture content.

When the optimum moisture content is known, the placement of backfill on a construction site to a specified unit weight can be controlled easily. Specifications for such a backfill might read "Backfill shall be compacted to a unit weight of 90% of that given by ASTM D1557-70." Periodic field checks on the unit weight will assure that the backfill is, in fact, placed at the specified unit weight.

The property of optimum moisture content applies only to plastic soils. There is no comparable property in nonplastic soils, or sands, involving moisture content. Sands are best compacted by vibration, with possibly some surface rolling, to shake the sand particles into their densest possible configuration.

The two extreme cases for configuration of particles in a sand mass are shown in Fig. 3-35, greatly exaggerated for the sake of clarity. The void ratio is minimum, e_{min}, when the particles are in their densest possible configuration, and is maximum, e_{max}, when the particles are in their loosest possible configuration. Similarly, the dry unit weight of the sand in the densest and loosest configurations are $\gamma_{d\,max}$ and $\gamma_{d\,min}$, respectively.

Obviously, the unit weight of any particular deposit of this sand must fall between the values of $\gamma_{d\,max}$ and $\gamma_{d\,min}$ and its void ratio must fall between the

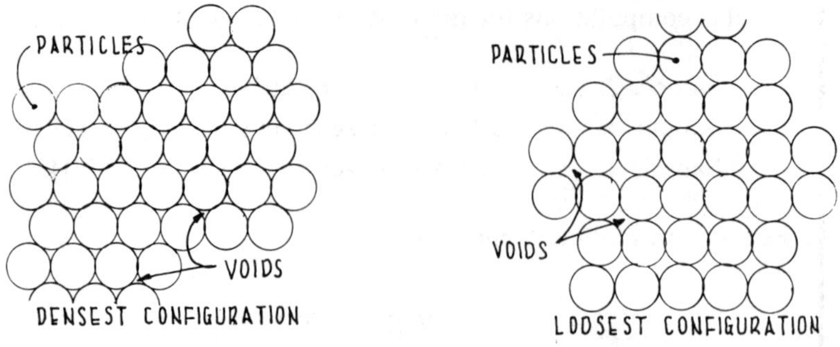

Figure 3-35 Configuration of sand particles.

values of e_{max} and e_{min}. The *relative* density D_r of a sand having a measured dry unit weight of γ_d or a measured void ratio of e is therefore given by

$$D_r = \frac{e_{max} - e}{e_{max} - e_{min}} \times 100\% = \frac{\gamma_{d\,max}}{\gamma_d}\left(\frac{\gamma_d - \gamma_{d\,min}}{\gamma_{d\,max} - \gamma_{d\,min}}\right) \times 100\% \qquad (3\text{-}44)$$

The process of obtaining $\gamma_{d\,max}$ and $\gamma_{d\,min}$ in the lab are quite straightforward. For $\gamma_{d\,max}$, the sand is placed in a graduated cylinder and shaken or tamped until its volume no longer decreases; it is then weighed and its unit weight computed. For $\gamma_{d\,min}$, the sand is placed very gently by a funnel into the cylinder to get the maximum possible volume; sometimes several trials are necessary before the minimum unit weight is obtained.

When a sand is to be compacted on a construction site, its compacted density is commonly specified as a percent of maximum relative density. A typical specification might read "Sand backfill shall be compacted to 90% of maximum relative density." The unit weight corresponding to $D_r = 90\%$ can be computed from Eq. (3-44); the inspectors will take field density tests during the placement of backfill to assure that this unit weight is achieved.

EXAMPLE 3-32

A sample of sand classified as an SM soil is tested and found to have a maximum unit weight of 128 pcf and a minimum unit weight of 103 pcf. Determine the unit weight at 95% maximum relative density.

SOLUTION The relative density is determined from Eq. (3-44):

$$D_r = \frac{\gamma_{d\,max}}{\gamma_d}\left(\frac{\gamma_d - \gamma_{d\,min}}{\gamma_{d\,max} - \gamma_{d\,min}}\right) \times 100$$

$$95 = \frac{128}{\gamma_d}\left(\frac{\gamma_d - 103}{128 - 103}\right) \times 100$$

Solve for γ_d:

$$\gamma_d = 126 \text{ pcf at 95\% relative density}$$

EXAMPLE 3-33

A sample of gravel classified as GP has a maximum unit weight of 140 pcf and a minimum unit weight of 114 pcf. What is its relative density if its in-place unit weight is 136 pcf?

SOLUTION The relative density is determined from Eq. (3-44):

$$D_r = \frac{\gamma_{d\max}}{\gamma_d}\left(\frac{\gamma_d - \gamma_{d\min}}{\gamma_{d\max} - \gamma_{d\min}}\right) \times 100$$

$$= \frac{140}{136}\left(\frac{136 - 114}{140 - 114}\right) \times 100$$

$$= 87\% \text{ relative density at 136 pcf}$$

SUMMARY

The following concepts are developed in the discussions of this chapter.

- The dispersion of a footing load into the supporting soil mass can be assumed to follow general concepts from the mathematical theory of elasticity; the Boussinesq solution for a concentrated point load at the surface of a semi-infinite elastic solid is widely accepted.
- Pressure dispersion into a soil mass from a point load at the surface can be roughly approximated as a cone having a side slope of 2 vertical to 1 horizontal.
- Larger loads produce larger pressure bulbs and therefore encompass and compress larger volumes of soil; settlements are proportional to total load rather than to contact pressure.
- A soil that is free of outside disturbances over a long period will eventually reach its at-rest condition; in the at-rest condition, lateral pressure at any point is about 40 to 50% of the static pressure due to overburden.
- Soil fails in shear at an angle $45° + \frac{1}{2}\phi$ from the plane of maximum principal stress, where ϕ is the angle of internal friction in degrees.
- Failure strength in soil due to bearing pressure is developed from three semiempirical bearing capacity factors: N_γ, a factor related to size of footing; N_q, a factor related to confining pressures; N_c, a factor related to cohesion.
- Size and shape of a footing has a slight effect on bearing capacity, increasing the bearing capacity in clays by as much as 20% above that of standard square shapes and decreasing the bearing capacity of sands by as much as 30% below that of standard square shapes.
- Lateral loads on a footing cause a distinct loss in strength in the supporting soil, producing as much as a 40% loss in clays and as much as a 75% loss in sands.

• When the strength of a soil is the limiting consideration and settlements are of no concern, a factor of safety of 2 to bearing capacity failure is commonly used.

• In clays, it has been found that using a factor of safety of 3 to bearing capacity failure will generally limit settlements to about 1 in.; this approach to limiting settlements in clay is empirical but has a long history of success.

• In sands, settlements are also controlled by empirical methods, but largely through the use of design graphs and charts whose validity has been established over many years of use.

• Presumptive values for allowable bearing pressures are in common use in the industry; the designer who uses presumptive values in foundation design remains personally responsible for their adequacy.

• Contact pressures under individual spread footings are not uniform as is usually assumed, but for smaller structures the variation is usually ignored.

• Consolidation settlements in clays are somewhat subject to calculations, based largely on assumed parabolic variations; calculations can in some cases offer a prediction of the time required for consolidation as well as the absolute amounts of settlement.

• Degradation settlements in sands are also somewhat subject to calculations, based on a finite evaluation of the general settlement integral; results depend heavily on being able to establish an accurate value for the modulus of elasticity (or modulus of settlement) of the soil throughout the affected zone. Settlement times are not included nor can they be determined in this approach.

• The modulus of subgrade reaction is a constant that appears in some of the older equations for settlement; the wishful concept of a modulus of subgrade reaction that is a constant throughout the soil mass is slowly disappearing in contemporary practice.

• Methods in current use for computing the absolute values for settlements of foundations may be of questionable accuracy. Methods that only compare two settlements within a soil mass are generally more believable.

• There is an optimum moisture content for plastic soils at which a given amount of comparative effort will produce the maximum reduction in volume.

• Plastic soils can be compacted to a given reduction in volume at a water content other than the optimum moisture content by increasing the amount (and cost) of the comparative effort.

• The density of sand is usually given as a relative density, stated as some percentage of its most dense possible particle configuration.

• The response of a clay to a foundation load can often be diametrically opposite to that of a sand; the concepts and implications listed in Table 3-4 are recommended for continuing study by foundation designers.

REVIEW QUESTIONS

3-1. The Boussinesq solution for pressure distribution applies to materials that are homogeneous, elastic, and isotropic. On what basis can it be applied to soils?

3-2. About what level of pressure is used for the outermost pressure contour of the Boussinesq bulb when it is used to depict pressure distribution under spread footings? Strip footings?

3-3. How far below a square spread footing of width *B* would the vertical pressure in the soil be reduced to about 10% of the contact pressure?

3-4. How far below a strip footing of width *B* would the vertical pressure in the soil be reduced to about 15% of the contact pressure?

3-5. Approximately, what is the equivalent fluid weight of a soil that weighs 124 pcf?

3-6. What is the theoretical at-rest vertical pressure at 15 ft below the surface in a soil weighing 115 pcf?

3-7. What would be the range of lateral pressures to be expected in the soil of Question 3-6?

3-8. What is the failure mode in soil when it is loaded by a bearing load such as a footing?

3-9. What is the shear stress at failure in a soil mass?

3-10. What is the failure angle in a soil, measured from a horizontal surface?

3-11. What are the bearing capacity factors for computing bearing failure under spread footings?

3-12. Is the bearing capacity for a spread footing on a clay soil higher or lower than that for a strip footing? On a sandy soil?

3-13. What maximum reduction in soil strength can occur due to application of lateral loads on a footing?

3-14. How do allowable soil pressures in a sandy soil usually incorporate limitations on settlements?

3-15. The size of a footing has a marked effect on the ultimate strength of a sandy soil. What effect does size have on the settlements?

3-16. What effect does submergence have on ultimate strength of sands? On maximum settlement in sands?

3-17. What effect does short-term submergence have on ultimate strength of clays? On maximum settlement in clays?

3-18. What is the usual factor of safety in soils to bearing capacity failure if settlements are not a major concern?

3-19. Where does one usually find the values for the presumptive bearing pressures that have been adopted for a particular locale?

3-20. In the event of a foundation failure due to the use of legally adopted presumptive soil pressures, who is liable?

3-21. Is the contact pressure at the edges of a footing higher or lower than that toward the center in a sandy soil? In a clay soil?

3-22. What causes the pressure under a spread footing to become more nearly uniform (with time) in a sandy soil? In a clay soil?

3-23. What is normally consolidated clay? An overconsolidated clay?

3-24. Why can a constant coefficient of consolidation be found for normally consolidated clays but not for overconsolidated clays?

3-25. Write the Skempton equation for computing the coefficient of consolidation in clays. On what single index property does it depend?

3-26. What size is the sample of clay that is usually used in a consolidation test?

3-27. Why is extreme care required in running a consolidation test?

3-28. What is an *e*–log time curve? Why must an *e*–log time curve be plotted for each increment of load?

3-29. What percent consolidation is used in drawing the final consolidation curve?

3-30. What is the difference between primary consolidation and secondary consolidation?

3-31. How is the level of zero consolidation found for an *e*–log time curve?

3-32. How is the level of 100% primary consolidation found for an *e*–log time curve?

3-33. Why is it necessary to compute settlements in clay stratum by using a series of thin layers rather than the entire stratum as one layer?

3-34. What is the relationship between consolidation time and the travel distance for the extruded water?

3-35. What is the relationships between consolidation time and the degree of consolidation?

3-36. What is the relationship between consolidation time and settlement?

3-37. What does preloading a building site accomplish?

3-38. What is the danger in building over a lightly preconsolidated clay? A sensitive clay?

3-39. In the Schmertmann method of computing the settlements of a footing on sand, how is the modulus of elasticity determined?

3-40. In the Schmertmann method, where does the idea of a variable influence coefficient come from?

3-41. In the Schmertmann method, how deep is the influence of the pressure bulb for a square footing? For a strip footing?

3-42. In the Schmertmann method, what is the maximum value of the influence coefficient, and at what depth does it occur for a square footing? For a strip footing?

3-43. Define the modulus of subgrade reaction. What are its units?

3-44. Why is the modulus of subgrade reaction not used more often in current practice?

3-45. How is it that comparisons of settlements in soil can be more believable than the settlement calculations themselves?

3-46. How is it that elastic theory is generally accepted for an inelastic material such as soil?

3-47. In what stage of loading on a soil is the elastic theory likely to be most accurate?

3-48. About how much more does a strip footing settle than does a spread footing of the same width and contact pressure? Why? Does the Schmertmann method bear out this difference in settlements?

3-49. Name a few circumstances under which a foundation might have to be founded on recompacted soil?

3-50. What is the optimum moisture content of a soil?

3-51. Which types of soil will have an optimum moisture content?

3-52. What is the zero air voids curve?

3-53. What is the minimum number of compaction samples that must be run to establish the optimum moisture content?

3-54. How can a soil be compacted to a higher density than that shown by the standard optimum moisture curve?

3-55. What is relative density?

3-56. To what kinds of soil is relative density applicable?

3-57. What is meant by "90% of relative density?"

3-58. How is maximum density determined?

3-59. How is minimum density determined?

3-60. How does optimum moisture content apply to sand?

PROBLEMS

3-1. A building 100 ft square exerts an average bearing pressure of 425 psf on its footprint area. At what depth can the increase in pressure be expected to drop to 100 psf or less?

3-2. At a depth of 12 ft below an isolated footing 6 ft square, the increase in soil pressure was measured and found to be 200 psf. Determine the approximate load being carried by the footing at the time.

3-3. A strip footing 6 ft wide has a contact pressure of 3000 psf. At what depth will the increase in soil pressure diminish to one-half the contact pressure? To one-fourth the contact pressure?

3-4. Solve Problem 3-3 for a square footing 6 ft on a side and compare the results.

3-5. The backfill against a basement wall is a silty sand, classified SM, weighing 111 pcf. After a long period of time, what will be the pressure against the wall at a depth of 3 ft? 6 ft? 9 ft?

3-6. For the basement of Problem 3-5, what will be the lateral pressures if the area becomes flooded? What is the percent increase in pressure at flood stage?

3-7. What is the equivalent fluid weight from the theory of elasticity for a soil weighing 109 pcf and having a Poisson's ratio of 0.20?

3-8. A soil has an actual weight of 121 pcf and an equivalent fluid weight of 59 pcf. What are the vertical and lateral pressures in this soil at a depth of 9 ft?

3-9. A sand weighing 126 pcf has an angle of internal friction of 28°. Assuming that cohesion is negligible, what will be the equivalent fluid weight at failure?

3-10. A soil weighing 109 pcf has an angle of internal friction of 22° and a cohesion of 650 psf. What is the lateral pressure at failure in this soil when the vertical pressure is 9300 psf?

3-11. A soil has an angle of internal friction of 29°. Compute the bearing capacity factors and check your results against the graphs of Fig. 3-13.

3-12. A soil weighing 109 pcf has an angle of internal friction of 22° and a cohesion of 220 psf. What is the contact pressure under a strip footing 6 ft wide founded 3 ft deep in this soil when failure occurs?

3-13. A silty gravel, GM, has an angle of internal friction of 34°. What would be the allowable bearing capacity of this gravel if settlements are not a consideration, where the D_f/B ratio is 0.25 for a strip footing with $B = 6$ ft?

3-14. What range of allowable bearing values should be expected for a medium sand in an area that is not subject to flooding if settlements are to be limited to 1 in.?

3-15. An isolated footing supporting a water tank in an area not subject to flooding is 10 ft in diameter. The SPT blow count on the supporting SM soil is 12; the soil has a unit weight of 124 pcf. Lateral loads can be as much as 15% of the vertical load. What is the allowable bearing pressure if settlements are to be limited to 2 in.? Estimate $D_f = 1$ ft.

3-16. A fine sandy soil, SM, is located in an area subject to flooding and has a unit weight of 119 pcf and a Dutch cone resistance of 110 ksf. What is the allowable bearing pressure on a 6-ft spread footing buried 3 ft into this soil if settlements are to be limited to 1 in.? If settlements are not limited?

3-17. An MH soil weighing 111 pcf has an SPT blow count of 9. The area is subject to flooding. What is the allowable bearing pressure for a square footing on this soil if settlements are to be limited to 1 in.? If settlements are of no concern? For reference, $B = 8$ ft, $D_f = 2$ ft.

3-18. A clay soil weighing 109 pcf has an unconfined compression strength of 2800 psf. What is the allowable bearing pressure under a strip footing on this soil if settlements are limited to $1\frac{1}{2}$ in. and lateral loads can be as much as 10% of the vertical loads? Use average $D_f/B = 0.50$.

3-19. What range of allowable bearing pressures should be expected under a square footing on a medium clay if settlements are limited to 1 in.? What would be the values for a round footing? A strip footing?

3-20. An isolated granary tank is to be carried on a circular foundation on a plastic silt weighing 112 pcf. The SPT blow count is 7. What is the allowable bearing pressure on this foundation?

3-21. A clay soil has a liquid limit of 58 and a plastic limit of 27. What is its coefficient of consolidation? Is this clay a soil of high compressibility or of low compressibility?

3-22. A building site is underlain by a stratum of normally consolidated clay averaging 60 ft deep and weighing 110 pcf. It is proposed to add 10 ft of sand fill over the area to partially preconsolidate the clay; the sand weighs 116 pcf. The clay at the surface has a void ratio of 1.06 and at 60 ft has a void ratio of 0.74. What settlement should be anticipated under this surcharge?

3-23. A sample of a saturated clay is subjected to a load of 3000 psf in a laboratory consolidation test. What is the porewater pressure at the beginning of the test and at 80% consolidation?

3-24. At a point 16 ft below the surface in a fully saturated clay, the porewater pressure is measured and found to be 6.933 psi. Is the clay layer fully consolidated under the existing loading?

3-25. Rework Example 3-22 using layers of clay 18 in. thick rather than 3 ft as used in the example. Compare the difference in results.

3-26. In Example 3-22, what would the settlement be if the contact pressure were increased to 4200 psf? Prove your conclusion, recognizing that the e–log p curve is not linear.

3-27. A clay stratum is 18 ft thick, drained two sides. If the stratum reaches 41% consolidation in 2 months under a particular load increase, how long will it take to reach 100% consolidation?

3-28. A consolidation sample $\frac{3}{4}$ in. thick, drained two sides, reached 100% consolidation in 296 sec under a particular increase in pressure. The actual field stratum from which the sample was taken is 7 ft thick and drained only at one side. How long will it take the field stratum to reach 80% consolidation under this same increase in pressure?

3-29. A building is expected to settle about 2 in. in the 2 years following its 12-month construction period. Assume the full building load is essentially in place for the last four months of construction and find the total settlement.

3-30. A soil is expected to settle $4\frac{1}{2}$ in. under a particular load. Three months after being loaded, the settlement is $2\frac{1}{2}$ in. How long will it take to settle the remaining amount?

3-31. The load on a 10-ft square footing founded near the surface of a stratum of saturated clay is increased by 112 kips, at which time a pressure gauge located 10 ft below the founding line indicates an increase in the porewater pressure of 336 psf. At the end of 39 days, the settlement is 1.02 in. and the increase in porewater pressure has diminished to 101 psf. How much more settlement can be expected? How long will it take for the remaining settlement to occur?

3-32. For the soil of Problem 3-31, estimate the porewater pressure at the founding line at the time the load is applied, at the end of 39 days, and when all settlements are complete.

3-33. For the footing of Example 3-28, compute the settlement if the founding line is placed 18 in. below surface.

3-34. For the footing of Example 3-28, compute the settlement if the founding line is placed 4 ft 6 in. below surface.

3-35. Using Equations (3-41), estimate the settlement of the footing of Example 3-28 if the pressure is reduced to 2000 psf and the size is increased to 8 ft.

3-36. Estimate the settlement of the footing in Example 3-28 if it were a strip footing rather than a square footing.

Problems 3-37 through 3-40. Standard tests for optimum moisture content produced the indicated results. Plot the moisture content–dry density curve for each set of data, find the optimum moisture content and plot the zero air voids curve.

3-37. $G_S = 2.59$
Sample 1: γ = 105.1 pcf, w = 7.9
Sample 2: γ = 118.2 pcf, w = 10.1
Sample 3: γ = 127.2 pcf, w = 12.5
Sample 4: γ = 128.1 pcf, w = 15.3

3-38. $G_S = 2.51$
Sample 1: γ = 109.4 pcf, w = 14.2
Sample 2: γ = 115.1 pcf, w = 17.3
Sample 3: γ = 118.0 pcf, w = 19.8
Sample 4: γ = 118.9 pcf, w = 21.5

3-39. $G_S = 2.68$
Sample 1: γ = 120.1 pcf, w = 11.4
Sample 2: γ = 128.5 pcf, w = 13.1
Sample 3: γ = 131.2 pcf, w = 17.3

3-40. $G_S = 2.63$
Sample 1: $\gamma = 117.2$ pcf, $w = 10.6$
Sample 2: $\gamma = 123.1$ pcf, $w = 14.1$
Sample 3: $\gamma = 123.6$ pcf, $w = 17.1$
Sample 4: $\gamma = 123.8$ pcf, $w = 19.6$

3-41. A sample of dry sand has a maximum unit weight of 136 pcf and a minimum unit weight of 109 pcf. What is its dry unit weight at a relative density of 95%?

3-42. A sample of sand has a maximum void ratio of 0.91 and a minimum void ratio of 0.49. Its void ratio in-place in a stratum 12 ft thick is 0.72. How much settlement would occur in the stratum if the stratum were to be vibrated to 95% relative density? [Use Eq. (2-19) for settlements.]

4

Field Tests and the Soils Report

Typically, the field investigation for a small-to-medium-size construction project will be subcontracted to a soils engineering contractor who specializes in such work. The field crews that actually conduct the field investigations will follow precisely a work order prepared in advance for the particular site; nothing more will be added and nothing will be deleted. The results of the field work will be transmitted to the contractor's geotechnical (soils) engineer, who will then supervise the related lab tests, evaluate the final data, and prepare the soils report.

The exact scope of work to be followed in conducting the field investigations and the questions to be addressed later in the soils report are supplied in theory by the client, but in practice, the entire soils investigation will probably be worked out between the client's structural engineer and the contractor's geotechnical engineer during the contract negotiations. In normal circumstances, the final submission of the soils report will mark the end of the soils engineer's work. When the soils report is accepted and the fee paid, the soils engineer withdraws from the project. The soils engineer cannot be called upon again (except possibly for clarifications) without renegotiation of fees.

Using the information supplied in the soils report, the structural engineer for the project will then prepare the foundation design. It is this structural engineer who earlier prescribed (in theory at least) the content of the soils investigation. It is also this structural engineer who will eventually sign the design drawings, thereby accepting full responsibility under the law both for the adequacy of the soils investigation and of the resulting foundation design.

It is thus essential to the success of the project that the structural engineer has prescribed a course of field investigations that is pertinent, complete, and relevant to the project. In practice, the structural engineer will depend heavily on the advice of the geotechnical engineer in developing the scope and content of the soils investigation. Informal advice from the geotechnical engineer will also bear strongly on the final selection of the type of foundation system, although such advice may never appear in the written records. A close working relationship between these two engineers is the best insurance possible that a workable, economical, and satisfactory foundation system will be selected.

Typically, the budget for the site investigation and soils report will not be generous; more likely, it will be marginal. As the reputation of soil mechanics has improved over recent years, budgets for site investigation have improved, but few commercial builders will allow much money to be spent poking holes in the ground. The structural engineer will have to plan the soils investigation very carefully to extract maximum information at minimum cost.

One purpose of this chapter is to outline some of the means to plan an economical soils investigation that is still pertinent, complete, and relevant to the design. A second purpose is to introduce some of the common field and lab tests that might be used in the investigation and to categorize their complexity, usefulness, and relative cost. A third purpose is to outline the required content of the soils report itself, in order that every relevant issue concerning the foundation design is addressed before the geotechnical engineer withdraws from the project.

The primary prerequisite for those who plan and conduct such an investigation is a sense of perspective. Perspective in turn is developed through experience and familiarity. Those in responsible charge of foundation work will have served a long apprenticeship in developing such perspective.

PRELIMINARY ASSESSMENT OF SITE

A soils investigation intended for use in a large horizontal earthworks project such as a highway will be much different from a soils investigation intended for an isolated building with its vertical punching load. While some of the final information will be similar or even identical, the primary concern in a soils investigation for a highway is the definition of properties laterally within a strip of right-of-way; for a building, the concern is on definition of properties vertically within the Boussinesq pressure bulb. A significantly different approach is used for these two types of investigations. The methods and equipment described in subsequent sections are those commonly used to investigate building foundations; they may or may not be appropriate for other types of projects.

The information to be gained from the soils investigation will include at least the following items.

1. Identification of the extent of stratification of the soils within the influence of the Boussinesq pressure bulb, to include depth, thickness, strength, compressibility, and classification of each soil stratum
2. Identification and classification of any nonuniform "wedges" of soil within the influence of the pressure bulb, to include their sizes, strength, and compressibility
3. Identification of any rock formations within the limits of the pressure bulb, to include thickness, strike, dip, state of weathering, existence of bedding planes, and presence of any fault lines or outcroppings
4. Location of the groundwater table and its annual variations, to include artesian pressures, if any
5. Extent of flooding due either to natural terrain or to proximity of major drainage channels

Preliminary information concerning the foregoing items may be gained indirectly by a study of topographic maps, aerial photographs, geologic maps, and soils surveys of the area. One of the more useful sources of information is the county soils survey, prepared by the Department of Agriculture; it is usually available at the county level through the Soil Conservation Service. Another valuable source of current information as well as design data is the city/county engineer, who can usually provide up-to-date information about the area surrounding the site as well as background information about potential problem soils in the vicinity. A third very valuable source of information are the local residents, who are probably the best source of unpublished information on cracked plaster, flooded basements, cracked masonry, and other symptoms of problem soils (or poor foundation design). Local residents can also provide solid information on the extent of local flooding. Just a few telephone calls to government officials such as those named above and to local residents in the general area can often provide a wealth of preliminary information concerning a particular site or a general area.

Having assimilated the available preliminary information from maps and soil surveys, the structural engineer should personally visit the site and make an assessment. Such a visit is not always feasible in today's practice, where the design office may be several hundred miles from the building site; travel budgets on smaller projects will not always permit such travel. Admittedly, the part of the site that the structural engineer wants to assess is below grade and inaccessible, but a general appraisal of the site and the topography will be a valuable asset later when prescribing the on-site soils investigation.

SCOPE OF THE SITE INVESTIGATION

It would be impractical to attempt to do the conceptual architectural designs for a large building without consulting the structural engineer. The structural engineer, in turn, could not make any authoritative recommendations on the structural system until he or she is reasonably sure what the foundation conditions are. As the conceptual design is progressing, therefore, one of the first responsibilities of the structural engineer is always to determine the foundation conditions, which usually will include awarding the contract for the soils investigation and report.

On smaller projects, however, the structural engineer may not even be engaged (or assigned to the project) until the conceptual architectural designs have already been completed and approved by the owners. Such a procedure is not unusual. Many, if not most, architects are perfectly capable of preparing a preliminary design, to include basic structure, on small-to-medium-size projects. With experience, the architect will have learned to allow for a considerable amount of latitude within the structural system in order that options remain open. In such circumstances, one of the first jobs of the structural engineer when he or she is finally assigned to the project is to determine exactly the foundation conditions as they affect the structure.

For almost all imaginable circumstances, then, one of the first problems to be resolved by the structural engineer is always to determine the foundation conditions. At these early stages, it would seem that the selection of the final structural system would have to await verification of suitable foundation conditions. With experience, however, and with a general knowledge of the soils and geology in a given area, the structural engineer can usually proceed with a tentative structural design with some degree of confidence that it will be suitable.

The soils engineering contractor who is to conduct the soils investigation will have to be supplied with preliminary design information and background information. Such information will include:

1. Site plan, to include grading, earthworks, and drainage plans as available
2. Building footprints for every structure in the project
3. Overall average uniform pressure exerted by each building in the project
4. Type of structural system used in each building
5. Foundation layouts, showing spread footings, strip footings, and grade beams
6. Location and estimated magnitude of lateral loads to be carried by each footing
7. Maximum and minimum column loads
8. Location and height of retaining walls and basement walls
9. Anticipated limits on settlements expected to be used in the structural calculations
10. Anticipated shearwall drift (angle of rotation) expected to be used in the structural calculations
11. Any isolated concentrated loads that are expected but are not included as a part of the column loads
12. Whether surcharge loading might be used to reduce settlements

There are very few ways to correct a foundation failure. (Courtesy of Acker Drill Co., Scranton, PA)

Once the foregoing list of items has been supplied to the soils contractor, a program of sampling and testing that relates specifically to the proposed design can be established. The purpose, of course, is to reduce contingency; as contingency is reduced, costs are reduced. The program of testing can then include every test that is relevant to the particular project design and exclude all those that are not.

Failure to provide the soils contractor with a definitive description of the project will require that the soils contractor extend the investigation to cover all reasonably possible cases, thereby increasing the work (and cost) considerably. Even then, there may be items in the final design that should have been verified by soils tests but simply were not anticipated. For minimum cost and maximum relevance, the soils investigation must be based on a definitive preliminary design.

There is little point, for example, in performing a soils investigation over an entire site if it is known that the building footprint covers only a small portion of the site. In such a case it is necessary only to perform the soils investigation on the soils within the general area of influence of the Boussinesq pressure bulb. A subsurface investigation of soils elsewhere on the site would have no relevance to the design of this particular building.

Similarly, having the in-place dry density of the soil would not seem to be necessary for the sake of the structural design. However, in the event the foundation material is disturbed during construction operations, it will be necessary to know its original dry density in order to control the inevitable recompaction. Determination of dry density during the soils investigation is convenient and inexpensive; later, it could produce an exasperating and expensive delay.

To establish the size of the area that must be investigated, it is necessary only to establish the extent of the pressure bulb. A typical case is shown in Fig.

4-1, where the three-dimensional pressure bulb is shown for the length and width of a rectangular building. It does not matter that the load is actually delivered to the soil by discrete footing loads rather than as an overall uniform pressure; the overall dissemination of pressure and the overall size of the bulb of influence are the same for either case.

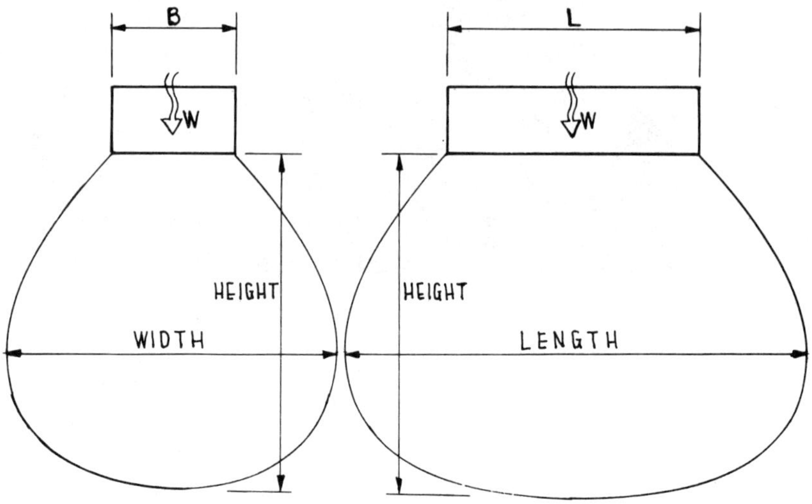

Figure 4-1 Size of pressure bulbs.

The outer pressure contour shown in Fig. 4-1 is the contour representing an increase in vertical pressure amounting to 10% of the average contact pressure p_{avg}. Within the usual limits of accuracy in soils, the influence of the building load outside this 10% pressure contour is minimal and can be ignored. (The depth and size of these pressure bulbs were developed in Chapter 3 and shown in Fig. 3-2 for an infinitely long strip and for a square. The solution shown in Fig. 3-2 for the size of the bulb is valid regardless of scale; B can be a footing width or a building width, but p_{avg} must be the *average* pressure over the width B.)

The Boussinesq solution in Chapter 3 is not extended to rectangles, which is the usual shape of a building in plan. An approximate solution for rectangles, accurate enough for planning the soils investigation, can be developed by taking the side slopes of the pressure bulb at approximately 2 : 1, as indicated in Fig. 4-2. The result is a truncated pyramid with a pressure increase at its base taken to be 5% above the at-rest pressures [34].

The total weight of the building in Fig. 4-2 is denoted $p_{avg}BL$. At the base of the pressure bulb, the at-rest vertical pressure is γH; hence, at 5% increase,

$$0.05\gamma H = \frac{p_{avg}BL}{(B + H)(L + H)} \tag{4-1}$$

Note that the 10% increase in vertical pressure used in Chapter 3 is based on the contact pressure p_a at the top of the bulb. In Eq. (4-1), however, the 5% increase

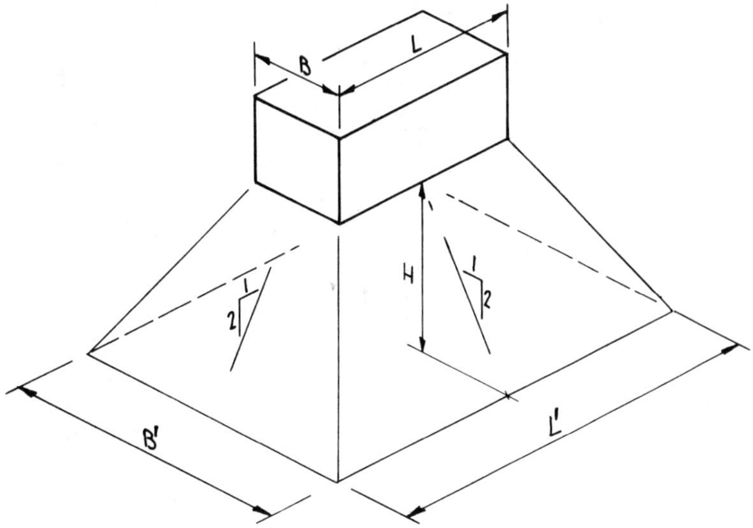

Figure 4-2 Approximate extent of influence.

refers to an increase above the vertical at-rest soil pressure at the bottom of the bulb, a significantly different concept. [The overall average pressure p_{avg} is so small compared to p_a that using 10% of p_{avg} in Eq. (4-1) would produce an insignificant increase in soil pressure at the base of the bulb and a disproportionately large pressure bulb.]

The ratio L/B is now denoted as n, and γ is taken at 100 pcf. With these additions, Eq. (4-1) is solved for H/B, yielding a cubic equation

$$\left(\frac{H}{B}\right)^3 + (n + 1)\left(\frac{H}{B}\right)^2 + n\frac{H}{B} - n\frac{p_{avg}}{5B} = 0 \tag{4-2}$$

The solution of this cubic equation is given in Table 4-1 for typical values of L/B and p_{avg}/B. The resulting values of L', B', and H are the outermost dimensions of the pressure bulb that would occur under a rectangular building having a width B and a length L.

The value of p_{avg} in Table 4-1 can best be estimated from the uniform dead and live loads at each floor and at the roof. Dead load per floor on a concrete building can be estimated at about 75 to 150 psf. For steel or timber, the weight is about 60 to 80 psf. Live load is commonly between 50 to 150 psf for floors and 20 to 80 psf for roofs. As a first estimate for a concrete building, the total average pressure p_{avg} might be estimated at about 175 psf per floor plus 150 psf for the roof; for steel or timber, deduct about 50 psf from each of these.

Some examples will illustrate the use of Table 4-1. Remember that the results are intended only to define a general volume that will be subject to an increase in pressure. Considerable latitude may be exercised in applying the numbers.

TABLE 4-1 OUTER LIMITS OF APPROXIMATE PRESSURE BULB EXPRESSED AS MULTIPLES OF THE LEAST WIDTH B[a]

	p_{avg}/B																				
	2			4			6			8			10			15			20		
L/B	L'/B	B'/B	H/B	L'/B	B'/B	H/B	L'/B	B'/B	H/B	L'/B	B'/B	H/B	L'/B	B'/B	H/B	L'/B	B'/B	H/B	L'/B	B'/B	H/B
1	1.3	1.3	0.3	1.4	1.4	0.4	1.5	1.5	0.5	1.6	1.6	0.6	1.7	1.7	0.7	1.9	1.9	0.9	2.0	2.0	1.0
2	2.3	1.3	0.3	2.5	1.5	0.5	2.6	1.6	0.6	2.7	1.7	0.7	2.8	1.8	0.8	3.0	2.0	1.0	3.2	2.2	1.2
3	3.3	1.3	0.3	3.5	1.5	0.5	3.6	1.6	0.6	3.7	1.7	0.7	3.8	1.8	0.8	4.1	2.1	1.1	4.3	2.3	1.3
4	4.3	1.3	0.3	4.5	1.5	0.5	4.6	1.6	0.6	4.8	1.8	0.8	4.9	1.9	0.9	5.1	2.1	1.1	5.3	2.3	1.3
5	5.3	1.3	0.3	5.5	1.5	0.5	5.6	1.6	0.6	5.8	1.8	0.8	5.9	1.9	0.9	6.1	2.1	1.1	6.3	2.3	1.3
6	6.3	1.3	0.3	6.5	1.5	0.5	6.7	1.7	0.7	6.8	1.8	0.8	6.9	1.9	0.9	7.2	2.2	1.2	7.4	2.4	1.4
7	7.3	1.3	0.3	7.5	1.5	0.5	7.7	1.7	0.7	7.8	1.8	0.8	7.9	1.9	0.9	8.2	2.2	1.2	8.4	2.4	1.4
8	8.3	1.3	0.3	8.5	1.5	0.5	8.7	1.7	0.7	8.8	1.8	0.8	8.9	1.9	0.9	9.2	2.2	1.2	9.4	2.4	1.4
9	9.3	1.3	0.3	9.5	1.5	0.5	9.7	1.7	0.7	9.8	1.8	0.8	9.9	1.9	0.9	10.	2.2	1.2	10.	2.4	1.4
10	10.	1.3	0.3	11.	1.5	0.5	11.	1.7	0.7	11.	1.8	0.8	11.	1.9	0.9	11.	2.2	1.2	11.	2.4	1.4

[a] L and B, length and width of the building; L', outermost length of the pressure bulb; B', outermost width of the pressure bulb; H, depth at 5% increase in at-rest pressure; p_{avg}, average uniform pressure exerted by the building on its overall footprint area.

EXAMPLE 4-1

A soils investigation is being planned for a concrete building 80 ft × 200 ft, five stories high. Estimate the size of the pressure bulb that should be used for the soils investigation.

SOLUTION Since the exact floor pressures are apparently unknown, use 175 psf for each floor plus 150 psf for the roof. The estimated uniform pressure over the building footprint is then

$$p_{avg} = 5 \times 175 + 150 = 1025 \text{ psf}$$

For entry into Table 4-1 compute L/B and p_{avg}/B:

$$\frac{L}{B} = \frac{200}{80} = 2.5$$

$$\frac{p_{avg}}{B} = \frac{1025}{80} = 13$$

Interpolate in Table 4-1 to find

$$\frac{L'}{B} = 3.45$$

$$\frac{B'}{B} = 1.95$$

$$\frac{H}{B} = 0.95$$

The size of the bulb which should be included in the investigation is then

$$\text{length} = 3.45 \times 80 = 280 \text{ ft}$$

$$\text{width} = 1.95 \times 80 = 160 \text{ ft}$$

$$\text{depth} = 0.95 \times 80 = 80 \text{ ft}$$

EXAMPLE 4-2

A soils investigation is to be conducted for an L-shaped building as shown. The building is six stories high, of steel construction. How large a volume should be included in the pressure bulb for the soils investigation?

SOLUTION Consider the building in two pieces, *ABGC* and *EFBH*. For a three-story steel building use a contribution of 125 psf per floor plus 100 psf from the roof.

$$p_{avg} = 3 \times 125 + 100 = 475 \text{ psf}$$

For entry into Table 4-1 for the portion *ABGC*,

$$\frac{L}{B} = \frac{240}{60} = 4.0 \qquad \frac{p_{avg}}{B} = \frac{475}{60} = 8.0$$

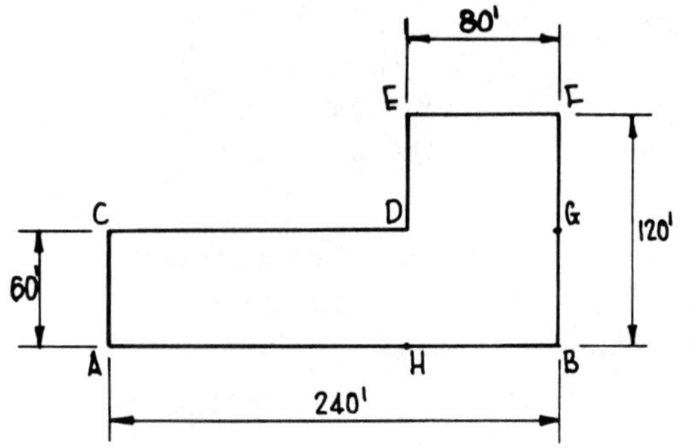

From Table 4-1, find

$$\frac{L'}{B} = 4.8,\ L' = 290\text{ ft}$$

$$\frac{B'}{B} = 1.8,\ B' = 110\text{ ft}$$

$$\frac{H}{B} = 0.8,\ H = 50\text{ ft}$$

Similarly for the portion *EFBH*,

$$\frac{L}{B} = \frac{120}{80} = 1.5 \qquad \frac{p_{\text{avg}}}{B} = \frac{475}{80} = 6.0$$

From Table 4-1 find

$$\frac{L'}{B} = 2.05,\ L' = 165\text{ ft}$$

$$\frac{B'}{B} = 1.55,\ B' = 125\text{ ft}$$

$$\frac{H}{B} = 0.55,\ H = 45\text{ ft}$$

Use the overall area shown in the sketch.

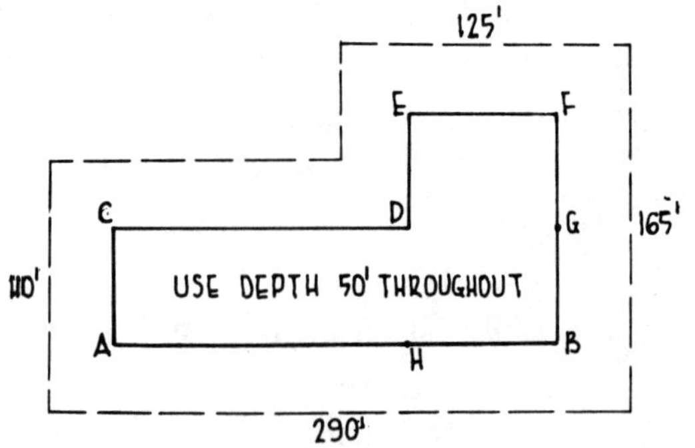

Having an approximate size for the pressure bulb and a preliminary design of the structure with its foundation loads, the structural engineer will be able to outline exactly the information on soils that will be needed for the final design calculations. During the contract negotiations with the geotechnical engineer, the program of soils investigation can be specifically tailored to suit those needs. Also at these negotiations, additional items may be added to the investigation by the project architect and the project civil engineer; many such items of information that are not related to foundations might possibly be added at this time.

At this point, the general knowledge and experience of the geotechnical engineer concerning the soils in the vicinity will be a valuable contribution to the proceedings. Using the same information that the structural engineer uses but with a different point of view, the geotechnical engineer may recognize problem areas in the foundation that would be beyond the specialized knowledge of the structural engineer. After suitable study, he may recommend the addition of specific tests or investigations to verify that the soil is in fact adequate for the proposed design or, on occasion, to show that particular features of the design are not well suited to the site conditions.

It is also at this point that the geotechnical engineer can advise the structural engineer, the project civil engineer and the project architect of any potential dangers that have been found in the overall site plan. Large cuts, fills, retaining walls, and steep slopes can sometimes contribute to soil slides that could later intrude into the building foundation. Desiccated clays in the area may be subject to annual cycles of shrink and swell that could affect the integrity of the foundation. Nearby rock outcroppings can cause perched water tables and consequent water problems not only in the foundation but also in other aspects of the project design. All such site conditions outside the immediate vicinity of the pressure bulb should be investigated and evaluated by the geotechnical engineer, and his or her appraisal included in the final soils report.

The contract should always list the specific problem areas that are to be discussed and evaluated in the soils report. Items other than these specific items may be left to the discretion of the geotechnical engineer; he or she should be given considerable latitude in formulating and expressing an evaluation. Wherever specific problem areas are to be addressed that will require a formal evaluation and recommendation, it is always well to name such requirements as a part of the scope of work.

Once the exact programs of soils investigation and testing (and cost) have been agreed upon, the entire scope of work can be specified and the contract can be executed. Where such a precise scope of work is specified, there will be little room for contingencies. As a result, it will be necessary to include special provisions in the contract to allow for renegotiations; renegotiations could become necessary if the field work reveals surprise conditions previously unknown or unsuspected by anyone on the project.

Such surprises are more common than one cares to admit, especially in areas where there has been little previous development. For example, the only limestone cave in the area always seems to be located directly under the proposed building rather than elsewhere on the site, or the only pocket of sensitive clay on

Access to a drilling site is often the most difficult problem to be solved. (Courtesy of Acker Drill Co., Scranton, PA)

the site will probably intrude into the pressure bulb, or the only old forgotten sanitary landfill within 30 miles will probably be located directly under this particular building site. Those who write the contract for the soils investigation should expect such complications and allow for a suitable extension of services within the terms of the contract.

Barring such surprises revealed by the field investigation, work on the structural design can usually be continued while the soils investigation is under way. Undoubtedly, however, there will be frequent exchanges during this period between the soils contractor's staff and the structural design staff, with the design sometimes proceeding on a "best guess" basis. Due to such exchanges, the final results of the soils investigation probably will be known to the design staff long before the actual soils report is finally received.

FIELD SAMPLING AND TESTING

Having agreed on the soils information to be provided and the cost to provide this information, the geotechnical engineer will set up the field and lab operations to perform the investigation. Almost certainly, he or she will be pressed for time. In American practice, the engineer will probably have at his or her disposal several major pieces of drilling and sampling equipment, a small staff of qualified and competent soils technicians, and a lab equipped to conduct all the more common soils tests. In less developed areas, the ultimate capability will probably be comparable, but there will be very little of the timesaving and laborsaving equipment; the test program will probably require a great deal more time than it would in American practice.

One of the requirements of the test program will likely be to sample and classify all soil strata that fall within the Boussinesq pressure bulb. Inevitably, at least one hole will have to be drilled down through those strata in order that samples can be extracted. These holes will necessarily be comparatively deep, located within the limits of the bulb. These holes serve no purpose other than access; the cheapest possible means to drill these holes or otherwise provide access to a stratum will of course be used.

Wash boring is a very economical way to gain access to a stratum for testing. (Courtesy of Acker Drill Co., Scranton, PA)

It may not always be necessary to drill to the full depth of the pressure bulb (as indicated by H/B in Table 4-1). For example, if it has repeatedly been verified that a deep stratum of a particular sand underlies the entire geologic area, there is little point in again drilling deep into the sand stratum. Once the top surface of that stratum is reached and the stratum identified, the stratum can be sampled for verification and drilling discontinued.

In addition to at least one deep hole drilled near the center of the pressure bulb, there may sometimes be two shallower holes drilled nearby. In plan, the three holes should form a large triangle; the shallower holes may even be located outside the limits of the pressure bulb. At each of the three holes, the depth of the first few soil strata are determined and logged. An evaluation of these depths will permit an evaluation of the levelness of the stratification, as well as identification of possible wedges of soil that do not extend across the entire building site.

Sampling is not usually performed from the shallower holes; their only purpose is to permit identification of depths and thicknesses of the upper two or three soil strata.

At shallow elevations, a *probe* may be useful in locating the depth of subsurface rock or even in finding changes in soil type, if such changes are distinctive. A probe is little more than a pointed steel rod or hollow pipe, driven into the ground until it hits something that feels different. A water *jet* is sometimes used to advance the probe in hard or gravelly ground. The applications of a probe are obviously quite limited.

A *test pit* may be useful at shallow excavations; a test pit is a hole that is actually dug (or bulldozed) down to the founding line or even lower in order that samples can be extracted. Insofar as the sampling itself is concerned, a test pit is probably the best means of access to the stratum. Its cost is usually too high to permit depths more than about 10 ft, although some backhoes can dig up to about 15 ft.

Augers provide a low-cost means of access, useful up to a maximum depth of about 100 ft; hole size can be anywhere between 4 and 12 in. in diameter. They may be used to bring the material up for examination, but the material will be disturbed and remixed to such an extent that it is usually suitable only for classifying the sample. Once the hole is drilled and cleaned, however, undisturbed samples can be extracted from the bottom of the hole as in any other method of drilling. Manual augers can go to depths of 6 to 8 ft, small (short-flight) powered augers to about 20 ft, and large (continuous-flight) powered augers to about 100 ft.

Wash boring is a means of drilling using a water jet and agitation to advance the hole. The effluent water is a soil-water slurry and is almost useless in identify-

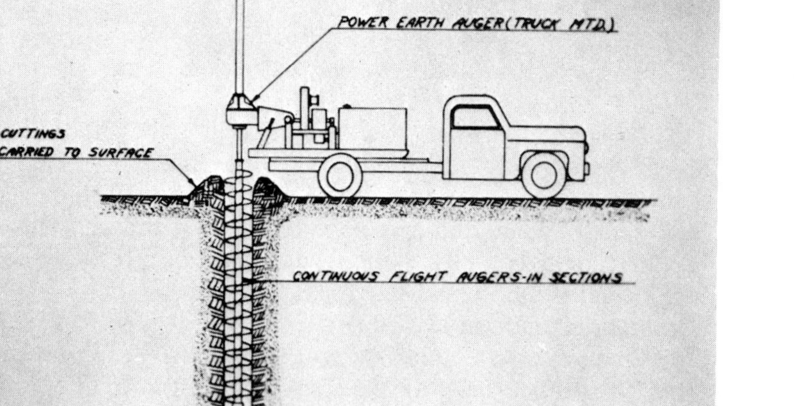

Truck-mounted earth augers are one of the more common drilling devices used in today's practice. (Courtesy of Acker Drill Co., Scranton, PA)

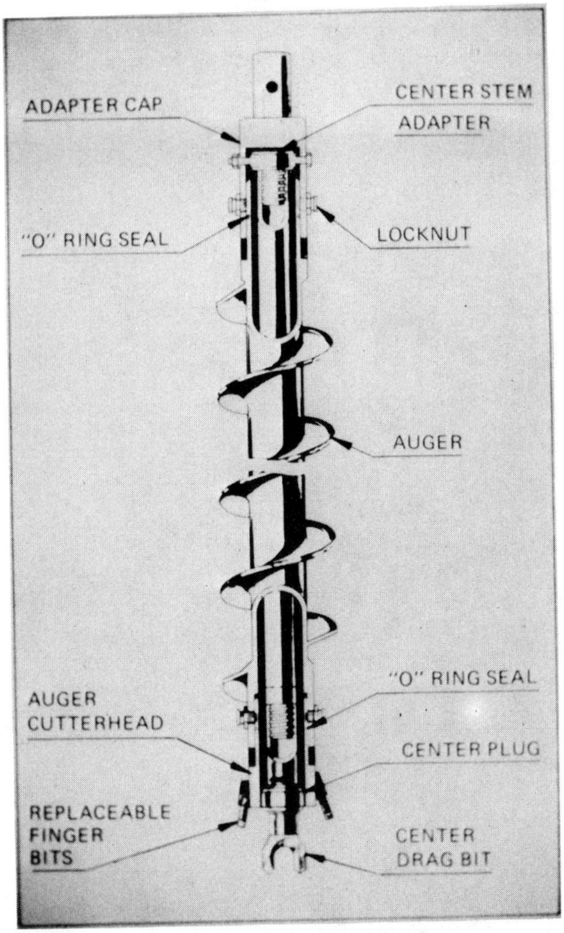

Nomenclature of the auger cutterhead. (Courtesy of Acker Drill Co., Scranton, PA)

ing the soil. The driller locates changes in soil type by noting changes in color and texture of the wash water effluent. At each level where a sample is required, the hole is cleaned and conventional sampling equipment is used to extract the sample.

Percussion drills, or *spudders,* literally pound a hole in the ground, using a long heavy steel bar suspended from a steel cable, having some type of drill bit attached to its base. The "tools" are repeatedly picked up and dropped about 2 ft, punching a hole in the ground. A water slurry is maintained at the bottom of the hole; the slurry water is periodically bailed out as it thickens, replaced with clean water, and spudding is then resumed. The method is cheap and can be used to drill holes up to 200 or 300 ft deep. It includes the capability of placing steel casing (a steel liner) to stabilize the sides of the hole. As with wash borings, at each level where sampling is required the hole is cleaned and conventional sampling equipment is used to extract the sample.

Rotary drills are often used when one is "pioneering" an area, that is, the geology of the locality is not well known and the soil samples must be identified throughout the entire depth of the boring. Rotary drilling utilizes a series of hollow drill rods with a drill bit attached at the bottom. Continuous sampling is possible since the undisturbed core of soil enters the hollow barrel and can then be retrieved and examined. Rotary drills have been adapted into such a variety of applications that they have become by far the most widely used type of drill in soils explorations in today's practice.

Rotary drills are the most adaptable of all drilling machines in performing soils investigations. (Courtesy of Acker Drill Co., Scranton, PA)

As a general rule, the larger the hole diameter, the greater the cost of drilling it. Unusually small holes, however, prohibit extracting a useful-size sample. In general, a hole size 4 to 6 in. in diameter will accommodate the more common sampling equipment.

Often, a hole will have to be "cased," that is, lined with steel pipe, to keep it from caving in; the problem is almost certain to occur in sands lying below the water table. Lightweight thin steel casing is commonly used to case the hole. The casing can be extracted when sampling is completed and used again in later operations.

Regardless how the hole was advanced to a particular stratum, the system for sampling or testing the soil in the stratum must be carried out at the bottom of a small hole by an operator located at the surface. Over the years, techniques and

Sampling the soil in a subsurface stratum is performed readily from a rotary drill. (Courtesy of Acker Drill Co., Scranton, PA)

equipment have evolved that permit such sampling and testing to be done with a minimum of unwanted disturbance to the soil.

A very common sampling device is the split-barrel sampler, shown in Fig. 4-3. The sampler is fitted to the end of a driving rod and lowered to the bottom of a thoroughly cleaned hole. The sampler is driven into the soil using a 140-lb hammer falling 30 in.; the sample undergoes considerable disturbance during this operation. The sampler is recovered and taken to the lab, where, upon disassembly of the sampler, the soil sample is recovered for subsequent testing. The number of blows required to drive a 2-in. sampler 1 ft into the soil is the standard

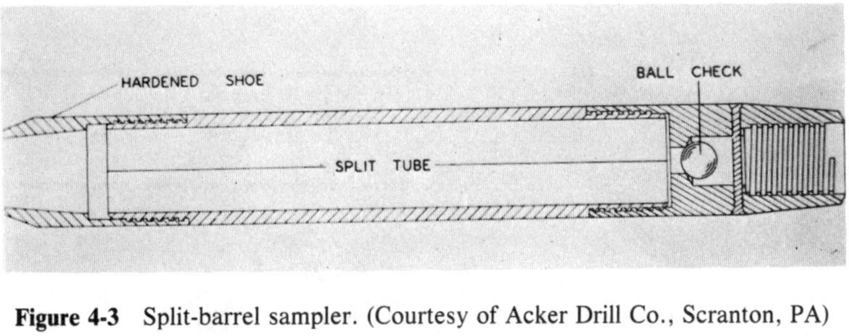

Figure 4-3 Split-barrel sampler. (Courtesy of Acker Drill Co., Scranton, PA)

penetration number (SPT) for that soil; the standard penetration test is described in more detail in Chapter 2.

Another common sampling device is the thin-walled (Shelby) tube sampler. The thin-walled tube is commonly pressed into the soil rather than driven, usually by a hydraulic jack on the drill rig. The thin-walled sampler does not produce the disturbance in the sample that the split-barrel sampler does, but it is somewhat more expensive. Samples of any length can be recovered, the tube and sample together are simply cut into convenient lengths after recovery, and these yet undisturbed test samples can then be distributed in the lab as needed. A typical thin-walled sampler is shown in Fig. 4-4.

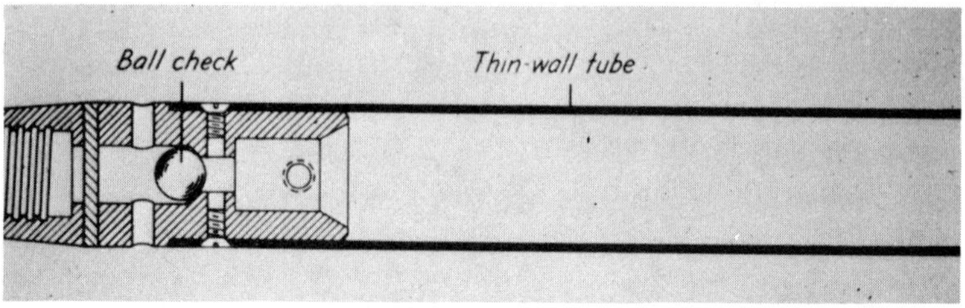

Figure 4-4 Thin-walled sampler. (Courtesy of Acker Drill Co., Scranton, PA)

There are other samplers in common use, usually modeled after the split-barrel sampler or the thin-walled sampler, many being nothing more than a variation of some kind to solve a particular sampling problem. An example of such a variation is a retainer mechanism fitted to the split-barrel sampler. In these devices a flap or spring-steel retainer is located near the mouth of the sampler, which closes and retains the sample as it is being taken to the surface. Such a retainer is usually necessary in sands located below the water table; without it, the sample would simply fall out of the sampler as it is being withdrawn.

Another variation is that of the piston sampler, in which a thin-walled tube is fitted with an internal piston which keeps the soil under pressure as the tube is pressed into the soil. A piston sampler is shown in Figure 4-5. The recovery of undisturbed samples in a piston sampler is much improved over those in an ordinary thin-walled sampler.

There are a variety of other types of sampling devices in existence, each with its own adherents, each with its own set of advantages and disadvantages.

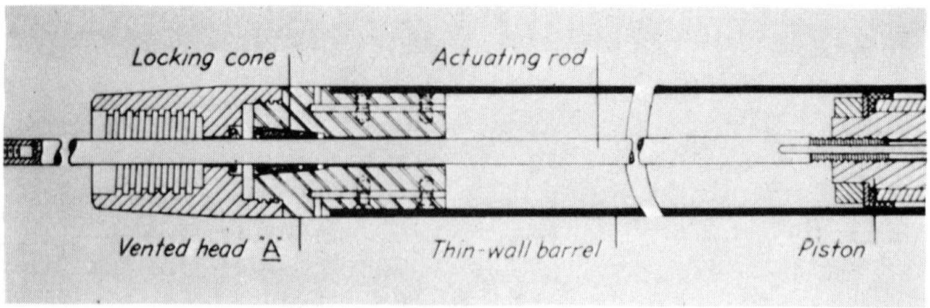

Figure 4-5 Piston sampler. (Courtesy of Acker Drill Co., Scranton, PA)

As a rule, when complexity increases, cost increases. The user must eventually accept whatever sampling system gives acceptable results within a reasonable cost. It must be remembered, however, that the quality of sampling which might be acceptable in a foundation investigation for a small building might be completely unacceptable in a soils investigation for a dam.

For foundations, the split-barrel sampler is very commonly used in those cases where the need is simply to determine stratification, density, and classification of the soil. The blow count on the sampler yields a measure of the angle of internal friction and strength, thus permitting a reasonably complete appraisal of the soil to be made at the particular depth using only a single test procedure. Unfortunately, the SPT blow count is most accurate in sands; its accuracy diminishes rapidly as cohesion (clay content) increases. In addition, samples taken in a split-barrel sampler are classified as disturbed samples and may not be used in sensitive lab tests such as the consolidation test or the unconfined compression test.

In recent years, the Dutch cone test has grown considerably in popularity. The test is discussed in Chapter 2; it is a penetration test, directly parallel to the SPT. It may also be conducted at the bottom of a borehole, but its use does require another operation in addition to the sampling operation. For the SPT, the sampling operation and the blow count are accomplished at the same time, reducing time and cost.

As clay content increases, however, the vane shear test described in Chapter 2 becomes more accurate than the penetration tests such as SPT and Dutch cone. The vane shear test is readily conducted at the bottom of a borehole, provided that the vane is driven to a depth such that disturbances from the drilling, washing, and cleaning operations are no longer present. Having the results of both the penetration test and the vane shear test at a particular level, along with a soil classification and a knowledge of the stratification, the geotechnical engineer can usually make an evaluation that is accurate enough for a small-to-medium-size construction project.

It should be emphasized at this point that these field tests reveal only the total shear strength of the soil. They do not provide a means to distinguish between the amount of strength that is derived from cohesion or that due to internal friction. Nor is there any direct assessment of settlements; the appraisals of Chapter 3 (Figs. 3-16 and 3-18) coupled with the judgment of the geotechnical engineer will provide the primary evaluation of settlements when the test program is limited to these two field tests.

Where additional test information is required, test samples would be taken. These samples would be carefully identified and taken back to the laboratory for further tests. Typical of such lab tests are the direct shear test in sands, the unconfined compression test in clays, and the triaxial compression test in mixtures of sands and clays. Even where the field tests are the only tests to be performed regarding the soil strength, however, laboratory testing will still be required for determination of water content, specific gravity of solids, and in-place void ratio.

Except for the consolidation test, none of these field or lab tests reveal any settlement properties of the soil. Since the consolidation test applies only to heavy clays, there is little information that can be developed regarding settlements in sands or in mixtures of sands and clays. Reliable and inexpensive methods to determine actual long-term settlements in the field have not yet been developed.

For determination of short-term settlements, however, a type of test called a load–deflection test has been in use for quite some time. A sketch showing a typical setup for such a load–deflection test is shown schematically in Fig. 4-6. There is no standard configuration or size or loading for the load–deflection test; each investigator conducts the test in his or her own fashion to suit his or her own needs.

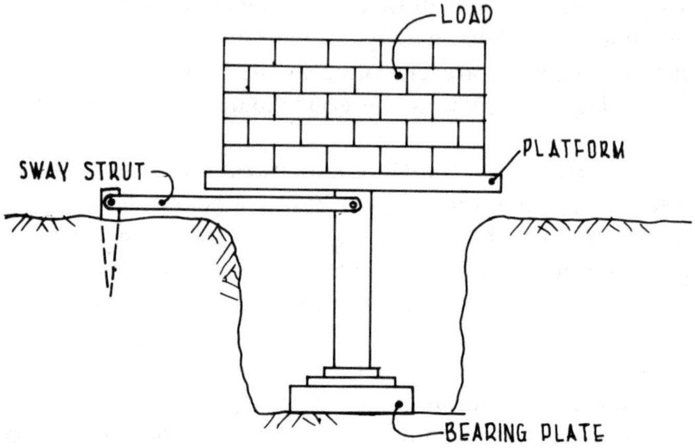

Figure 4-6 Typical setup for load-deflection test.

In concept, the load–deflection test is quite simple. A bearing plate representing a footing is placed at a depth comparable to the founding line for the actual foundation. It is then loaded in increments, one increment being in the range of about one-tenth the estimated failure load. When an increment of load is added, deflection will be quite rapid at first, slowing markedly after only a few minutes or hours. When the deflection rate has decreased to less than about 0.005 in. in 30 min, another increment is added and the procedure repeated (24 hours per increment is also used as a convenient time interval). The test is complete when deflections no longer cease in a reasonable length of time.

The results of the test are plotted very similarly to those of the consolidation test as discussed earlier. A settlement–log time curve is plotted for each increment of load as indicated in Fig. 4-7. From each of these curves, a point is chosen which best indicates the maximum short-term deflection under the particular load; this point is usually chosen as the beginning point of the final straight-line portion of the curve, designated Δ in the curves of Fig. 4-7a.

The final load–deflection curve for the test is then drawn; the total load at any time is plotted against its maximum short-term deflection as indicated in Fig. 4-7b. It is emphasized that this curve represents only short-term deflections and does not include any effects of long-term consolidation or degradation. Even so, it probably does include a major portion of the total settlement.

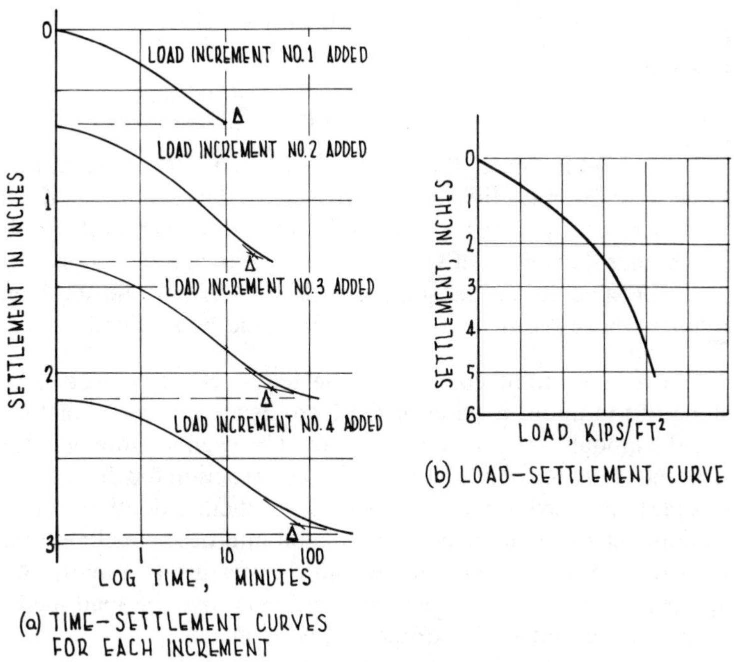

Figure 4-7 Load test curves.

The load-deflection curve of Fig. 4-7b represents the behavior of only one size of plate. The behavior of a smaller plate or a larger plate could be much different. Extrapolation of results into larger size of foundations is quite commonly done, however, especially when comparing settlements between adjacent footings (see Chapter 7).

The field load–deflection test is obviously quite slow, expensive, and cumbersome. If results are to be meaningful, the bearing plates should be large, but for large plates the required load becomes upward of 30, 40, or 50 tons. For smaller sizes of plates, say 2 ft square, loads become more manageable, but the Boussinesq pressure bulb extends such a short distance into the stratum that the results mean little. The following example illustrates this problem with "scale" in a load–deflection test.

EXAMPLE 4-3

A footing 10 ft square having an average uniform pressure p_{ftg} is assumed to have a pressure bulb extending roughly $2B$ or 20 ft below the depth of founding, producing a pressure of $0.10p_{\text{ftg}}$ at that depth. How much pressure would have to be exerted on a test plate 2 ft square to produce the same increase in pressure at the same depth?

SOLUTION The depth of influence of 20 ft for the footing amounts to $10B$ for the 2-ft plate. From the Boussinesq pressure bulb of Fig. 3-2, the pressure at that depth is roughly $0.005p_{\text{pl}}$. This pressure is equated to that for the footing,

$$0.005p_{\text{pl}} = 0.10p_{\text{ftg}}$$

Hence

$$p_{\text{pl}} = 20p_{\text{ftg}}$$

This result indicates that the pressure on a 2-ft square test plate would have to be some 20 times the contact pressure of the real footing if it is to reproduce the Boussinesq pressure bulb and provide meaningful results. It is doubtful whether the soil itself would be strong enough to sustain such a load, since the factor of safety on the soil is usually less than 5; the soil would most likely fail long before the necessary pressure could be attained.

One last point concerning the load–deflection test merits comment. The slope of the initial portion of the load–deflection curve of Fig. 4-6 is something like the modulus of elasticity that would be used in more elastic materials. It is, in fact, the short-term modulus of subgrade reaction for this particular soil loaded by this particular size plate located at this particular depth of founding. Whether the modulus can or cannot be extrapolated into other loading rates, into other plate sizes or into other elevations within the stratum is a matter for each foundation designer to decide; the design methods that use the load–deflection test as their basis require that such extrapolations be made.

Concerning field tests in general, one of the more notable shortcomings in field testing is the lack of a suitable test to determine field densities at in-place pressures that can be conducted at the bottom or at the sides of the borehole. The current tests for field densities require an operator; they can only be conducted therefore at atmospheric pressures in a place accessible to the operator. There are specialized density tests that can be performed in the borehole on particular kinds of soils, but an inexpensive general test to measure in-place densities (or void ratios) at in-place pressures anywhere along the borehole remains an elusive point in the technology.

COMMON LABORATORY TESTS

Throughout the preceding chapters, various laboratory tests for soils have been introduced and discussed. For the sake of convenience, these tests, along with other common lab tests, are summarized in the following list. The standard procedure for conducting these tests, when such a standard procedure exists, is indicated by its test designation as given by the American Society for Testing and Materials (ASTM). In some cases there are particular requirements for field sampling that must be observed in conjunction with a particular test. Such requirements are indicated, together with the approximate size of the sample that is required for the test.

TESTS FOR INDEX PROPERTIES AND SOIL CLASSIFICATION

1. *Water content:* ASTM D2216, natural moisture content protected while awaiting test, 10- to 20-g sample required for test

2. *Specific gravity of solids:* ASTM D854 and ASTM C127, 150 to 500 g of sample required, depending on coarseness of grain sizes
3. *In-place density:* ASTM D1556 or D2167, usually performed in-place as the moisture sample is being extracted, but may be made on a large (10 kg) undisturbed sample
4. *Liquid limit:* ASTM D423, performed on fine-grained soils, requires 100 to 500 g of that portion of the sample passing the No. 40 sieve
5. *Plastic limit:* ASTM D424, performed on fine-grained soils, requires about 20 to 30 g of that portion of the sample passing the No. 40 sieve
6. *Sieve analysis:* ASTM D422, performed on dried whole sample, requires up to 500 g of sample with grain sizes up to $\frac{3}{8}$ in., up to 5000 g for larger sizes

TESTS FOR FRICTION, COHESION, AND COMPRESSION STRENGTH

7. *Direct shear:* ASTM D3080, performed on disturbed cohesionless sample about 4 in. by 4 in. square, up to 1 in. thick
8. *Unconfined compression:* ASTM D2166, performed on undisturbed cohesive soils, sample cylindrical in shape, diameter 2.8 in., length two to three times diameter
9. *Triaxial compression:* ASTM D2850, sample size and shape same as that for unconfined compression test, but sample may be cohesionless if undisturbed
10. *Vane shear:* not a standard test, usually performed in-place in the field but may be performed in the lab on a large sample having dimensions at least three times the vane size

TEST FOR COMPRESSIBILITY

11. *Consolidation:* ASTM D2435, performed on undisturbed sample of cohesive soil at least 2.5 in. in diameter, with thickness one-third to one-fourth the diameter, usually $\frac{3}{4}$ in.

TESTS FOR COMPACTION AND RECOMPACTION

12. *Modified proctor:* ASTM D1557, disturbed sample of cohesive soil up to 50 kg; permitted to reuse sample in successive compactions but preferable not to do so
13. *Maximum and minimum densities:* ASTM D2049, performed on disturbed samples of cohesionless soil, from 10 to 100 kg of sample required, depending on grain size

There are, of course, many other laboratory tests for soils than these, covering a variety of such properties as soil chemistry, soil permeability, electrical resistivity, shrink–swell potential, response to cyclical loading, and others. Such tests are not usually appropriate for foundations for small buildings and are not presented here. Under special conditions, however, such tests may reveal particular items of information relevant to a particular foundation design and may be justified even for small buildings; the advice of the geotechnical engineer should be heeded in such special cases.

THE SOILS REPORT

The end product of the soils investigation is the soils report. It forms the basis of all decisions to be made regarding the site throughout the remaining design stages of the project, and often is used by the construction contractor throughout the construction stages as well. The soils report must stand alone; it cannot be assumed that those who originated the investigation or who carried out the work will be available for further discussions over the next few months or years while the report is being used.

Under ideal circumstances, the soils report would be simply a statement of facts, indicating exactly the tests performed, the results obtained, and the conclusions derived. In soils, however, the conclusions must often be shaded by judgment and opinion. The soils report must point out those areas where conclusions have been subject to such judgment or bias, and must include the relevant test or background data that justify the conclusions drawn.

There will be little room for subjective narrative in the soils report. Every statement must be based on a test result or a field observation that was included in the program of investigation. Similarly, every conclusion must indicate clearly its source in order that months or years later, any new personnel assigned to the project can fully understand the basis of the design they are expected to take over and to complete.

Over the years, the soils reports prepared for smaller projects by a particular soils firm usually will have evolved into a rather rigid, condensed format. The recommendations made in the report will probably be limited to those listed in the "scope of work" in the contract. Where the contract specifies a subjective analysis on a particular subject, the narrative will usually be brief and succinct to the point of being an outline. Such is the legacy of least-cost soils investigations; the structural engineer must understand implicitly all the ramifications of the conclusions stated so briefly in the soils report.

In general, the soils report will include at least the following items, stated in narrative form:

1. Identification of the contract and its date as the basis for the work.
2. Description of the scope of work, stating notable contract limitations, if any, in certain areas of the work.
3. An executive summary of the conclusions and recommendations that were developed as a result of the investigations, to include all major findings that will affect the design.
4. Description of the project, with a reduced-size site plan, noting access routes, survey markers, or other identification and orientation points on site.

5. Summary of foundation types, column loads, lateral loads, basement walls, retaining walls, shearwalls, and other structural and architectural features as they affect the investigation.
6. Notable site features such as streams, ponds, cuts, fills, rock outcroppings, and other topographic and geologic features as they affect both the investigation and the project as a whole.
7. Description of the field soils investigation, type and manufacturer of equipment used, type, size, and manufacturer of sampling and testing devices utilized, listing of field tests run and data gathered, with a map locating all boreholes, sampling locations, and test locations. (A machinery and equipment list may be included in the appendix to shorten the narrative.)
8. Listing of the lab tests run in conjunction with the soils investigation, naming the type, size, and manufacturer of test equipment where appropriate. (A lab equipment list used in this contract may be included in the appendix to shorten the narrative.)
9. Findings and conclusions derived from the investigation, to include sketches showing the pressure bulbs, the extent of stratification of the soil, areas in which the investigation did not extend, index properties of each stratum as appropriate, locations of any formations that could adversely affect the design, location of water table at minimum and maximum levels, extent of reported flooding, and other relevant findings derived throughout the course of the investigation. Pertinent test data should also be reproduced, to include *e*–log time curves, consolidation curves, load–deflection curves, filter curves, or other commonly used graphic solutions corresponding to a particular line of investigation. The interpretation of the test data and the conclusions drawn as a result of the test should always be stated specifically.
10. Recommendations based on the findings and conclusions, to include allowable soil pressures under both vertical and lateral loads and the anticipated extent of settlement under these allowable soil pressures, limitations in the strength of the soil, limitations in the estimated settlements, and means to include the effects of ground water as appropriate. Recommendations will also cover problem areas specified in the contract as well as possible remedies for other problem areas revealed by the investigation subsequent to execution of the contract. Where appropriate, recommendations may also be made to add or delete certain design features that are inappropriate to the site conditions.
11. Report of field operations, to include all boring logs, test data, and other information prescribed for the field investigation and paid for under this contract. This report and related data are usually included as an appendix to the report.
12. Report of lab tests, include backup lab information such as *e*–log time curves, major test items (not to include index properties), triaxial tests, unconfined compression, direct shear, and other significant test items paid for under the contract. This report and related data are also usually included in the appendix.

SUMMARY

The following topics have been presented in this chapter with the indicated conclusions.

- The scope of work to be followed in a typical soils investigation for a small-to-medium-size building will probably be developed jointly by the project design firm and the soils engineering contractor.

- The budget for the soils investigation will probably be limited; a maximum of information must be obtained at an absolute minimum of cost.

- The soils engineer will not be accessible after the soils investigation is completed and the soils report is accepted; the soils report must therefore contain every specific item of information about the site that will be needed throughout the final design.

- Few things are more effective in reducing the cost of the soils investigation and in assuring that all the necessary information is obtained than is producing a workable conceptual design for the project and conforming the soils investigation to the needs of that design.

- Preliminary information concerning the site is almost always available; such information can often be used to reduce the scope of the soils investigation or to make the soils investigation more relevant to the proposed project design.

- Typically, the project design will continue while the soils investigation is under way; on occasion, the design may have to be based on interim information provided by the soils contractor during the course of the investigation.

- The contract for the soils investigation should always include both the detailed scope of work and the specific items to be answered in the soils report.

- The contract for the soils investigation should allow for renegotiating the scope of work in the event the investigation reveals unexpected conditions on the site.

- Conducting the field tests is left to the contractor, who may use whatever means are at his or her disposal to complete the specified scope of work within the agreed time.

- Samples taken by driving a split-barrel sampler with an impact hammer are classified as disturbed samples and may not be used for sensitive tests; samples taken by advancing a thin-walled sampler with a jack are classified as undisturbed and may be used in all general test procedures.

- No field tests are currently in general use for determination of long-term settlements; short-term deflections can be determined by a load–deflection test, but the results are limited in their applications.

- Measurement of field densities at various elevations along the borehole at the natural in-place pressures is possible only for certain types of soil.

• The soils report is the end product of the soils investigation; it will serve throughout the remainder of the design stages and will probably be used throughout the subsequent bidding and construction stages.

• The soils report must be written in such a form that it will stand alone; it cannot be assumed that those who produced it will be accessible at a later date for consultations and guidance.

REVIEW QUESTIONS

4-1. If the geotechnical engineer is not given a preliminary design by which he or she can plan a soils investigation on a building site, what is the recourse?

4-2. Name the primary items of information to be obtained in the soils investigation.

4-3. What are some of the sources of general information that can be used to make a preliminary evaluation of a construction site?

4-4. On ordinary small-to-medium-size construction projects, who is responsible to determine the foundation conditions?

4-5. Name some of the typical items to be included in the preliminary design information that could produce an impact on the nature of the soils investigation.

4-6. In conducting a soils investigation for a building, why limit the subsurface investigation to the soil enclosed by the Boussinesq pressure bulb?

4-7. Why limit the investigation to those locations where the increase in pressure will be about 5% of the existing at-rest pressures?

4-8. When estimating the overall average pressure exerted by a building over its footprint area, what is a nominal average pressure attributable to each floor? The roof?

4-9. There will almost always be at least one deep boring in the program of soils investigation for a building foundation. If only one such deep boring is to be drilled, where should that deep boring be located?

4-10. Where only one deep boring is to be drilled in an investigation, there will often be shallower borings around the site in support of this one deep boring. Should these be in a straight line? Why?

4-11. What are the advantages and disadvantages of a probe in sounding out the changes in soil type under a building?

4-12. What are the advantages and disadvantages of a test pit when conducting a soils investigation?

4-13. About how far can a powered auger drill a hole?

4-14. What is a wash boring? What is its primary disadvantage when used in soils investigations?

4-15. What is the most common type of drilling rig used in conducting soils investigations. Why?

4-16. Does a split-barrel sampler produce a disturbed sample or an undisturbed sample?

4-17. Does a thin-walled sampler produce a disturbed sample or an undisturbed sample?

4-18. Name the two penetration tests that are commonly conducted at the bottom of a borehole.

4-19. Name the common shear test that can be conducted at the bottom of a borehole.

4-20. In what type of soil are the penetration tests more accurate?

4-21. In what type of soil is the vane shear test more accurate?

4-22. What lab test is used to provide settlement properties in clay soils?

4-23. What is the purpose of the field load–deflection test?

4-24. What is the primary disadvantage in the use of the smaller plate sizes in the load–deflection test?

4-25. What is the primary disadvantage in the use of the larger plate sizes in the load–deflection test?

4-26. Why is so much emphasis placed on having all the information from the soils investigation written down?

4-27. Why is it necessary to have a complete list of the lab and field equipment that was used in conducting the soils investigation?

4-28. Who prescribes the information to be included in the soils report?

4-29. Who writes the soils report?

4-30. Why is the geotechnical engineer not available for consultations throughout the design stages of the project?

PROBLEMS

Determine the size of the pressure bulb that should be used for the soils investigations for each of the given building shapes.

4-1. Concrete construction, four stories

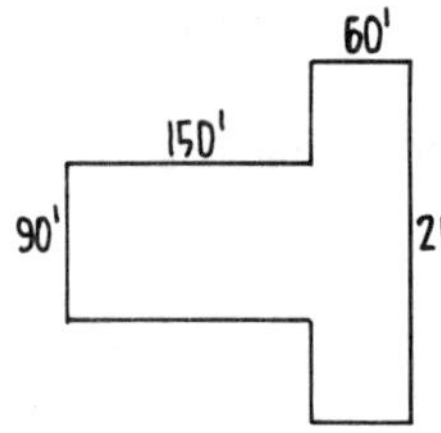

4-2. Steel construction, six stories

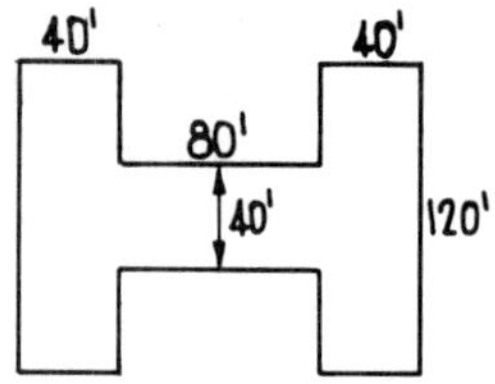

4-3. Timber construction, two stories

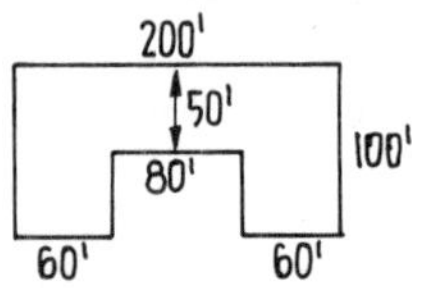

4-4. Concrete construction, five stories

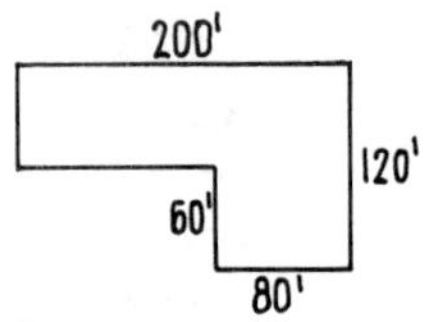

4-5. Timber construction, three stories

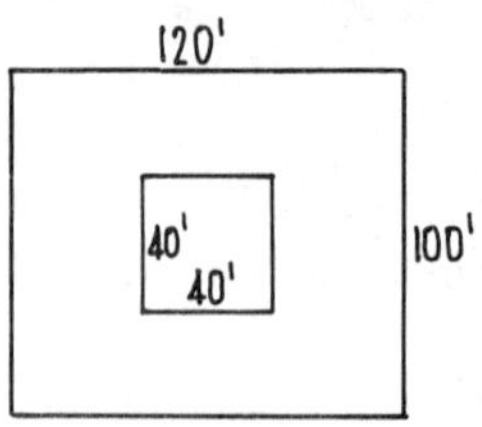

4-6. Steel construction, five stories

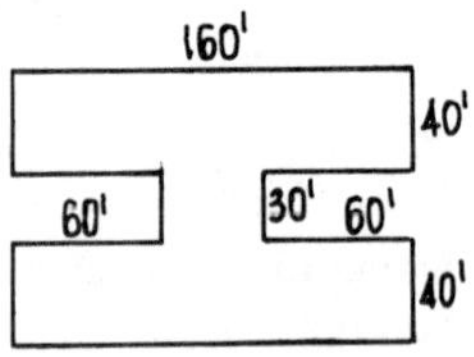

5

Loads on Building Foundations

Loads to foundations are generally broken into two broad categories, gravity loads (dead and live) and lateral loads (wind and earthquake). All loads to foundations are treated as static loads. Live loads, wind loads, and earthquake loads may actually be highly dynamic, but in practice such loads are applied as equivalent static loads rather than as dynamic loads.

Environmental loads are a special type of loading that may occur on structures. Typical of such loads are snow, ice, sand accumulation, rain, and other regional environmental hazards. The proper application of such loads is highly dependent on local practices; no attempt is made here to account for all such special applications.

In Chapter 6, an accurate analytical method is developed for selecting the size and shape of the contact area of a footing. In that method, only the vertical design load is used; moments are not needed. The vertical design load used in the method may be taken directly from the structural analysis, or, for preliminary designs, may be computed in an independent analysis.

Since there will be no need later to know the moments on the structure, a simplified but accurate analysis for the remaining unknowns, the vertical loads, becomes readily possible. The techniques and methods developed in this chapter present an accurate but simplified analysis to determine these vertical design loads.

One of the many uses for an accurate preliminary analysis is the use made by the soils engineer who conducts the soils investigation. The engineer's investigation can be made considerably more relevant if he or she knows the locations and magnitudes of the loads to be applied, from the largest to the smallest, as well as any requirements for settlements or for lateral forces.

More important, however, if the smallest and largest column loads are known, the Boussinesq pressure distribution can be more accurately estimated and differential settlements can be predicted much more accurately. Further, potential hazards at weaker underlying strata can be identified accurately without undue guesswork. The techniques and methods presented in subsequent discussions permit just such a preliminary determination of foundation loads to be made.

GRAVITY LOADS

As one may suppose, the gravity loads acting on a building are those due to the fixed mass of the building itself (dead load) and those due to transient masses (live load). Gravity loads usually act vertically, but in some cases they may act horizontally (e.g., soil loads against a basement wall). Gravity loads are usually of much longer term than the lateral loads (wind and earthquake), which, by comparison, are extremely transient and erratic.

The distinction between dead loads and live loads is an important one. Concrete structures and components, including the concrete footings, are designed for a different factor of safety for dead load than for live load. Also, earthquake loads are computed using only the dead load of the building; live load is excluded. Reasonable care must therefore be exercised in distinguishing between the two loads throughout the structural analysis in order that wasteful overdesign or dangerous underdesign does not occur.

Dead loads are those due to the weight of the building components, whether constructed as a part of the structure (beams, frames, columns) or as appurtenances (firewalls, parapets, equipment housings). Dead loads cannot be changed during the life of the building except by additional construction or remodeling of the building. A brief list of dead loads for some of the more common building materials is given in Table 5-1.

Mechanical equipment, ductwork, cable trays, carpets, and other such loads are not classified as dead loads. Although they are indeed permanently present, they are subject to renewal, replacement, or alteration during the life of the building. Since their weights are subject to change during the life of the building, they are not classified as dead loads. In addition, they are not rigidly attached to the building frame; when the building undergoes earthquake motion, these items simply shift around without resisting the motion of the building.

Those loads in a building that are movable are called live loads. Or alternatively, live loads are those that occur due to the usage or occupancy of the building. Over the life of the building, however, the usage of the building can

TABLE 5-1 DEAD LOADS: WEIGHTS OF COMMON BUILDING MATERIALS

Material	psf	Material	psf
Ceilings		Partitions	
Furred channel system	1	Clay tile	
Acoustic fiber tile	1	3 in.	17
Floors		4 in.	18
Concrete, per inch		6 in.	28
Stone	12½	8 in.	34
Slag	11½	10 in.	40
Lightweight	6 to 10	Gypsum block	
Fills, per inch		2 in.	9½
Gypsum	6	3 in.	10½
Sand	8	4 in.	12½
Cinders	4	5 in.	14
Finishes, per inch		6 in.	18½
Terrazzo	13	Plaster, per inch	
Quarried tile	12½	Cement	10
Mastic	11½	Gypsum	5
Hardwood	5	Lathing	
Softwood	4	Expanded metal	½
Roofs		Gypsum board ½ in.	2
Copper	1	Walls	
3-ply felt and gravel	5½	Brick	
5-ply felt and gravel	6	4 in.	40
Shingles		8 in.	80
Wood	2	12 in.	120
Asphalt	3	Hollow concrete block	
Clay tile	9 to 15	Heavy aggregate	
Slate, ¼ in.	10	4 in.	30
Sheathing, per inch		6 in.	44
Wood	4	8 in.	56
Gypsum	4	12 in.	80
Insulation, per inch		Light aggregate	
Loose	½	4 in.	21
Poured in place	2	6 in.	30
Rigid	1½	8 in.	38
Corrugated asbestos, ¼ in. thickness	3	12 in.	56
		Clay tile	
		4 in.	25
		6 in.	30
		8 in.	33
		12 in.	45
		Structural glass, 1 in.	15

change sharply as tenants change or as new owners subject the building to new functions. Any new loads must, of course, be kept within the original design values, regardless of what the new function is.

For a starting point, ranges of live load which include a reasonable latitude for future changes are prescribed by the various codes. These ranges of live load are based on rather broad categories for the intended usage of the building; over the years, they have been found by the industry to be generally satisfactory. A brief list of some of the more commonly used live loads is given in Table 5-2.

The magnitude of the vertical gravity load to be supported by each footing can readily be established, based on known and accepted methods of structural analysis. Since there is no need to find moments, the complexity of the analysis is reduced considerably. It should be apparent that the overall accuracy of such a simplified analysis is limited to the accuracy of the gravity loads themselves, which in the case of live load has been shown to be quite arbitrary.

The major concepts of the simplified analysis are:

1. The vertical gravity load that is tributary to a particular footing can be computed with reasonable accuracy as the sum of all gravity loads halfway to the next vertical support in any direction.
2. The vertical gravity load that is tributary to a footing includes the loads at all floor levels, and includes the entire load from any cantilevers that are tributary to the support.
3. Any redistribution of gravity loads within a building due to flexural moments can be neglected.

For proof of the foregoing conclusions, one need look no further than the ACI coefficients [23], a commonly applied method of analysis for regular structures of any material. This widely accepted method does not require any redistribution of vertical load due to flexure. The ACI method is derived analytically and is accurate within its prescribed limits; the actual proof of its accuracy is readily available in the literature [24] and is not repeated here.

The following limitations apply to the simplified analysis:

1. The building must be reasonably regular; that is, adjacent span lengths may not differ by more than 20%.
2. Loads must be relatively uniformly distributed.
3. Live load may not exceed three times the dead load.
4. The building must be framed from prismatic members, although shear panels and bearing walls may of course be used.

In using this simplified approach, intuitive methods are usually reliable. A concentrated load, for example, can be distributed to the supporting members in proportion to its proximity to each support. Similarly, a floor slab on grade makes no contribution to footing loads.

Municipal building codes sometimes require that a beam be capable of taking a randomly placed load of 2000 lb at any point along its span, presumably to represent a safe or other heavy concentrated load. The end result of this provision is to require an increase in the capacity of the beam in shear when the load is placed very close to the end of the beam. The load itself is a part of the floor live load and is not intended to be a separate load to be added to each beam; it should not be included when computing column loads.

TABLE 5-2 MINIMUM DESIGN LIVE LOADS

a. Uniformly distributed live loads based on occupancy

Occupancy	psf	Occupancy	psf
Armories and drill rooms	150	Residential	
Assembly halls, theatres		Multifamily units	
Fixed seats	60	Private apartments	40
Movable seats	100	Public rooms	100
Balconies, exterior	100	Corridors	60
Bowling alleys	75	Single-family units	
Dance halls	100	First floor	40
Dining areas	100	Habitable attics	30
Gymnasium floors	100	Storage areas	30
Hospitals		Hotels	
Operating rooms	60	Guest rooms	40
Private rooms	40	Public rooms	100
Wards	40	Primary corridors	100
Libraries		Public corridors	60
Reading rooms	60	Private corridors	40
Stack rooms	150	Schools	
Manufacturing	125	Classrooms	40
Marquees	75	Corridors	100
Office buildings		Stairs and fire escapes	100
Offices	80	Stands and bleachers	100
Lobbies	100	Stores, retail	
Penal institutions		First-floor rooms	100
Cell blocks	40	Upper floors	75
Corridors	100	Stores, wholesale	125

b. Uniformly distributed live loads based on use

Use	psf	Use	psf
Air-conditioning equipment	200	Garages, ramps, drives	
Amusement-park structures	100	Trucks, 3 to 10 tons	150
Bakeries	150	Trucks above 10 tons	200
Boathouse floors	100	Greenhouses	150
Boiler rooms, framed	300	Hangars	150
Broadcasting studios	100	Incinerator floors	100
Catwalks	25	Kitchens, commercial	150
Dormitories		Laboratories, science	100
Partitioned	40	Laundries	150
Nonpartitioned	80	Libraries, corridors	100
Elevator machine rooms	150	Morgues	100
Fan rooms	150	Public rooms	100
File rooms		Rest rooms	60
Letter	80	Rinks, ice skating	250
Card	125	Storage, hay or grain	300
Addressograph	150	Telephone exchanges	150
Foundries	600	Toilet rooms	60
Fuel rooms, framed	400	Transformer rooms	200

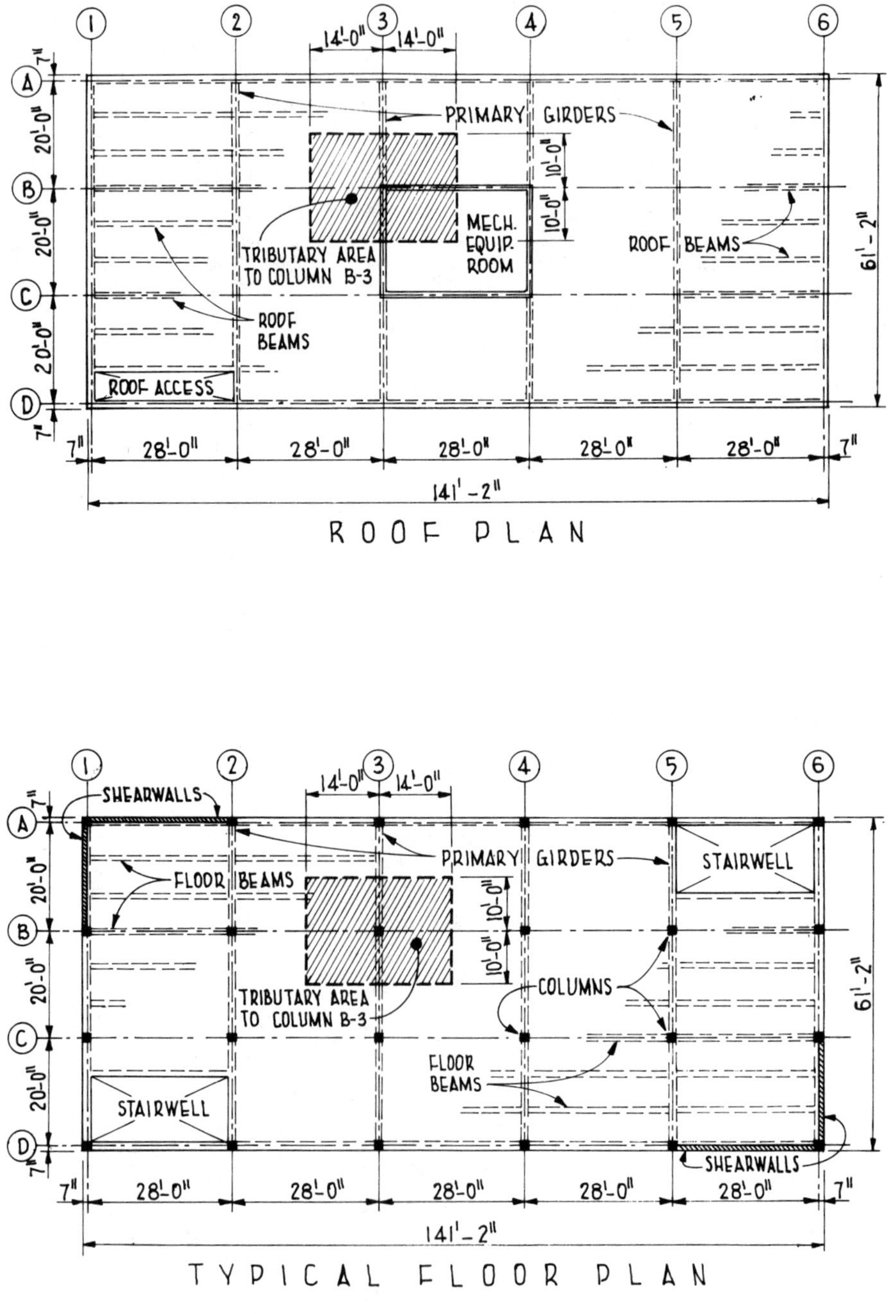

Figure 5-1 General configuration and sections.

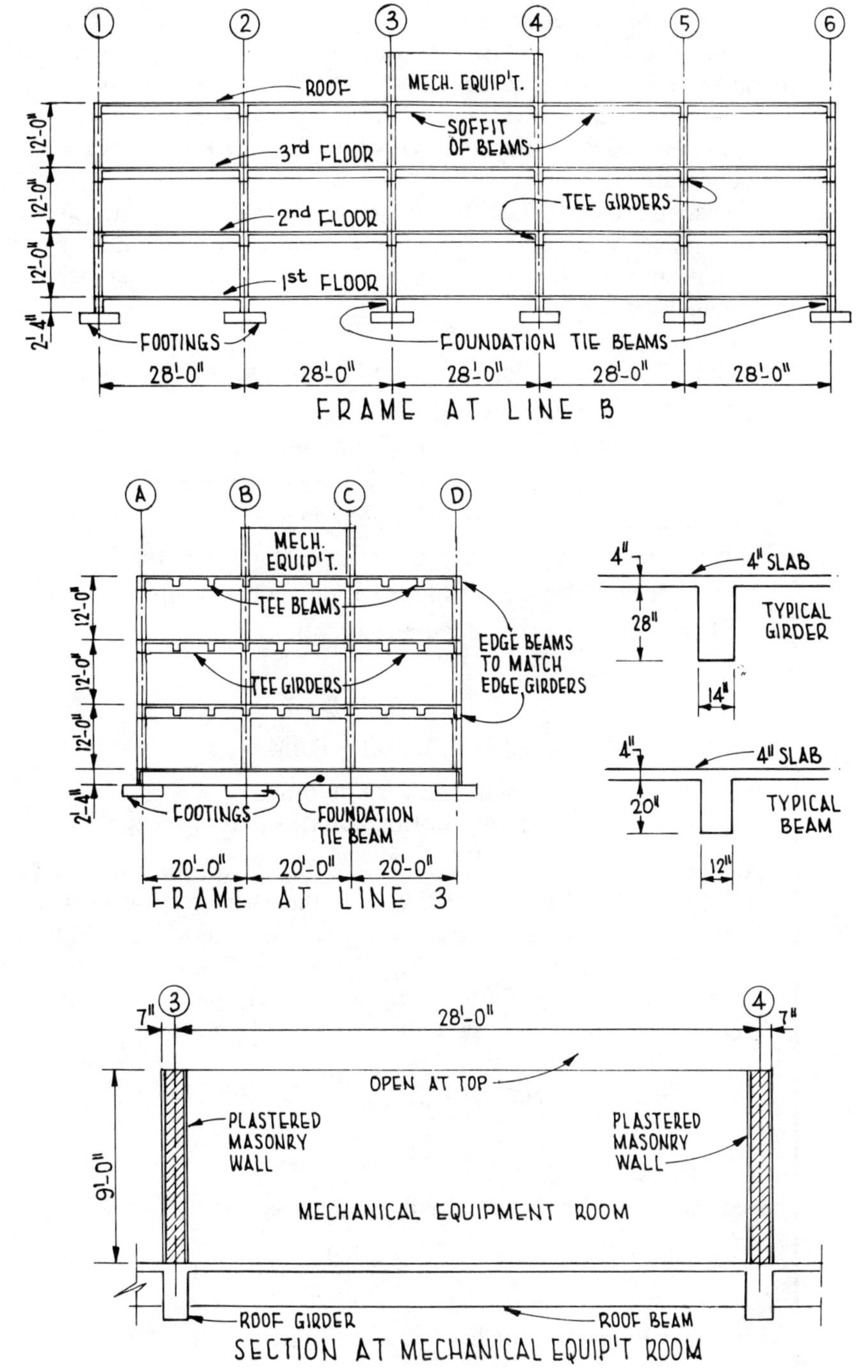

Figure 5-1 cont.

It should be recognized at this point that municipal building codes may sometimes permit a reduction in live loads where tributary areas are large. Additional reductions are sometimes permitted in multistory building due to the improbability that all stories will be loaded to maximum loads at the same time. Such refinements are indeed worthwhile, but for the sake of simplicity are not included here.

Example calculations follow, showing the computation of vertical design loads both for dead load and for live load. A simple building is used for this computation; the layout is shown in Fig. 5-1. The following loads will be used in subsequent calculations:

Roof live load 30 psf; floor live load 60 psf; stairwell live load 100 psf
Roofing: built-up roofing 15 psf
Wall around mechanical equipment housing 600 plf
Rooftop equipment room: allow 150 psf for equipment
Floor covering: terrazzo concrete, 20 psf
Perimeter walls: masonry and glass, 650 plf per story
Interior partitions: movable partitions, allow 12 psf
Shearwalls: concrete, 1400 plf per story

EXAMPLE 5-1: Loads on Building Foundations
Vertical dead loads only.
Typical interior column and footing.

The dead load to the footing at grid point B3 will be computed. It should be noted that lateral loads are taken by the shearwalls at corners A1 and D6. Consequently, footing B3 does not carry any lateral load. For the sake of the computations, the weight of the floor and roof beams will be distributed uniformly over the floor. Within one 20 ft × 28 ft bay, there are three such beams, each 26 ft 10 in. long, with a stem 20 in. × 12 in.

$$\text{Stem wt} = \left(\frac{12 \times 20}{144}\right)(26.83 \text{ ft})(145 \text{ pcf})(3)$$

$$= 19.5 \text{ kips}$$

Distributed uniformly, the stem weight becomes

$$W = \frac{19{,}500}{20 \times 28} = 35 \text{ psf}$$

For the main girders the load per foot for the stem, 28 in. × 14 in., is given by

$$\text{wt per ft} = \left(\frac{28 \times 14}{144}\right)(145) = 395 \text{ lb/ft}$$

The tributary areas to column B3 are shown as the hatched areas in Fig. 5-1. The following calculations include all dead loads within those areas.

At roof level (including mechanical equipment room):

Wt of wall = 600 plf × 28 ft/2 + 600 plf × 20/2	=	14.4 kips
Wt of roofing = 15 psf × 20 × 28	=	8.4 kips
Wt of slab = (4/12)(145 pcf)(20 × 28)	=	27.1 kips
Wt of beams = (35 psf)(20 × 28)	=	19.6 kips
Wt of girder = 395 plf (18.83 ft)	=	7.4 kips

Column from roof to third-floor level:

Wt of column = (14 × 14/144)(9.33 ft)(145 pcf)	=	1.8 kips

At third-floor level:

Wt of terrazzo = (20 psf)(20 ×28)	=	11.2 kips
Wt of slab = (4/12)(145 pcf)(20 × 28)	=	27.1 kips
Wt of beams = (35 psf)(20 × 28)	=	19.6 kips
Wt of girder = 395 plf (18.83 ft)	=	7.4 kips

Column from third floor to second floor:

Wt of column = (14 × 14/144)(9.33 ft)(145 pcf)	=	1.8 kips

At second-floor level:

Wt of terrazzo = (20 psf)(20 ×28)	=	11.2 kips
Wt of slab = (4/12)(145 pcf)(20 × 28)	=	27.1 kips
Wt of beams = (35 psf)(20 × 28)	=	19.6 kips
Wt of girder = (395 plf)(18.83 ft)	=	7.4 kips

Column from second floor to top of footing:

Wt of column = (14 × 14/144)(11.67 ft)(145 pcf)	=	2.3 kips
Total dead load to footing	=	214 kips

It should be observed in Example 5-1 that door openings, floor recesses, and minor deviations in dead load are ignored. Since the method itself is approximate, any undue refinement of loads is hardly warranted. In fact, many designers would add an arbitrary 5% or even 10% to the final sum to account for any items that might have been overlooked in the analysis or to allow for items that may be added in the future. The next example shows the calculation of live loads tributary to the same footing.

EXAMPLE 5-2: Loads on Building Foundations
Vertical live loads only.
Typical interior column and footing.

The live load to the footing at grid lines B3 for the building shown in Fig. 5-1 will now be calculated. The live loads are stated in the problem data accompanying Fig. 5-1 but are repeated here for immediate reference.

At roof:	equipment load	=	150 psf
	live load	=	30 psf
At floor:	live load	=	60 psf
	partitions	=	12 psf

The tributary areas to column B3 for live load are identical to those for dead load, as shown by the dashed lines in Fig. 5-1. The following calculations include all live loads within those boundaries.

At roof level:

mech. equip't load = (150)(20 × 28)/4 = 21.0 kips

roof life load = (30)(20 × 28)(3/4) = 12.6 kips

At third-floor level:

partition load = (12)(20 × 28) = 6.7 kips

live load = (60)(20 × 28) = 33.6 kips

At second-floor level:

partition load = (12)(20 × 28) = 6.7 kips

live load = (60)(20 × 28) = 33.6 kips

At ground level: all live load carried on grade

Total live load to footing = 114.2 kips

The tributary dead loads and live loads calculated in Examples 5-1 and 5-2 are the gravity loads to be carried by the footing at grid point B3. It should be noted that such calculations are direct and simple, with little refinement. Their accuracy, however, is within the same range of accuracy as the loads themselves.

Where transient horizontal loads such as wind or earthquake occur on a building, the building codes permit an increase (usually 33%) in the allowable vertical bearing pressures on the soil. The circumstances whereby the horizontal loads on a building produce both horizontal and vertical loads on the foundation is presented in the following sections.

LATERAL LOADS

Lateral loads come from two sources, wind and earthquake; each has a distinctly different effect on the structure. At the foundation, however, the end result in both cases is a lateral force on the foundation system and, in some circumstances, an increase or decrease in vertical load.

Winds try to move the building laterally across the earth's surface. Earthquakes try to move the earth's surface laterally underneath the building. In both cases, the result is a base shear as shown in Fig. 5-2. The effects of these two cases of load on the building are quite different. Such effects within the building itself are not of concern here, however. The only concern in this study is the effect on the foundation.

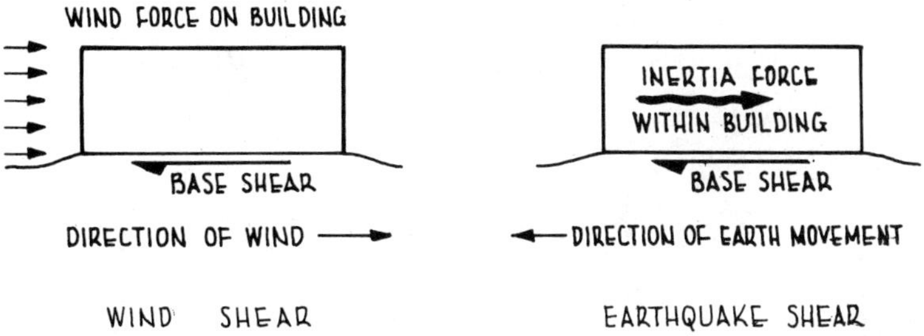

Figure 5-2 Base shears.

In general, wind and earthquake loads are of such short duration that they do not produce settlements. They are therefore of primary concern where the allowable bearing capacity of the soil is governed by the strength of the soil rather than by its susceptibility to settlements. There is of course a possibility that the vibration of sandy soils produced by wind or earthquake loads will produce settlements, but such vibrations are more likely to produce area-wide settlements than differential settlements between footings; it is only these differential settlements that are of concern here.

BASE SHEAR DUE TO WIND

The wind force on a building is obviously proportional to wind velocity. Wind velocity, in turn, is highly dependent on latitude, geography, terrain, and local obstructions. The maximum wind velocity to be expected on a particular building site is therefore quite difficult to determine; it is very sensitive to particular site conditions.

On a broad geographical basis, wind-velocity charts have been created to show generally the wind velocities that may be expected within various regions. Figure 5-3 shows one such chart for the contiguous United States; the given wind

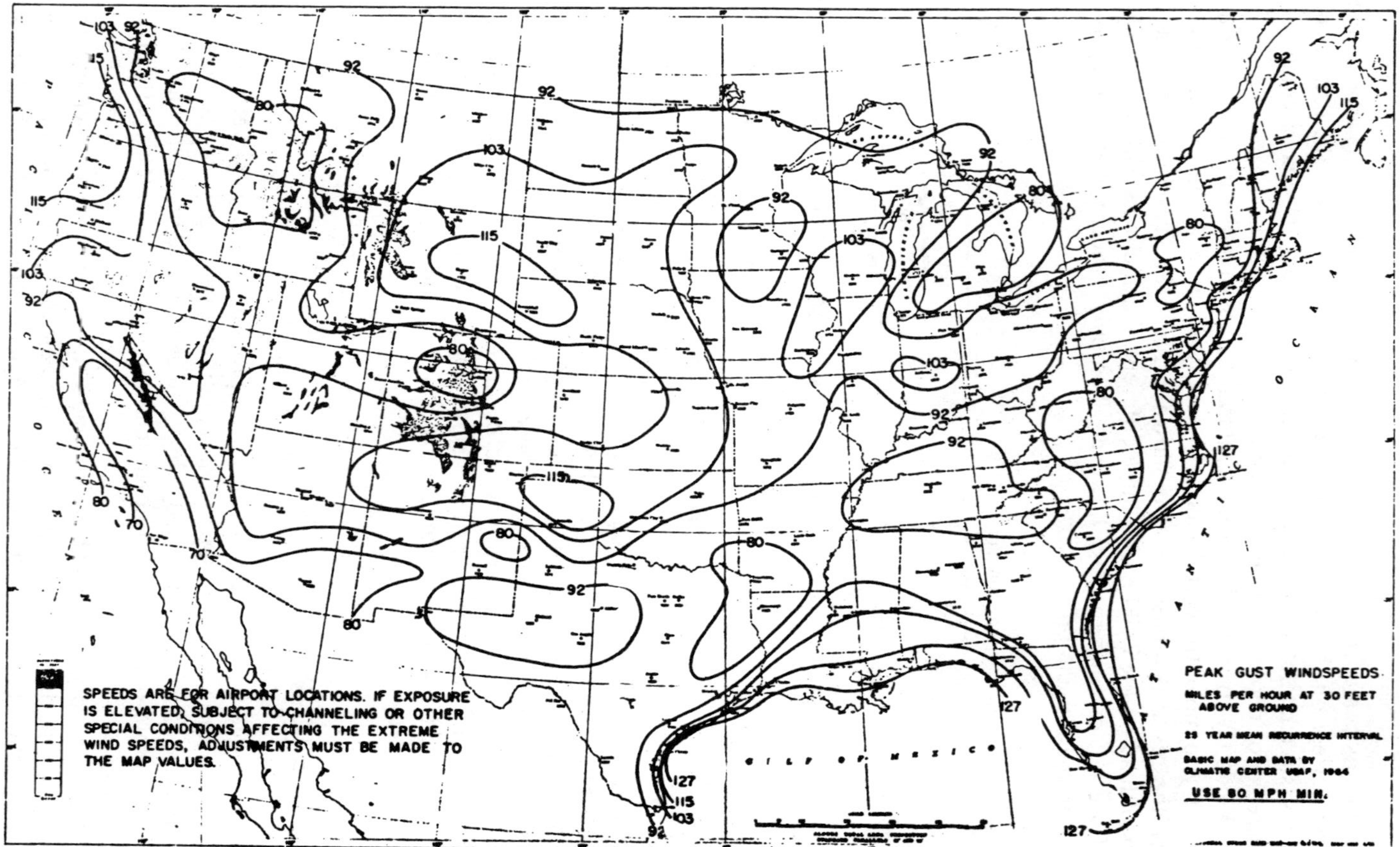

Figure 5-3 Wind velocity map. (From Uniform Building Code. Reprint by permission of International Conference of Building Officials.)

velocities are those at a standard height of 30 ft above the surface. For general design use, the accuracy of the chart is about as accurate as anything else available, considering the highly approximate nature of any such forecast.

The effects of wind loads on low buildings (less than about 60 ft high) are reasonably predictable. The wind pressure on the building increases along the height of the building, being considerably less near the ground than at the top of the building. A typical wind pressure distribution is shown in Fig. 5-4.

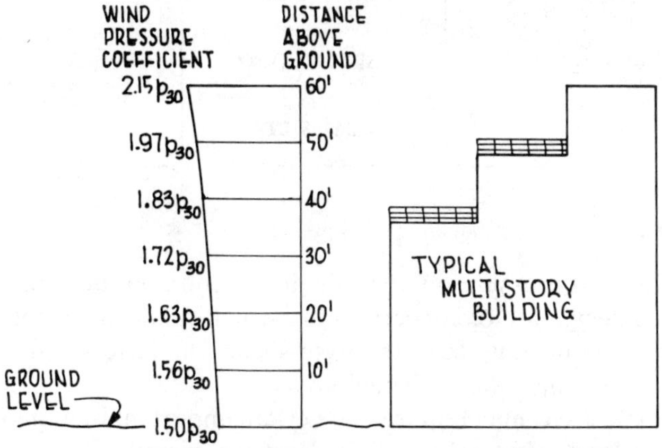

Figure 5-4 Typical wind pressure distribution.

Since the wind pressure varies along the height of the building, a reference "stagnation" wind pressure is chosen at some standard height above ground level, usually 30 ft. At that level, the reference stagnation wind pressure in pounds per square foot is given as [9]

$$p_{30} = 0.00256V^2 \tag{5-1}$$

where V is the wind velocity in miles per hour. The actual wind pressure at various levels is shown in 10-ft increments in Fig. 5-4, given as a factor times this reference stagnation pressure. The factors shown in Fig. 5-4 are those for the most severe conditions of terrain, with few or no obstructions to block or otherwise diffuse the wind pressure.

The wind "sees" the building as if it were a flat silhouette, projected horizontally onto a flat screen. Setbacks, reveals, recesses, and other surface features of the building do not significantly affect the wind loading. The total force on the building is thus the accumulation of forces due to the pressure at any level, p_i, times the projected area at that level, A_i:

$$\text{applied force} = \text{base shear} = \sum p_i A_i \tag{5-2}$$

To simplify calculations for the base shear, the wind pressure is taken in stepped increments rather than as a smooth curve, as shown in Fig. 5-5. The pressure at any level can be interpolated from the values given in Fig. 5-4. The sum of pressures times projected areas then yields the wind base shear, as defined by Eq. (5-2).

The wind pressure on certain components in and around a building may be much higher than on others, depending on the shape of the building. Similarly,

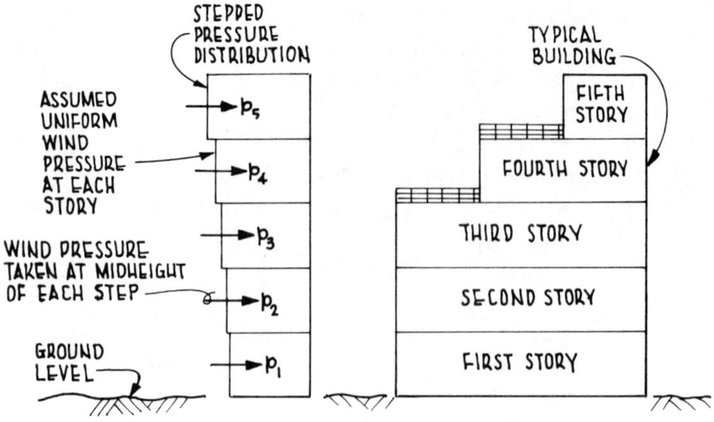

Figure 5-5 Typical application of wind pressure.

the building may experience wind pressure on one side and suction on the other. Such features of the building design are of concern to the structural designer, but they do not materially affect the magnitude of the base shear; the base shear is simply the sum of the pressures times the projected areas.

The building code in some areas may require that certain important buildings be designed for higher wind loads than other buildings in the vicinity. Typical buildings that might fall in this category are hospitals, fire stations, civil emergency centers, and police stations. The increased wind load is obtained by multiplying the wind base shear of Eq. (5-2) by an "importance factor" I, which can be as high as 1.25 for some buildings. In the examples presented in this book, the importance factor I is taken at its base value of 1.00. The following example illustrates the calculation of the base shear due to wind on a typical small building.

EXAMPLE 5-3: Wind Loads on a Small Building

Computation of base shear due to wind.
Design wind velocity 120 mph.

The base shear due to wind acting on the building of Fig. 5-1 will be computed for each of the two axes. The silhouettes of the two views of the building are shown in the following sketch.

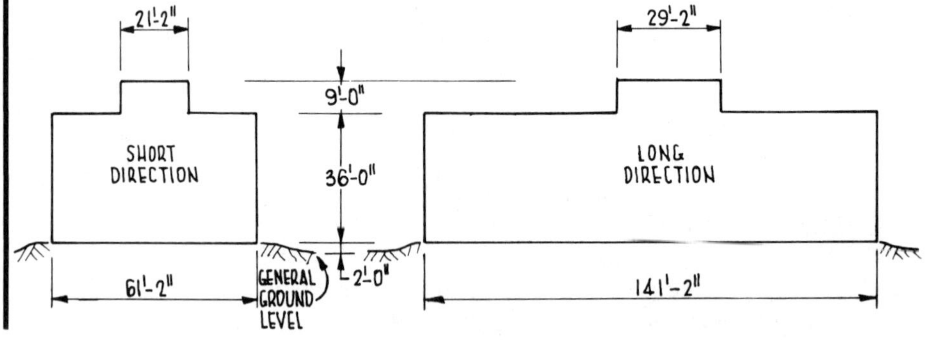

The variation of pressure along the height of the building is interpolated from Fig. 5-4. The interpolated values to the middle of each floor level are shown in the following sketch.

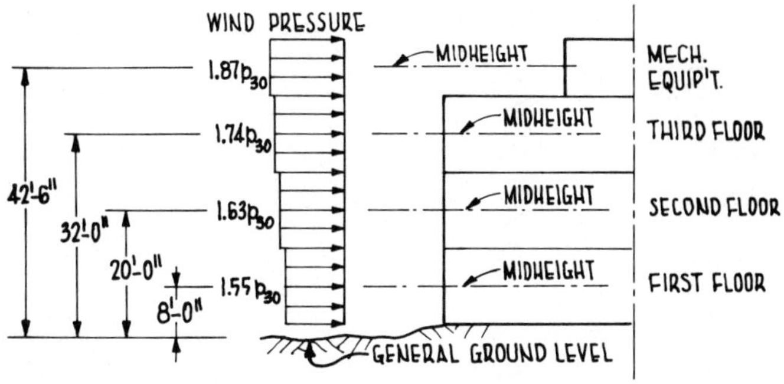

The reference wind pressure at 30 ft off the ground is computed from Eq. (5-1),

$$p_{30} = 0.00256\,V^2 = 0.00256(120)^2 = 37 \text{ psf}$$

The total force at various levels is computed simply as pressure times projected area. For wind normal to the short side of the building, the pressures are:

At mechanical room: $P = 1.87 \times 37 \times 9 \times 21.2 = 13.2$ kips
At third floor: $P = 1.74 \times 37 \times 61.2 \times 12 = 47.3$ kips
At second floor: $P = 1.63 \times 37 \times 61.2 \times 12 = 44.3$ kips
At ground floor: $P = 1.55 \times 37 \times 61.2 \times 12 = 42.1$ kips

Total base shear normal to the short side = 146.9 kips

The base shear for wind normal to the long side of the building is computed similarly:

At mechanical room: $P = 1.87 \times 37 \times 9 \times 29.2$ ft $= 18.2$ kips
At third floor: $P = 1.74 \times 37 \times 141.2 \times 12 = 109.1$ kips
At second floor: $P = 1.63 \times 37 \times 141.2 \times 12 = 102.2$ kips
At ground floor: $P = 1.55 \times 37 \times 141.2 \times 12 = 97.2$ kips

Total base shear normal to the long side = 326.7 kips

These two base shears, 146.9 kips normal to the short side and 326.7 kips normal to the long side, must be taken by the foundations at the two sets of shearwalls.

The distribution of this base shear to the two sets of shearwalls is discussed in a later section. At this point, only the two base shears due to wind load were to be determined; they have been computed and are shown in the following sketch. These forces will be used in combination with other loads later, in a continuation of this example.

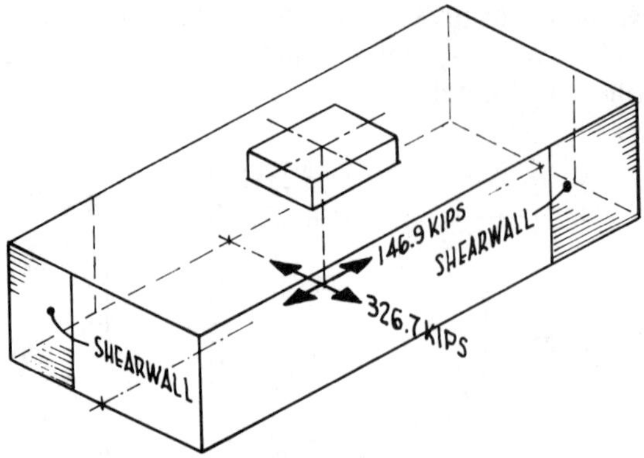

It should be noted that the forces shown in the sketch are shown in two directions, since the wind load can come from any direction.

The foregoing analysis for base shear due to wind load is seen to be relatively simple and straightforward, even though it recognizes that the wind force varies with the height of the building. The following analysis for base shear due to earthquake will be seen to be equally straightforward.

BASE SHEAR DUE TO EARTHQUAKE

Earthquake loads are distinctly different from wind loads. They are inertia loads, created when the earth takes a sharp lateral displacement underneath the building. The magnitude of this lateral displacement is quite high; under severe earthquakes, movements of 8 in. or even more should be expected. This high-velocity displacement occurring at the base of the building can have devastating effects on the building unless the structure is properly designed to accept such a displacement.

Since the building must also undergo the same lateral displacement as the earth, the forces created within the building are directly proportional to the mass that is being accelerated, $F = ma$. Much higher forces are therefore generated in a massive heavy building built of concrete than in a lighter building built of steel or timber. Also, much higher forces are generated within an extremely rigid building than in a flexible one, since accelerations of building components will be much higher.

The level of intensity of earthquake forces has been established for various regions. A map showing the earthquake risk zones of the contiguous United States is given in Fig. 5-6. The legend given with the map indicates the intensity to be expected in each zone.

The base shear created by earthquake forces on a building is given as a factor times the dead weight of the building. Some structural designers include as much as 25% of the live load along with the dead load when computing the inertia force. In this book, only the dead load is used.

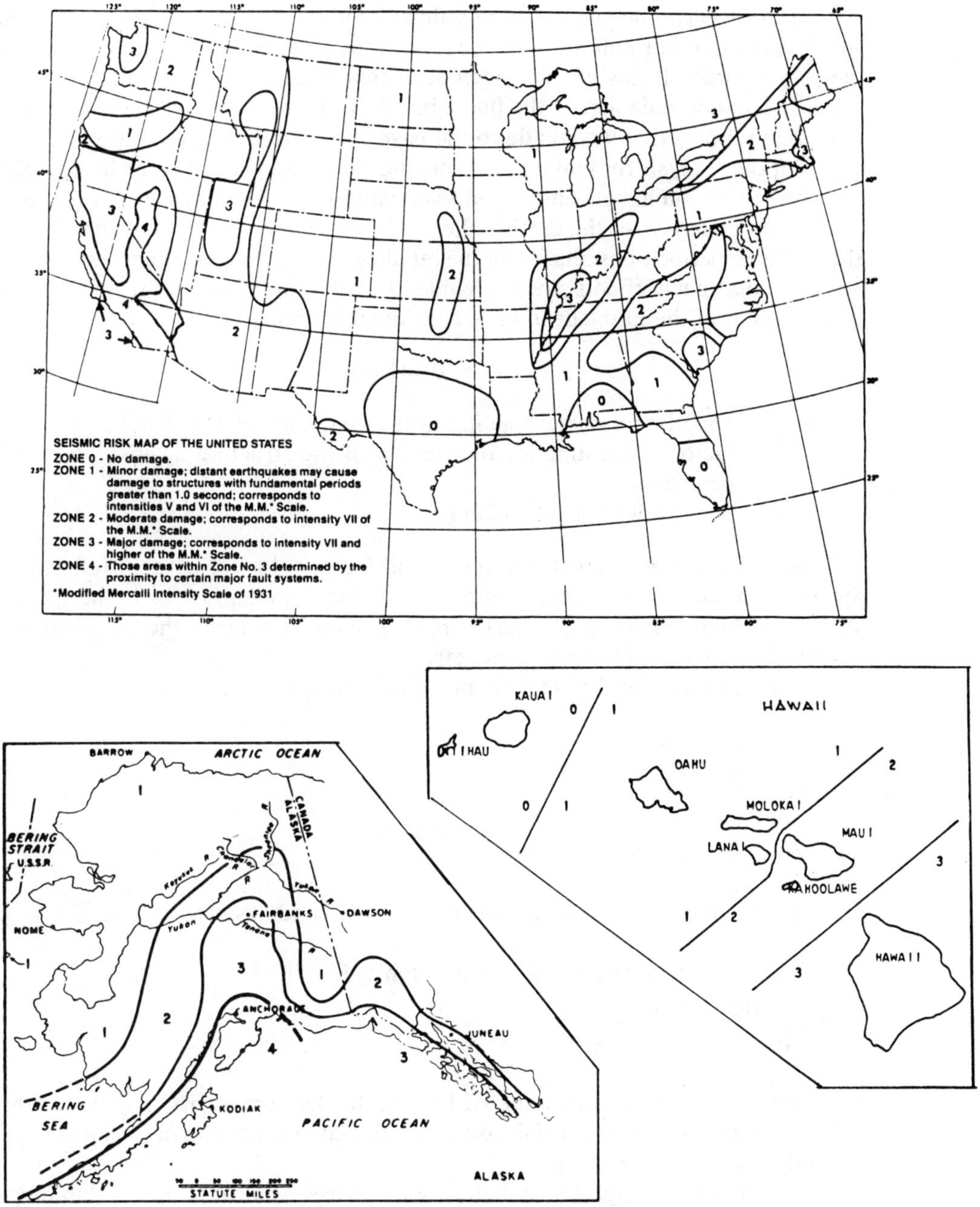

Figure 5-6 Seismic risk map. (From Uniform Building Code. Reprint by permission of International Conference of Building Officials.)

The rationale behind using only dead load deserves clarification. The only loads that can contribute to base shear are those which can be accelerated, thus causing a force. It has already been seen that live loads are loose; they might move around or slide across the floor, but they would not resist the acceleration and create forces that would add to the base shear. It is therefore doubly important that the analysis for gravity loads distinguish clearly between the loads rigidly attached to the building frame (dead loads) and loads free to move (live loads).

The magnitude of the earthquake force on a building can vary sharply with the dynamic period of oscillation of the building, but for low buildings the period is so short that such effects are negligible. For low buildings (70 ft or less), the computation of the base shear V_b is considerably simplified [9]:

$$V_b = 0.14ZKW \tag{5-3}$$

where Z = zone factor corresponding to the zone map of Fig. 5-6
K = factor accounting for the rigidity of the structure against lateral displacement
W = dead weight of the building

As a matter of interest, the numerical factor 0.14 in Eq. (5-3) reflects the dynamic characteristics of the building. For low buildings, 0.14 is the limiting value. For high, flexible buildings (more than about 70 ft high), the factor must be recomputed to account for dynamic effects.

The value of Z in Eq. (5-3) is prescribed by code [9]:

Zone 1: $Z = \frac{3}{16}$
Zone 2: $Z = \frac{3}{8}$
Zone 3: $Z = \frac{3}{4}$
Zone 4: $Z = 1.0$

The value of K in Eq. (5-3) depends on the structural system:

For diaphragm–shearwall construction: $K = 1.33$
For rigid frames: $K = 0.67$
For other structural systems: $K = 1.00$

The municipal building code adopted by a particular community may have other values of K permitted for special cases. A detailed examination of the local building code is of course essential.

In addition, an importance factor I may be assigned for particular buildings, as in wind design. This factor is assigned by the local building authority and for earthquake may be as high as 1.5. Analogous to wind load, the base shear given by Eq. (5-3) is multiplied by the importance factor I to obtain the final design value.

Examination of Eq. (5-3) reveals that in most areas, the maximum earthquake force on a low building is about 14% of its gravitational dead weight. In areas of light earthquake intensity, the force may be as low as 2.625% of the gravitational dead weight. These percentages of the gravitational dead weight of the building are commonly known as the "*g*-load" or "*g*-force" on the building.

The following example will demonstrate the calculation of the base shear due to earthquake on a typical small building.

EXAMPLE 5-4: Earthquake Loads on a Small Building

Computation of base shear due to earthquake.
Earthquake risk zone 2.

The base shear due to earthquake acting on the building of Fig. 5-1 will be computed. The base shear is the same along either major axis of the building. The total dead weight of the building has been computed by the methods of Example 5-1 and has been estimated to be 4300 kips. The building with its base shear is shown in the following sketch.

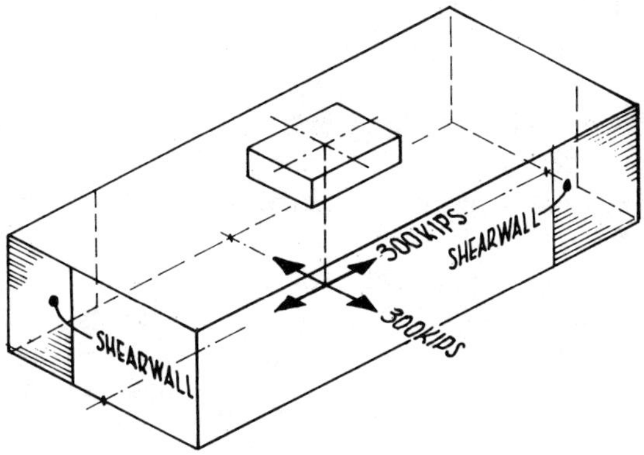

The base shear is computed readily from Eq. (5-3):

$$V_b = 0.14ZKW$$

where $Z = 0.375$ for risk zone 2, $K = 1.33$ for diaphragm–shearwall construction and the total dead weight is 4300 kips. Hence

$$V_b = 0.14 \times 0.375 \times 1.33 \times 4300 = 300 \text{ kips}$$

The distribution of this base shear to the two sets of shearwalls is presented in the next section. At this point, only the base shears due to earthquake were to be determined; the base shears are shown in the following sketch. The forces could occur in any direction, of course, since the earthquake could come from any direction.

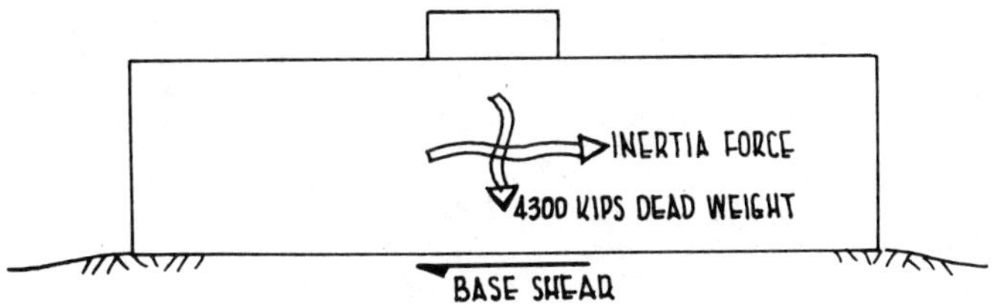

When the results of the earthquake analysis are compared to the results of the wind analysis given in Example 5-3, it is seen that the earthquake forces in this case are larger in one direction and the wind forces are larger in the other direction. There is almost no way to predict in advance which of these two cases will govern the design. In most cases, both analyses have to be done to determine which load is more severe.

DISTRIBUTION OF BASE SHEAR TO FOOTINGS IN A RIGID FRAME

Regardless of whether the highest base shear is generated by wind or by earthquake, the foundation system must sustain this lateral load. The overall design of the structural system itself must, of course, define which of the footings have been chosen to resist the lateral loads. Consequently, the general configuration of the structural system must be established before the lateral load on any one footing can be determined.

For rigid frames, the distribution of the base shear to individual columns and footings can be made according to the ''portal analysis'' for rigid frames. In this method, the rigidity of interior columns is assumed to be twice that of perimeter columns; interior columns therefore take twice as much shear force as the perimeter columns take. For a symmetrical case, the distribution is shown schematically in Fig. 5-7.

The determination of lateral load on any individual footing of a rigid frame becomes quite simple using the portal analysis. In the frame of Fig. 5-7, for example, there are four interior columns and two perimeter columns. The equivalent number of interior columns is therefore 4 + 2/2, or five equivalent interior

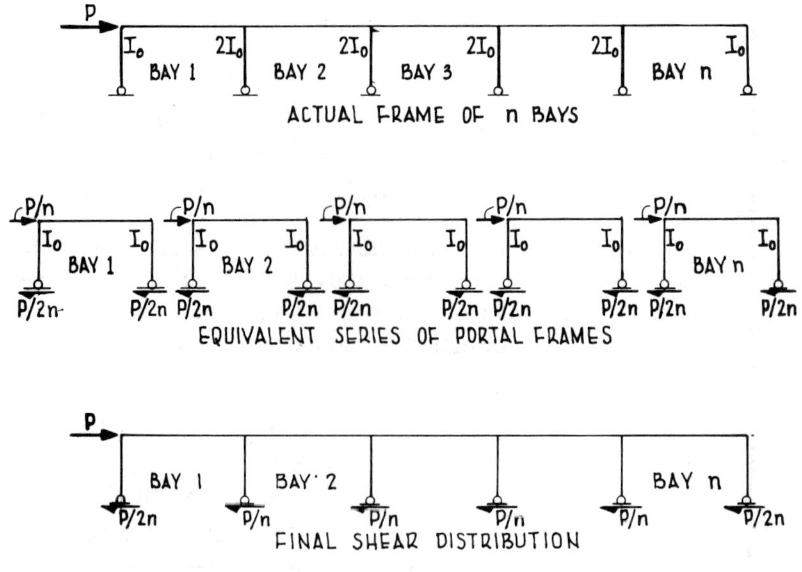

Figure 5-7 Portal method of shear distribution.

columns. The lateral load to each interior column is then the total base shear distributed among the five equivalent interior columns, or $P/5$ kips to each interior column. The load to a perimeter column is half that amount, or $P/10$ kips. A sketch showing the foregoing loads on a typical interior footing and on a typical perimeter footing is presented in Fig. 5-8.

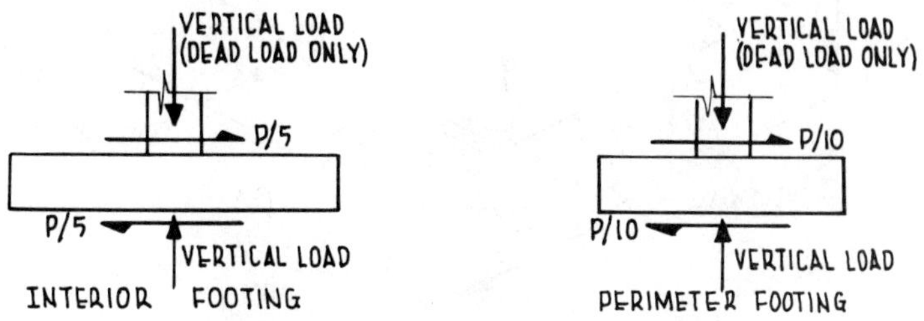

Figure 5-8 Typical rigid frame footings.

Sliding of a rigid frame footing due to lateral loads is not usually a problem, but it is nonetheless well to check the maximum friction load that the footing can develop. The possibility of sliding increases as the building becomes lighter and the lateral load stronger. The maximum friction load that a footing can develop is computed as the vertical design dead load times the coefficient of friction.

The coefficient of friction between a footing and the supporting soil is prescribed by code, normally about 0.30. For the interior footing of Fig. 5-8, the maximum lateral force the footing could take without sliding is then 0.3 times the vertical force. Note that the soil pressure and size of footing have no effect on the allowable friction force; the allowable friction force is simply 0.3 times the vertical force. Friction forces on footings are discussed further in Chapter 10.

DISTRIBUTION OF BASE SHEAR TO FOOTINGS IN A BRACED FRAME

The portal method for distributing the base shear among the footings can be applied only to rigid frame structures. For diaphragm–shearwall structures, the distribution of base shear is somewhat more involved than it is for rigid frames. A typical diaphragm–shearwall structure is shown in Fig. 5-9, with resisting forces shown at the base of each resisting shearwall.

It should be noted that insofar as the foundation is concerned, it makes no difference whether the "shearwall" is a solid concrete wall, a masonry panel, or a cross-braced steel frame. The load delivered to the foundation is identically the same for any of these rigid panels. In further discussions it should be understood that the terms "shearwall," "panel," and "braced panel" may be used interchangeably.

In Fig. 5-9, the lateral load is resisted entirely by the two panels parallel to it, shown as panels *AB* and *CD*. For the symmetrical case, the shear is distributed

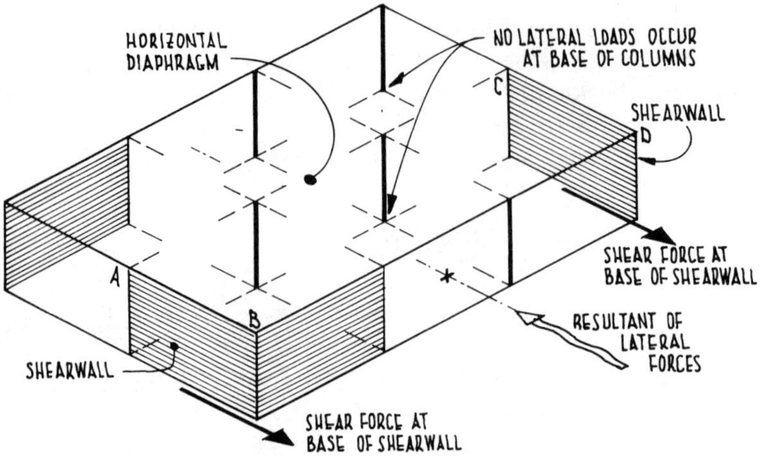

Figure 5-9 Loads on a shearwall.

evenly between the two panels, half being resisted by each panel. It is assumed, of course, that all columns are hinged top and bottom and therefore carry no part of the lateral loads.

The shearwall and its foundation are shown in Fig. 5-10 for two possible configurations. In the first configuration, the shearwall is assumed to carry only shear, with all vertical loads carried by columns at each end. In the second configuration, the shearwall is assumed to be a bearing wall as well, having a continuous strip footing as shown.

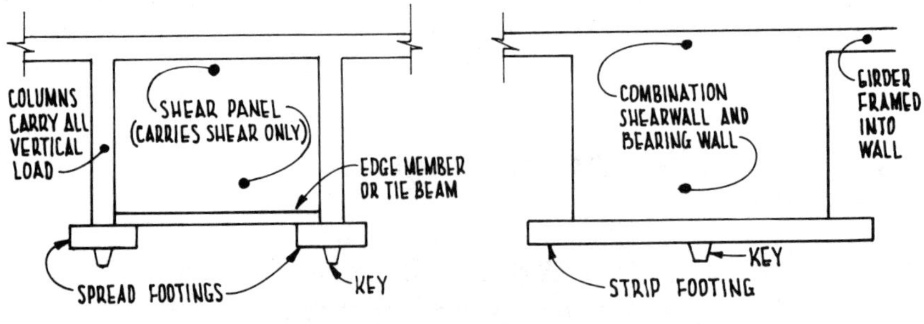

Figure 5-10 Shearwall foundations.

It should be evident that the lateral load on a footing supporting a shearwall is much higher than it was for any of the column footings in a rigid frame structure. Only a few of the footings in the entire building carry the total base shear for the diaphragm–shearwall structure, while every footing in a rigid frame carries its small proportionate part of the base shear. Problems with potential sliding of the foundation are therefore more common with diaphragm–shearwall structures than with rigid frames. Such a problem is encountered in Example 5-5.

"Keys" are sometimes provided at the base of shearwall foundations as shown in Fig. 5-10 to help resist the lateral load. Their effectiveness is questionable; this point is discussed further in Chapter 10. More often, ties are extended to adjacent footings; the additional frictional resistance of those footings can thereby be realized. In some cases, the shearwall may even be dowelled to the floor slab to add that additional friction force and thereby provide additional resistance to lateral loads.

It should be noted at this point that both the foregoing examples were for cases of symmetrical loading on symmetrical structures. Where the building is sharply unsymmetrical, the analysis becomes more complex. Unsymmetrical cases are not drastically different from those shown here; they are just more complex and consequently more tedious.

Since this book is devoted to the design of foundations, a detailed analysis of complex structures is outside the intended scope of this book. For complex or unusual structures, the foundation loads should be taken directly from the actual structural analysis; simplified or approximate methods in such cases should be used only with discretion.

VERTICAL LOADS INDUCED BY WIND AND EARTHQUAKE ON RIGID FRAMES

Lateral loads on a building induce an overturning effect on the building as a whole. On rigid frames this overturning moment must produce increased loading on the foundations on one side of the building and reduced loading on the other. A typical example is shown in Fig. 5-11, where the building is assumed to be rigid enough to act like a monolithic solid block.

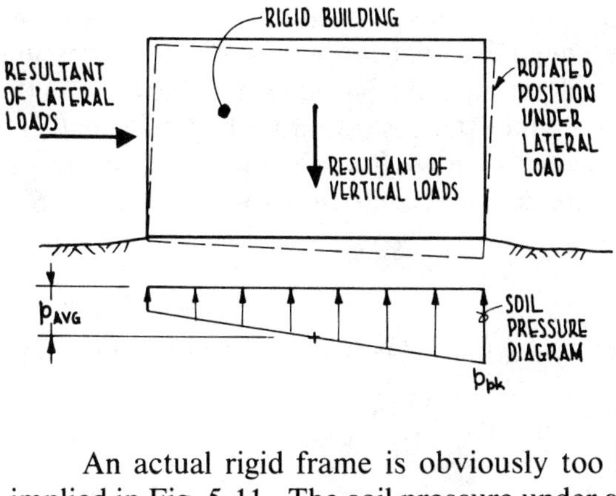

Figure 5-11 Overturning on a rigid building.

An actual rigid frame is obviously too flexible to act as a solid block, as implied in Fig. 5-11. The soil pressure under an actual rigid frame therefore is not likely to reach the peak p_{pk} shown at the extreme end of the building; this peak pressure will probably be relieved by an unknown amount as the flexible building deforms under the loads. Consequently, if the soil pressure diagram shown in Fig. 5-11 is used in the analysis of rigid frame structures, it will yield a result that is probably in error, but always on the "safe" side insofar as the foundation load is concerned.

For short, wide buildings, the soil pressures can be expected to increase only a nominal amount due to the incidence of lateral loads and the consequent overturning. A typical case is shown in Fig. 5-12, where the soil pressure has been deliberately limited to an increase of 33%. The sketch is valid for both wind and earthquake loads.

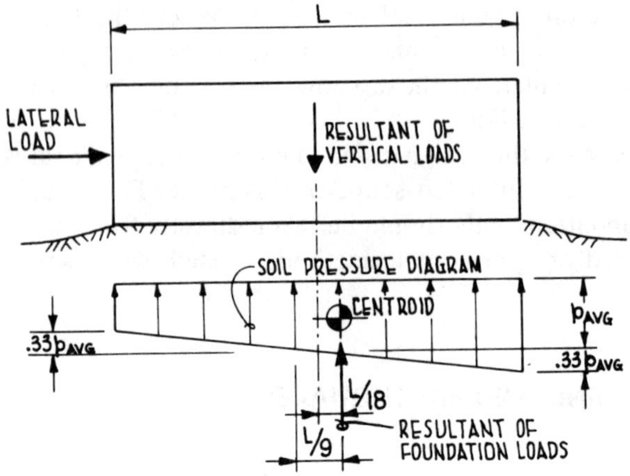

Figure 5-12 Soil pressures due to overturning.

Design codes [9,23] permit a 33% increase in the allowable soil pressure when maximum wind or earthquake loads occur. This 33% increase is shown on the soil pressure diagram of Fig. 5-12. The centroid of the resulting trapezoidal diagram corresponding to this 33% increase is therefore off center by an amount $L/18$, as shown. Since the load is reversible, the centroid must always fall within $L/9$ of the building footprint if the increase in soil pressure cannot exceed 33%.

Conversely, it can be concluded that the increase in soil pressure cannot exceed 33% as long as the resultant of vertical dead loads is kept within the middle ninth of the building footprint. Since the base shear for both wind and earthquake are known and the total dead load is known from the earthquake calculations, it is a simple matter to compute the actual eccentricity e of the resultant R:

$$Re = V_b \times \frac{h}{2} \tag{5-4a}$$

Hence for rigid frame structures, where R is taken as the dead load DL,

$$e = \frac{\frac{1}{2}V_b h}{\text{DL}} \tag{5-4b}$$

It is concluded that if the value of e computed by Eq. (5-4b) is within the middle ninth of the building footprint, the increase in soil pressure cannot exceed the allowable 33% increase in soil pressure. There is no particular need to know the exact amount of the increase. It is enough to know that the increase is within the allowable 33% margin.

Buildings having the resultant of dead loads within the middle ninth of the building footprint are defined herein as "low-rise" buildings, although the criteria admittedly have more to do with slenderness and lightness than with height. Almost all buildings small enough to utilize shallow footings should fall in this category of "low-rise" buildings. Unusually narrow or unusually light buildings could, however, fall outside this category; such exceptional cases, when found, might be better suited for a shearwall–diaphragm structural system than for a rigid frame system.

VERTICAL LOADS INDUCED BY WIND AND EARTHQUAKE ON BRACED FRAMES

In diaphragm–shearwall structures, the overturning effect of the lateral loads is resisted entirely by the shearwalls. Consequently, the shear panels will undergo an overturning rotation as shown in Fig. 5-13 and the entire building will "rack" or "drift" through an angle as indicated. The bottom surface of the shear panel must, of course, rotate through the same angle, displacing the foundations and producing sharply increased (or decreased) displacements.

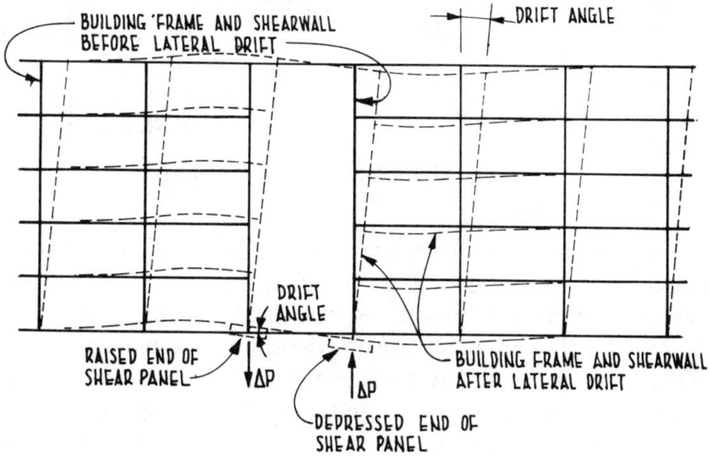

Figure 5-13 Drift angle in a building.

The drift angle shown in Fig. 5-13 is limited by code to 0.005 radian [9]. Excessive drift is highly damaging to glass and finishes as well as to the structure and must be tightly controlled in the structural design. An angle of 0.005 rad is about $\frac{5}{8}$ in. per story, or 1 in. in 16 ft; an angle of this size, if it occurs, can produce serious problems. About half this much drift is preferred.

The control over the drift angle must be exercised by the structural engineer when designing the structural system for the building. As noted, it is required by code that the drift be limited by the structural design to 0.005 rad. For the sake of the foundation design, it will be assumed herein that such control within the structure has, in fact, been accomplished and that the drift angle is never more than 0.005 rad, even when lateral load is maximum.

In addition to the foregoing limitations on drift, practical limitations on settlement place a further limit on the amount of rotation the shear panel can be allowed to undergo. In the foundation shown in Fig. 5-14, for example, the foundations would probably be designed with a maximum settlement S under gravity loading of about 1 in. maximum. Assuming the foundations to be properly designed, the shear panel will settle uniformly under its gravity loads such that long-term settlements at both ends are roughly equal before the lateral loads occur.

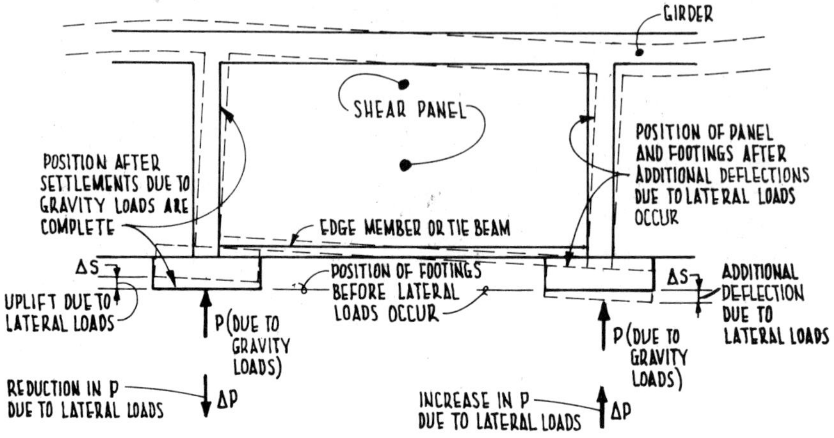

Figure 5-14 Shearwall footing pressures.

When the short-term lateral loads are applied, the panel of Fig. 5-14 will rotate until at some instant the uplift ΔS at the raised end will relieve all of the pressure at that end, producing zero pressure there. At that instant, the downward load at the other end must be the sum of the two loads, which is taken here to be the limiting case. Any further differential movement between the raised footing and the other footings around it could seriously damage the structure.

The actual magnitude of the differential movement ΔS would depend on the modulus of subgrade reaction for short-term loading, a highly uncertain property in soils. Depending on the source of data, the modulus for short-term loading may be estimated to be anywhere from 3 to 30 times the modulus for long-term loading. With such an estimate, any attempt to calculate the actual rotation is hopeless; such calculations are not pursued further here. In this book, limits on pressures rather than limits on deformations will be used to control the drift.

The limiting case, therefore, as shown in Fig. 5-14, occurs when the soil pressure at one end of the shear panel decreases to zero; the load at the other end then becomes the sum of all tributary vertical loads, or roughly doubled (assuming that the two loads are about equal). The restoring moment is then limited to the value of the couple $P_w s/2$, where s is the span between the footings and P_w is the sum of the tributary vertical loads. To increase the restoring moment, either a longer shear panel may be used or an additional number of shear panels may be used.

If a higher resisting moment is needed, several alternative configurations are readily possible. One of the possibilities is simply to extend a heavy grade beam outward from the base of the panel to the adjacent footings at either end, as shown in Fig. 5-15. The allowable restoring moment of this configuration is some three times as much as the configuration without these extensions.

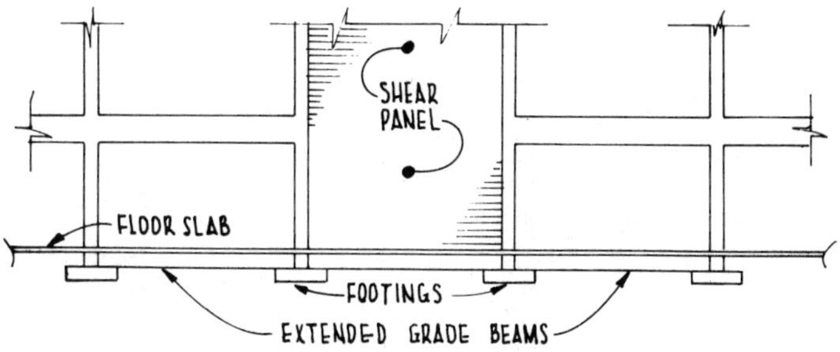

Figure 5-15 Extended grade beam.

The foregoing considerations must be borne in mind by the structural designer when selecting the overall structural system for the building. As far as the design of the foundation is concerned, however, the end result is that the soil pressures at the ends of a shear panel can be expected to be roughly doubled when lateral loads are applied. Only in exceptional circumstances should the rotations be allowed to increase further than this, in consideration of potential drift and settlement problems.

The limitation on the restoring moment produced by these restrictions is a serious drawback to the use of shear panels. Short shear panels (less than about 16 ft long) can produce excessively high soil pressures and will probably permit excessively high amounts of drift. Heavily loaded shearwalls are at their best when they can be founded on very rigid foundations that can take a load reversal; deep piles are one such foundation, although deep foundations are not treated in this book.

In this book the vertical design load on either of the two footings at the ends of a shear panel will be taken as the sum of all tributary vertical loads due to dead load plus live load, designated herein as P_W. The restoring moment offered by the shear panel is computed using this limiting value of P_W; it is assumed that the structure itself is properly designed to resist the lateral loads with this limit on the level of the restoring moment. The footing pressures and forces associated with this case are shown in Fig. 5-16.

If the foundation supporting a shearwall is a continuous strip footing rather than two spread footings, the same basic assumption still applies: the shearwall rotates until the soil pressure at one end of the strip footing becomes zero, as shown in Fig. 5-16. In this rotated position, the soil pressure diagram for a strip footing becomes triangular and the maximum soil pressure becomes double that in the unrotated position. The same end result therefore occurs—that the soil pressure at either end of the footing could double. The restoring moment for a strip

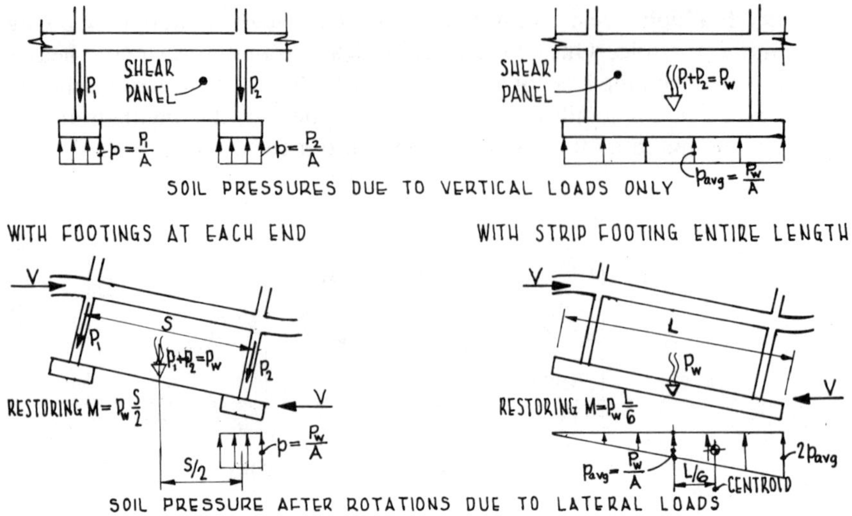

Figure 5-16 Comparison of shear panel footing pressures.

footing and its triangular pressure diagram is $P_W L/6$, where in this case L is the length of the strip footing; this restoring moment is seen to be only about one-third of that produced by the use of two spread footings.

LOAD COMBINATIONS FOR FINAL DESIGN

It is not reasonable to assume that all the potential loads that might occur on a building will occur simultaneously. It is unlikely, for example, that a building would be subjected to a peak earthquake load while the wind is blowing at maximum velocity. Nor is it likely that the roof live load will be in place during peak wind velocities.

Further, it is not reasonable to require that a structural material sustain transient rarely applied loads at the same stress level at which it carries continuous long-term loads. A higher stress level (and possibly even limited damage) might be allowed to occur, for example, for the short duration of an earthquake, recognizing that the maximum earthquake load (and limited damage) will probably occur only once in the life of the structure. Conversely, a lower stress level might be desirable under dead load, which occurs at 100% of its design value for 100% of the time.

In addition, it should be recognized that a lesser load rather than a higher load could, in some circumstances, cause "worst-case" stresses. For example, the restoring moment exerted by a shearwall is at its lowest value when the vertical load is minimum (i.e., when the vertical load consists of dead load only). For such a case, the existence of live load would actually improve the restoring moment.

In view of such possibilities, the design of a building is performed for several combinations of load that will produce worst-case stresses under reasonable load combinations. Also, stress levels are allowed to increase as much as 33% whenever transient loads such as wind or earthquake are being considered. Further, the structure is usually checked for stress reversals under minimum vertical load (dead load only) in combination with the maximum lateral load (wind or earthquake).

The following list of load combinations are among the more commonly used load cases. Special load cases could, of course, be required for unusual circumstances. Note that using 75% of a load at the usual allowable stress will produce a 33% increase in stress at 100% of the load.

1. Dead plus full live (or roof snow) (5-5a)
2. (0.75)[Dead plus full live plus wind (or earthquake)] (5-5b)
3. Dead plus wind (or earthquake) (5-5c)

FINAL DESIGN LOADS ON FOOTINGS

When loads are combined into Eqs. (5-5a) through (5-5c) for the final load case, the design loads degenerate into a few simple cases. For foundation design, it is usually much easier to use these cases than to go through a complete analysis of loads for each building.

For footings other than those supporting shearwalls, Eqs. (5-5a) through (5-5c) degenerate to:

Maximum vertical load = DL + LL.

Minimum vertical load = DL.

For low-rise rigid frame structures, the distribution of the horizontal base shear to each footing may be computed using the portal method of analysis; the increase in vertical load due to overturning for such structures is known to be within allowable limits and need not be computed.

In no case may the actual lateral load on any footing exceed 0.3 times the actual vertical load (limited by friction).

For diaphragm–shearwall structures, there is no lateral load on any footing other than those supporting the shearwalls.

For each of the two spread footings at the ends of a shear panel, the most severe case is that where the vertical load due to wind or earthquake is roughly equal to the (DL + LL) case, causing zero soil pressure at one end of the panel, for which case Eqs. (5-5b) and (5-5c) yield:

Maximum vertical load = $0.75(P_W)$, where P_W is the sum of all tributary vertical loads on the panel due to dead load plus live load. At the time this level of vertical load occurs, this one footing carrying $0.75P_W$ must also carry 100% of the lateral load that is to be carried by the shear panel.

Minimum vertical load = 0. At the time this level of vertical load occurs, this footing cannot resist any lateral loads.

In no case may the total lateral load V exceed 0.3 times the total vertical load $0.75P_W$ (limited by friction).

Restoring moment = $P_W s/2$, where P_W is the total vertical load on the shear panel as defined earlier and s is the spacing between footings.

For strip footings supporting shearwalls, Eqs. (5-5b) and (5-5c) degenerate to:

Maximum vertical load = $0.75P_W$, where P_W is the sum of all tributary vertical loads due to dead load plus live load.

The soil pressure diagram is triangular at maximum load, with pressure at one end being zero; the entire footing is assumed to remain in contact with the soil.

The base shear being carried by the shearwall is distributed along the length of the footing proportional to the vertical contact pressure at any point (also a triangular stress distribution).

In no case may the total lateral load V exceed 0.3 times the total vertical load $0.75\ \Sigma\ (P_{DL} + P_{LL})$ (limited by friction).

Restoring moment = $P_W s/6$, where P_W is the vertical load defined earlier and s is the length of the strip footing.

In the foregoing simplifications, it is assumed that the shearwall is placed within the regular column module and is not a multiple-span shearwall. Multiple-span shearwalls can be treated in much the same way, however, where the angle of drift is controlled such that the soil pressure at one end goes to zero. Such analyses are too lengthy for presentation here.

The final design loads to be carried by the footings of the building shown in Fig. 5-1 can now be determined. The foregoing simplifications will be applied to footings A2 and B3 as typical footings.

EXAMPLE 5-5: Final Design Loads on Footings
Computation of combined loads.
Dead plus live plus wind plus earthquake.

Compute the design loads to be sustained by footings A2 and B3 of the building shown in Fig. 5-1. The vertical loads on footing B3 are given in Examples 5-1 and 5-2. For footings A1 and A2, the vertical loads have been computed independently. They are, when no lateral loads are included,

	Footing A1	Footing A2
P_{DL}	183 kips	219 kips
P_{LL}	24 kips	49 kips

SOLUTION The loads on footing B3 are, from Examples 5-1 and 5-2,

$$P_{DL} = 214 \text{ kips}$$

$$P_{LL} = 114 \text{ kips}$$

The final vertical design load to footing B3 is computed as the sum of the dead and live loads,

$$P_W = P_{DL} + P_{LL} = 214 + 114 = 328 \text{ kips}$$

The final vertical design load to footing A2 (located at the end of a shearwall) is computed using the values of P_{DL} and P_{LL} given above at both ends of the shearwall,

$$P_W = 0.75 \sum P = 0.75(183 + 24 + 219 + 49) = 356 \text{ kips}$$

The foregoing loads are the final vertical loads to be used for the design of the footing.

Lateral loads are taken from the analyses given in Examples 5-3 and 5-4; the lateral load delivered to footing B3 is zero, since all lateral loads are taken by the shearwalls. The lateral load delivered to footing A2 is 100% of the lateral load to be carried by the shearwall, which in this case is half the total earthquake load of 300 kips computed in Example 5-4. The lateral loads on the shearwalls are shown in the following sketch. The load is distributed equally to the two shearwalls since the lines of action of the loads are symmetrical.

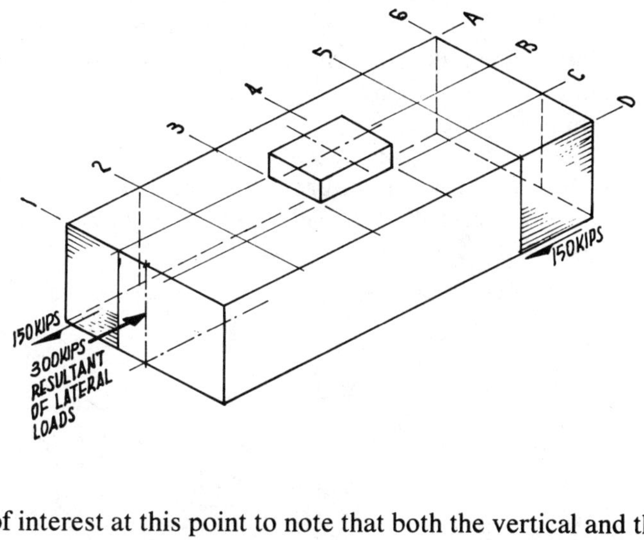

It is of interest at this point to note that both the vertical and the horizontal design loads for footing A1 are identical to those for footing A2, even though their tributary gravity loads are different. A moment's reflection will confirm that the restoring moment offered by the shearwall is therefore the same for either direction of wind, regardless of what the initial distribution is.

The final loads on the two footings are shown below:

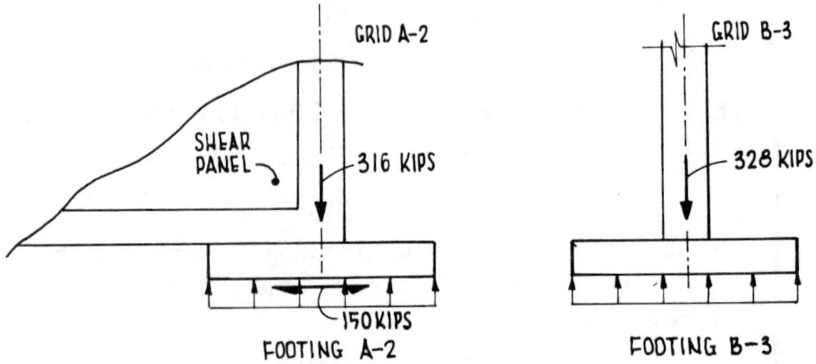

The footing at grid point A2 must now be checked to see if it can sustain the indicated lateral load. With a coefficient of friction of 0.30 to be applied to dead load only, the maximum lateral load on the footing is easily computed:

$$\begin{aligned} \text{allowable friction load} = 0.3 \times P &= 0.3(P_{DL}) \\ &= 0.3(219 + 183) \\ &= 121 \text{ kips} \end{aligned}$$

The applied lateral load of 150 kips that must be taken by footing A2 is seen to be some 24% larger than the allowable load that can be developed by friction. The footing therefore cannot sustain the design load.

A result such as this is often encountered in shearwall–diaphragm structures. The usual remedy is to run a tie beam around the perimeter of the building, tieing all the perimeter footings together. The total frictional resistance of the six footings along line A is then available to resist the 150-kip lateral load.

A typical section through such a tie beam is shown in the following sketch. The tie beam performs multiple functions in the design since it also acts as a cutoff wall at the sides of the building and, in addition, supports the exterior wall of the building.

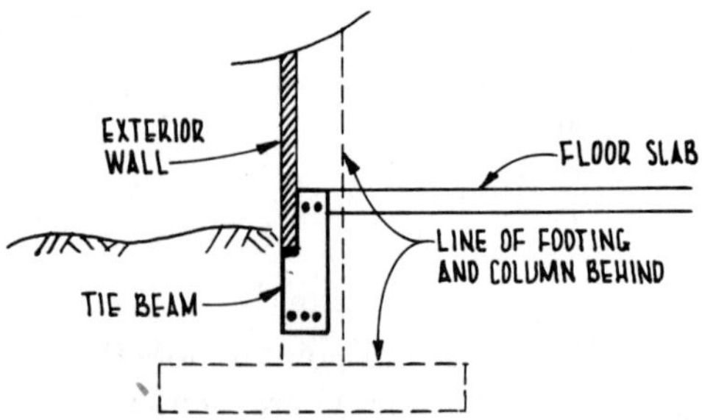

Such a tie beam is assumed to be adopted here. The tie beam is of course designed to be effective either in tension or in compression since the earthquake load could occur in either direction. The proof of the adequacy of the six footings along line A to sustain the 150-kip lateral load is assigned as Problem 5-29. (The lateral load is divided among the six footings in direct proportion to their vertical design loads.)

SUMMARY

A "low-rise" building has been defined herein as a building in which the resultant of dead loads remains within the middle ninth of the building footprint when the maximum lateral loading occurs. The following conclusions were developed in this chapter for such low-rise buildings.

- For foundations of prismatically framed, reasonably modular, uniformly loaded structures, the redistribution of vertical loads due to flexure on beams and columns can be neglected.
- The vertical gravity load that is tributary to a footing can be computed with reasonable accuracy as the sum of all gravity loads halfway to the next vertical support in any direction.
- The final vertical design load on any footing *not* carrying a shear panel may be taken simply as the sum of gravity loads tributary to the footing; lateral loads have no direct effect on these vertical design loads (applies both to rigid frames and to diaphragm–shearwall systems).
- For the design of footings carrying a shear panel when maximum lateral load occurs, vertical load on the shear panel should be taken as 75% of the sum of all gravity loads $\Sigma(DL + LL)$ tributary to the panel throughout its length. The panel is then assumed to rotate until the soil pressure at one end decreases to zero. If the foundation is a strip footing, the soil pressure diagram becomes triangular and the average soil pressure doubles. If the foundation is two spread footings, either of the two footings could be loaded by the entire panel load; the design load on each footing could therefore be effectively doubled.
- In addition to carrying the vertical design load, each of the two footings carrying the ends of a shear wall must be designed to sustain the base shears due to wind or earthquake; the base shear is resisted by friction between the foundation and the soil.
- For computation of maximum frictional resistance to the base shear, a coefficient of friction of no more than 0.30 between a footing and the supporting soil is commonly accepted.
- For rigid frames, the base shear is distributed to all the individual footings (usually by portal analysis).
- For diaphragm–shearwall structures, the base shear is distributed only to the foundations supporting the shearwalls (usually by simple statics).

• Where lateral loads are resisted by the action of a shearwall or braced panel, the angular rotation of the shearwall or braced panel due to the lateral loads may not exceed 0.005 rad (a building code limitation).

• The maximum restoring moment offered by a shearwall or braced panel occurs when the soil pressure under the raised end of the panel reaches zero; further uplift could have serious detrimental effects on the structure.

• The structural design of the building itself must accommodate the limitations on the restoring moment and must limit the drift of the building to the specified amount.

REVIEW QUESTIONS

5-1. Give typical examples of items in a building that would be classified as dead loads.

5-2. Give typical examples of items in a building that would be classified as live loads.

5-3. Of the four major types of load—dead load, live load, wind load, and earthquake load—which are treated as long-term loads and which are treated as transient loads?

5-4. Define the "tributary area" for a column loading.

5-5. Describe the configuration of a typical building that would be subject to the simplified ACI analysis.

5-6. How are the effects of wind and earthquake acting on a building similar?

5-7. How is the response of a building to a wind load different from its response to an earthquake load?

5-8. How is the reference wind pressure on a building computed?

5-9. How is the distribution of wind pressure over a building arrived at?

5-10. Define the "importance factor" in the design of buildings for wind load. To what kind of buildings might it apply?

5-11. Where does a building designer find the requirements for the use of importance factors in his or her locality?

5-12. What is the earthquake intensity zone number in the vicinity of the Missouri "bootheel"? The Georgia coast? The New Hampshire coast?

5-13. Where are the two regions in the United States that have a zone 4 earthquake damage risk? The four regions that have a zone 0 earthquake damage risk?

5-14. Estimate the percent of the contiguous United States that has some risk of damage due to earthquake, that is, lying in risk zone 1 or higher.

5-15. How is the earthquake base shear computed for low buildings?

5-16. Where does a building designer find the value of the zone factor Z for a particular locality?

5-17. How is live load treated in an earthquake analysis?

5-18. How is the base shear distributed to the individual footings in a rigid frame structure?

5-19. About what maximum coefficient of friction can be assumed for ordinary soils?

5-20. Does the coefficient of friction apply to loads or to contact pressures?

5-21. What happens to the coefficient of friction if the size of the footing is increased?

5-22. How is the base shear distributed to the shear panel foundations in a braced frame or a diaphragm–shearwall structure?

5-23. What happens to the footing pressures under the columns in a rigid frame when the building undergoes maximum lateral load?

5-24. What happens to the footing pressures under the shear panels when a braced frame undergoes maximum lateral load?

5-25. Compare the distribution and magnitudes of the soil pressure under a shearwall foundation when a continuous strip footing is used and when spread footings are used.

5-26. How is it that lateral loads produce vertical forces on the foundation?

5-27. What is a tie beam? What does it do for foundation loads?

5-28. List the three most common combinations of load in building design.

5-29. Where does a building designer find the required combinations of building loads in a particular locality?

5-30. What is a "low-rise" building insofar as lateral loading is concerned?

PROBLEMS

5-1. Determine the dead load on footing A-1 of Fig. 5-1, using the load criteria of Example 5-1.

5-2. Determine the dead load on footing A-2 of Fig. 5-1, using the load criteria of Example 5-1.

5-3. Determine the dead load on footing D-1 of Fig. 5-1, using the load criteria of Example 5-1. Use a dead load of 175 psf over the entire stairwell area to account for the stairwell structure.

5-4. Determine the dead load on footing B-1 of Fig. 5-1, using the load criteria of Example 5-1.

5-5. Determine the live load on footing A-1 of Fig. 5-1, using the load criteria of Example 5-2.

5-6. Determine the live load on footing A-2 of Fig. 5-1, using the load criteria of Example 5-2.

5-7. Determine the live load on footing D-1 of Fig. 5-1, using the load criteria of Example 5-2. Apply the live load of 100 psf over the entire stairwell area.

5-8. Determine the live load on footing B-1 of Fig. 5-1, using the load criteria of Example 5-2.

Problems 5-9 through 5-12. Determine the wind pressures against the building silhouettes shown. Find the load acting at each story and find the overall base shear.

5-9. Wind velocity 100 mph
Total dead load 10,800 kips
Earthquake risk zone 2
Rigid frame structure

5-10. Wind velocity 120 mph
Total dead load 5950 kips
Earthquake risk zone 3
Rigid frame structure

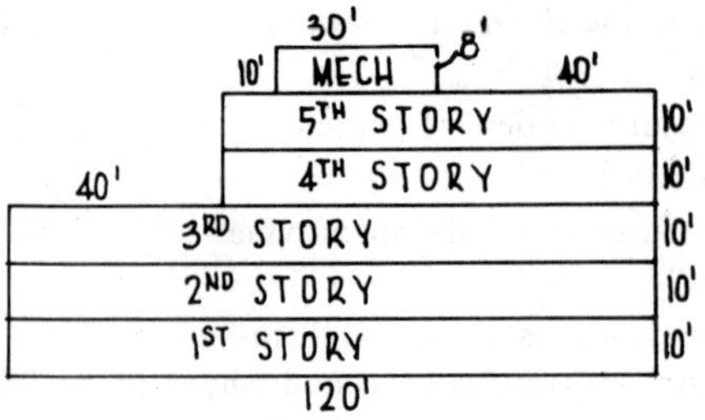

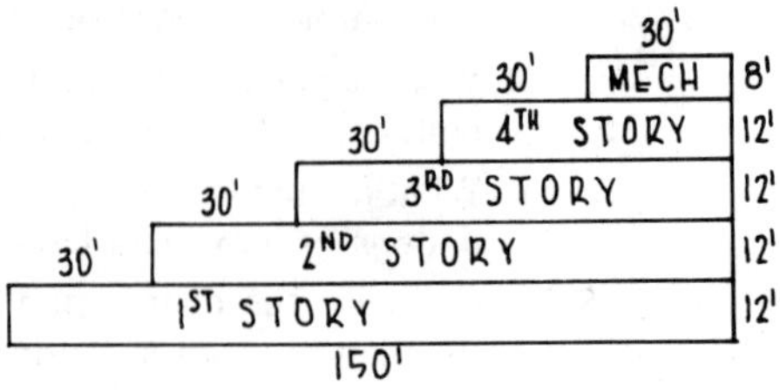

5-11. Wind velocity 150 mph
Total dead load 1920 kips
Earthquake risk zone 2
Diaphragm–shearwall structure

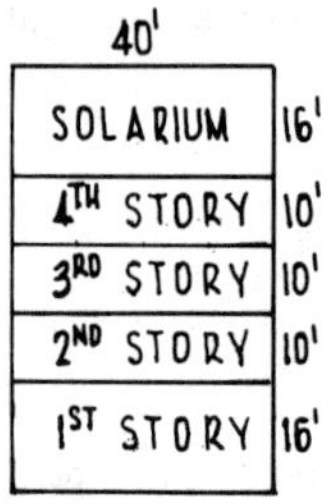

5-12. Wind velocity 90 mph
Total dead load 3100 kips
Earthquake risk zone 1
Diaphragm–shearwall structure

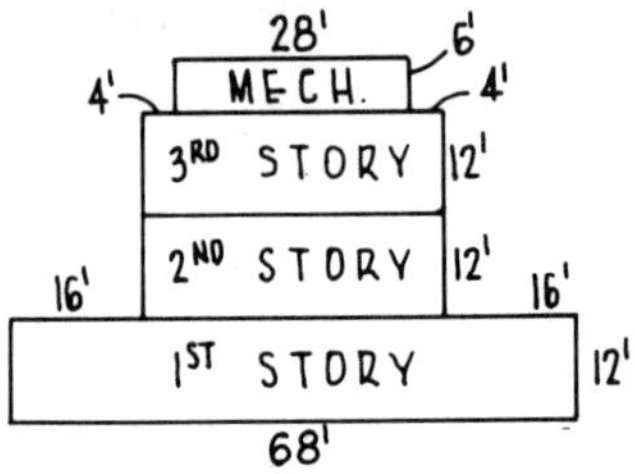

Problems 5-13 through 5-16. Determine the base shear due to earthquake for the structures shown with Problems 5-9 through 5-12.

5-17. A rigid frame structure is 120 ft × 192 ft in plan, with columns at 24 ft on center in both directions (6 columns × 9 columns). The base shear due to wind is 321 kips on the short side and 514 kips on the long side. Distribute the base shear to each column.

5-18. For the structure of Problem 5-17, find the base shear at each column if the six center columns are removed to create an open-air atrium measuring 72 ft × 96 ft as shown.

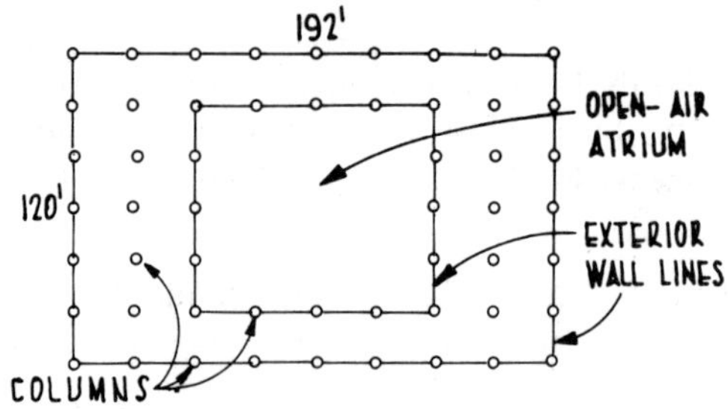

5-19. A diaphragm–shearwall structure is subject to a base shear of 391 kips in any direction due to earthquake. Location of the shearwalls in plan is shown in the sketch. Determine the maximum shear that could occur at the foundations of each shearwall.

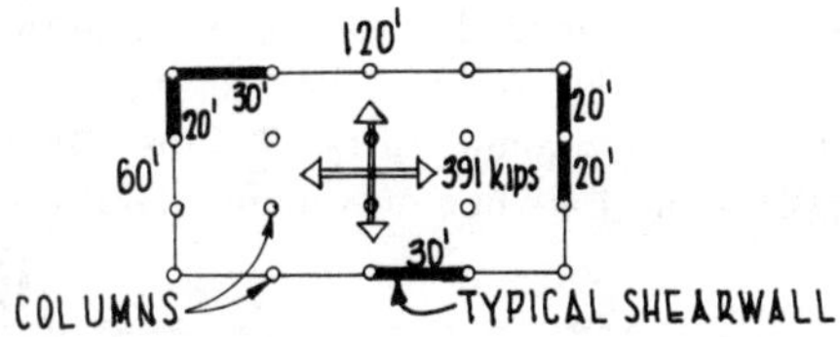

5-20. A braced panel structural system is subject to a base shear due to wind as indicated in the sketch. Determine the maximum base shear that could occur at the foundations of each braced panel.

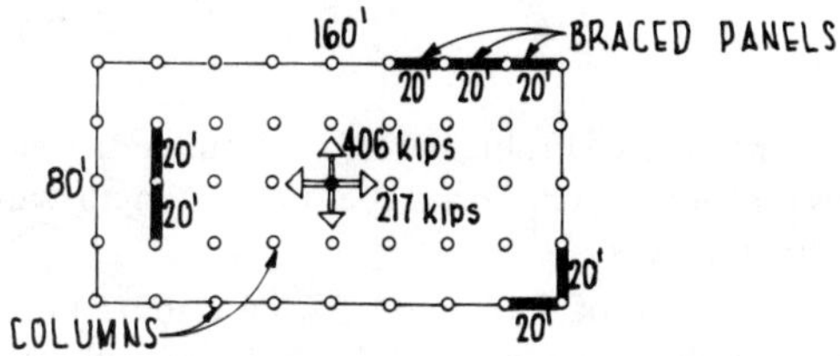

5-21. A braced panel structural system is subject to a base shear of 131 kips due to earthquake. Locations of the braced panels are shown in the sketch. Determine the maximum shear that could occur at the foundation of each braced panel.

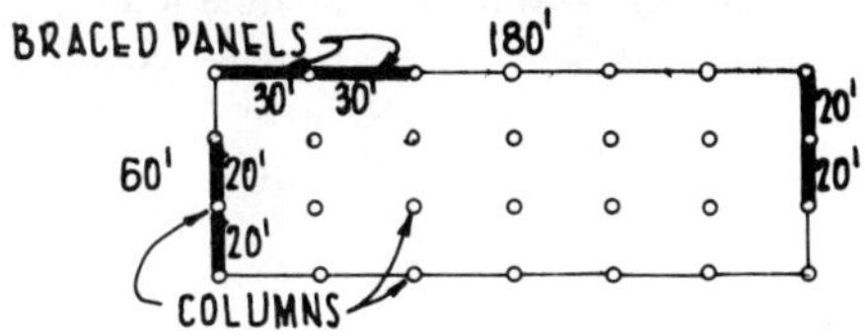

5-22. A diaphragm–shearwall structure could be subject to a wind load of 219 kips against its long side or a wind load of 152 kips against its short side as indicated. Determine the maximum shear that could occur on the foundations of each shearwall.

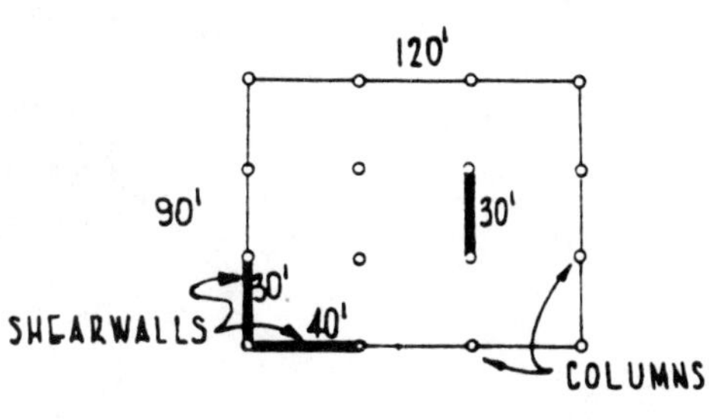

5-23. Using the results of Problems 5-1, 5-2, 5-5, and 5-6, determine the maximum vertical load that could occur on footings A1 and A2 when a maximum lateral load occurs.

5-24. Using the results of Problems 5-1, 5-4, 5-5, and 5-8, determine the maximum vertical load that could occur on footings A1 and B1 when a maximum lateral load occurs.

5-25. The interior columns in the braced frame structural system shown in the sketch carry a total vertical load of 340 kips. The perimeter columns (except corners) carry 241 kips and the corner columns carry 196 kips. Determine the maximum vertical load on footings E3 and E4 when maximum lateral load occurs.

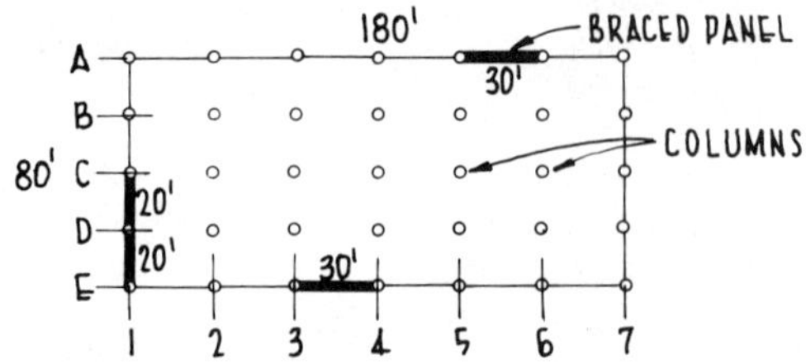

5-26. For the structure of Problem 5-25, determine the maximum vertical load on each of the three footings supporting the double-length panel C1–D1–E1 when maximum lateral load occurs.

5-27. For the structure of Problem 5-25, determine the required size of a strip footing centered under the panel E3–E4 if the allowable soil pressure is 3500 psf.

5-28. For the structure of Problem 5-25, determine the required size of a strip footing centered under the panel C1–D1–E1 if the allowable soil pressure is 3500 psf.

5-29. Complete the calculations of Example 5-5 by showing that the six footings along line A of the building will develop far more resistance to lateral load than is required. For the calculations, assume the vertical loads at A1 and A6 are roughly equal and those at A2, A3, A4, and A5 are roughly equal.

6

Determination of the Contact Area

Given the loads acting on a footing from above and the allowable pressure that can be supported by the soil below, calculations for the size of the contact area of a footing would seem to be very straightforward. The interaction between soil and structure, however, can have a significant influence on the final selection of the size of the footing. A means to account for such soil–structure interaction is developed in this chapter.

The effects of lateral loads acting on a footing were presented in Chapter 3. In that presentation, it was seen that the effects of lateral loading can be quite severe, in some cases reducing the strength of the soil by as much as 75%. Once the allowable bearing pressure is reduced by the appropriate percentage, however, the design load can be taken thereafter as the vertical load only, with no further consideration of the lateral load being necessary.

In addition to the loads acting from above and the allowable pressure in the soil below, the rotations of the footing due to structural deformations in the building above must also be accounted for. In the method developed in this chapter, it will be seen that these rotations (when they occur) can be accounted for independently and accurately, permitting the final design to be made using only a modified vertical load; the actual moments and rotations need not be found. The final selection of the required footing area then becomes a relatively simple calculation involving only a vertical load and an allowable pressure.

In these design calculations, the accuracy of the physical properties of soil and concrete become of primary importance. As noted earlier, however, the physical properties of soil can be determined only to about one significant figure and concrete to two. With such limited accuracy, highly refined theoretical approaches are subject to doubt; simple direct approaches are far more believable.

In recognition of the level of accuracy, simplified statically determinate approaches are used in subsequent discussions. The final calculations use pressures rather than settlements since the prediction of the magnitude of settlements is in itself questionable. However, even though the exact magnitude of one calculated settlement may be questionable, the thoughtful comparison of two such calculated settlements can often provide insight into the relative movements of the two foundations; such comparisons are utilized repeatedly in the following discussions.

It was shown in Chapter 3 that at equal contact pressures, larger footings will settle more than smaller footings. In all discussions in this chapter concerning a single isolated feeling, it is assumed that the footing under discussion is of a relatively large size, at least large enough that its settlement approaches the maximum allowable value for settlements.

DIFFERENTIAL SETTLEMENTS

When all the footings under a building settle the same amount, the resulting uniform settlement may have very destructive effects on sewer, water, and power connections to the building, but it has no effect on the structure. Only the differential settlement between adjacent footings or supports will affect the stress levels in the structure. The structural analysis is therefore concerned only with differential settlements between supports, not with total settlements.

In a group of footings, if the settlement of the footing that settles most is limited to 1 in., the differential settlement between any two footings in the group can be expected to be somewhat less than 1 in., say a maximum of $\frac{3}{4}$ in. It is this approach that is commonly used in practice for the design of shallow footings—that the settlement of the footing that settles most will be limited to about 1 in. and that differential settlements, if any occur, can then be expected to be less than about $\frac{3}{4}$ in.

There are myriads of combinations of differential settlements that might occur in a routine structure. If, however, the differential settlements are less than about $\frac{3}{4}$ in., the change in total stress in any of the structural members due to any reasonable combination of loads can be expected to be less than about 15% of the total. For this amount, a separate analysis for potential differential settlements is not usually considered to be necessary and is rarely performed in practice. Where circumstances are unusual, or where a key foundation is known to be subject to large settlements, a thorough study of the effects of differential settlements is necessary.

ALLOWABLE PRESSURES AND SETTLEMENTS

Unless specified otherwise, the allowable bearing pressure is defined as the *increase* in vertical pressure that may be applied to a soil within a certain band of depths. It can be assumed that the soil will accept the increase in pressure in addition to its regular overburden pressures. The band of depths in which the increase may be applied is that band where shallow foundations are normally founded, usually between 16 in. to 6 ft below the surface.

The allowable bearing pressure may be presumed to include all effects that could, within reason, affect the soil adversely. Soil moisture may be presumed to be at that level where the capacity of the soil is lowest. Effects of excavations in the near vicinity, however, should not be presumed to be included unless specifically stated.

The magnitude of the allowable bearing pressure may be obtained from any one of several sources. Such sources include the soils report, code maxima, personal evaluations, city and county standards, or other reliable sources. In addition to knowing the magnitude itself, it is necessary for the designer to know also whether the magnitude is limited by the strength of the soil or by its susceptibility to settlements.

Where the prescribed bearing pressure is limited by strength criteria, it may be presumed that settlement at that pressure is less than 1 in. In actual fact, it may be considerably less for higher-quality soils. Alternatively, where the prescribed bearing pressure is limited by susceptibility to settlements, it may be presumed that the prescribed soil pressure is that pressure estimated to produce a nominal 1-in. settlement on a relatively large footing; 8 ft square is commonly used as a reference size. The ultimate pressure that would produce a strength failure can be expected to be significantly higher than this pressure at 1 in. of settlement. Consequently, where bearing pressure is limited by settlement criteria, it may be presumed that strength is not a problem.

As discussed in Chapter 1, the soils treated in this text are those having an allowable bearing pressure less than about 5000 psf. In these soils it should be expected that settlements will be high. Even when the allowable bearing pressure of the soil is limited by strength criteria, a settlement at or near 1 in. for the larger footings should be anticipated.

In subsequent analyses, it will be assumed that settlements of the larger footings will approach the 1-in. limit as pressure approaches its allowable value, regardless of whether the allowable bearing pressure is limited by strength or by settlement criteria. This assumption is consistent with general pressure–settlement relations (modulus of subgrade reaction) given in standard references [13]. It applies reasonably well for the better soils, such as medium sands or very stiff clays, which are the better soils within the range of interest here.

In a stratified soil, it may occasionally happen that the capacity of one stratum is limited by strength and the capacity of another by settlements. Completely independent solutions will be necessary for each stratum. Within each stratum, however, the foregoing assumptions apply. Throughout all calculations involving soils, only working levels of load are used. The use of design concepts involving ultimate loads are not applied to soils in this text.

FOOTING WITH VERTICAL LOAD ONLY

The simplest case for footing design is that of a footing carrying vertical load only. Moments are prevented by providing a hinge between the column and footing, as shown in the example problem of Fig. 6-1. The structural analysis of the building above must, of course, include the effects of the hinged column base.

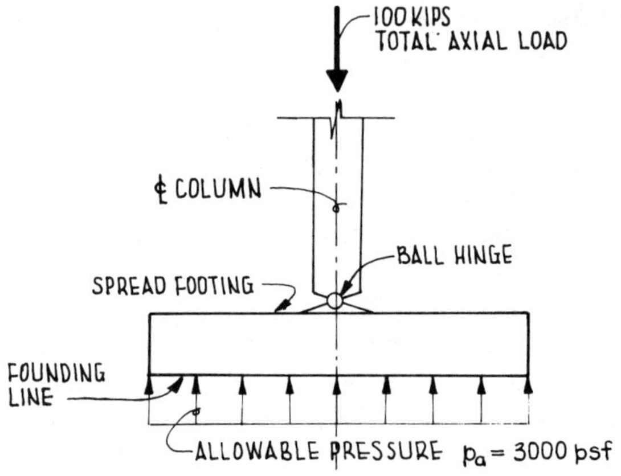

Figure 6-1 Direct loads on footings.

With no moment on the footing, the bearing pressure is assumed to be uniform, as shown in Fig. 6-1. The required contact area for the footing is found very simply:

$$A = \frac{P}{p}$$

where P = vertical load
A = contact area
p = bearing pressure

In the example of Fig. 6-1, the indicated column load is 100 kips and the allowable bearing pressure is 3000 psf. The required contact area is therefore found to be

$$A = \frac{P}{p} \qquad A = \frac{100{,}000 \text{ lb}}{3000 \text{ psf}} = 33 \text{ ft}^2$$

A square footing 6 ft × 6 ft would satisfy the requirements. In this case it does not matter whether the bearing pressure is limited by strength or by settlement criteria.

The use of hinged column bases as shown in Fig. 6-1 is much preferred insofar as foundation design is concerned. It has the advantage that the system is completely statically determinate and is therefore accurate. There are no statically indeterminate deformations with their dependence on unreliable pressure–settlement relations.

Rotations of a footing are a very common feature, however, occurring whenever the footing is fixed to the base at the column. The effects of column moments and the rotations of the footing produced by these moments are considered next.

EFFECTS OF MOMENT

The reason for limiting the settlement of a footing is to prevent damage to the structure above; settlements by themselves do no harm to the footing. The only settlement that is of interest, therefore, is that at the centerline of the column (or bearing wall). The magnitude of the settlements elsewhere on the footing is not usually of interest.

A typical concentric column rigidly attached to its footing is shown in Fig. 6-2. Exaggerated rotations due to moments on the column are shown. A strip footing with its bearing wall would show a similar response.

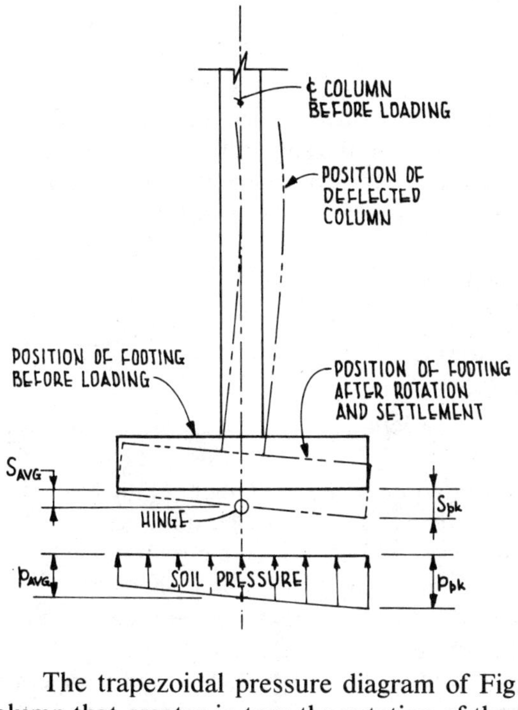

Figure 6-2 Typical column and footing.

The trapezoidal pressure diagram of Fig. 6-2 is a result of bending in the column that creates in turn the rotation of the footing. A minute's reflection will confirm that the average settlement as well as the average pressure will occur at the centerline of the symmetrical footing. The rotation due to the column moment is seen to have no effect on the average pressure p_{avg}, but it does have an effect on the peak pressure p_{pk}. Where strength of the soil is the limiting factor, it is the peak pressure p_{pk} in Fig. 6-2 that must not exceed the allowable pressure p_a.

Alternatively, where settlement of the column is the limiting factor, it is the average settlement of the footing, S_{avg}, that must be limited (Fig. 6-2). At the centerline, it is noted that settlement of the column will be proportional to the average pressure p_{avg}. Consequently, if the settlement of the column is not to exceed 1 in., it is the average pressure p_{avg} that must not exceed the allowable pressure p_a.

The foregoing conclusions for concentrically loaded footings are summarized:

Average pressure and average settlement occur at the centerline (centroid) of a footing.

Moments in the column have no effect on the average pressure or average settlement.

The peak pressure occurs at the edge of the footing and is produced by rotations of the footing.

The foregoing conclusions will be used repeatedly in subsequent discussions.

EFFECTS OF ROTATIONS

A typical concentric column and footing are shown in Fig. 6-3. Exaggerated rotation of the footing and the effect on the soil pressures are shown. The formulas shown for computing the magnitudes of moments and rotation are given in standard texts [26]. The following symbols appear in Fig. 6-3:

f'_c is the ultimate strength of concrete.

E_c is the modulus of elasticity of concrete.

I is the moment of inertia of the cross section.

b is the width of the cross section.

t is the depth of the cross section.

If the footing is not allowed to rotate at all, the moment at the base of the column will be $\frac{1}{2}M_c$, as shown. Alternatively, if the footing is allowed to rotate freely, the moment at the base of the column will be zero, with a rotation angle θ_f as shown. It should be evident that the amount of rotation that the footing is allowed to undergo (corresponding to the degree of restraint) will determine the magnitude of the moment that will exist at the base of the column.

Refer again to Fig. 6-3, where the soil pressures and their corresponding settlements are shown on the soil pressure diagram. It is assumed that within the range of pressures that may occur, the settlement is directly proportional to pressure.

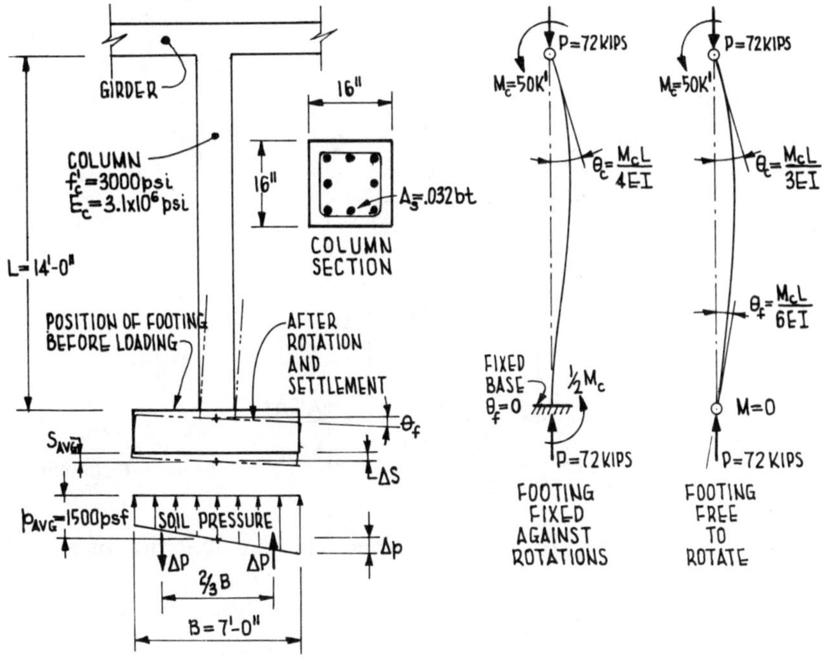

Figure 6-3 Typical footing rotations.

The large moment at the top of the column in Fig. 6-3 can be attributed to wind or earthquake loads; vertical loads do not usually produce such large column moments. The rotation at the top of the column is readily computed, using the approximation $I = bh^3/12 = 5461$ in. and $E_c = 3{,}100{,}000$ psi:

$$\theta_c = \frac{M_c L}{4E_c I} = \frac{50{,}000 \times 12 \times 14 \times 12}{4 \times 3{,}100{,}000 \times 5{,}461} = 0.0015 \text{ rad}$$

For the sake of reference, the stress f_c in the concrete at the top of the column was found in an outside calculation using conventional methods [25],

$$f_c = 827 \text{ psi}$$

This stress is in a comfortable range insofar as column stress is concerned.

To see the effect if the footing rotation is so small it is negligible, it is assumed that the base of the column is fixed to the footing and that the footing is highly resistant to rotations (but it does rotate slightly). The maximum value of moment that could occur at the base of the column is then

$$M_f = \frac{M_c}{2} = 25 \text{ kip-ft}$$

For this moment, the resultant trapezoidal soil pressure diagram is shown in Fig. 6-4, where the couple formed by the force ΔP must oppose the applied moment M_f.

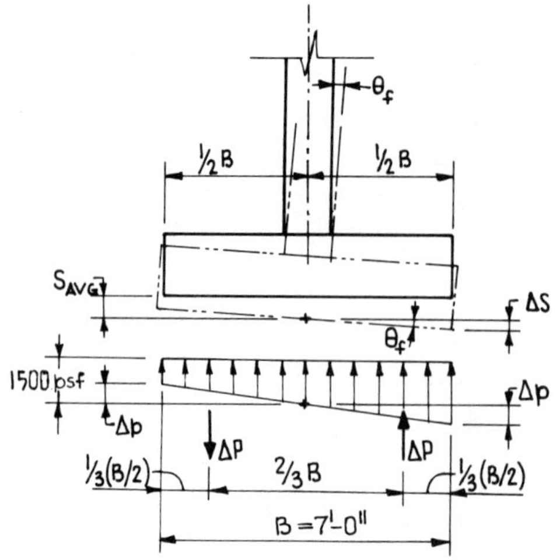

Figure 6-4 Footing rotations, first case.

The solution for ΔP is obtained from the definition of a couple,

$$M_f = \Delta P\left(\frac{2B}{3}\right) \qquad 25{,}000 \times 12 = \Delta P(7 \times 12 \times \tfrac{2}{3})$$

Hence

$$\Delta P = 5360 \text{ lb}$$

The solution for the increase in pressure Δp is obtained by statics,

$$\Delta P = (\tfrac{1}{2}\,\Delta p)(\tfrac{1}{2}B^2) \qquad 5360 = (\tfrac{1}{2}\,\Delta p)\tfrac{1}{2}(7 \times 7)$$

Hence

$$\Delta p = 438 \text{ psf}$$

The change in settlement ΔS at the edge corresponding to this change in pressure is found by ratios, where the indicated pressure of 1500 psf is assumed to produce a settlement in the range of 1 in.:

$$\frac{1}{1500} = \frac{\Delta S}{438}$$

Hence

$$\Delta S = 0.3 \text{ in.}$$

The corresponding footing rotation is then

$$\theta_f = \frac{\Delta S}{B/2} = \frac{0.3}{42} = 0.007 \text{ rad}$$

Rather than being nearly zero as assumed, this rotation is seen to be some $4\frac{1}{2}$ times the amount of elastic rotation (0.0015 rad) already computed at the top of the column. This rotation would cause a stress of more than 3800 psi, exceeding the ultimate strength of the concrete. It would also produce impossible conditions for compatibility of deformations at the base of the column.

It is concluded, therefore, that the assumption that produced this result (that the footing and soil could provide a relatively rigid base) was grossly incorrect. That assumption will therefore be discarded and the other extreme will be examined. It will now be assumed that the footing still provides adequate support for its vertical load but that it undergoes small rotations relatively freely, rather like a raft floating in water.

For this second case, the base of the column is assumed to be fixed to the footing as before, but it is now assumed that the soil offers very little resistance to rotations (but it does offer some). The result is that the footing rotates through the angle θ_f as shown in Fig. 6-5, again producing a trapezoidal pressure–settlement diagram as shown. The angle θ_f is computed as before,

$$\theta_f = \frac{M_c L}{6EI} = \frac{50{,}000 \times 12 \times 14 \times 12}{6 \times 3{,}100{,}000 \times 5461} = 0.0010 \text{ rad}$$

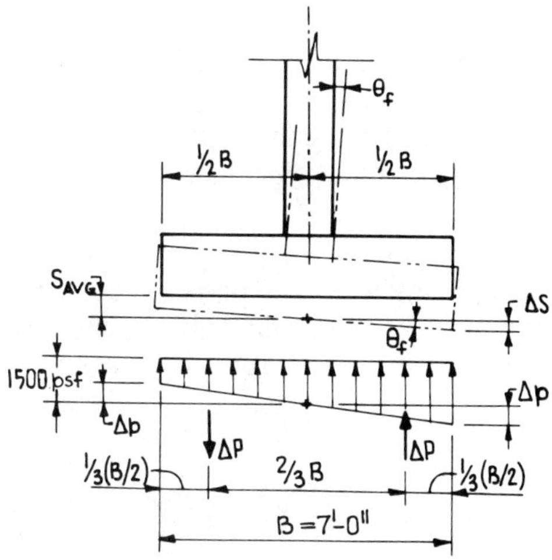

Figure 6-5 Footing rotations, second case.

The settlement ΔS at the edge of the footing is readily computed:

$$\Delta S = \left(\frac{B}{2}\right) \theta_f = 42 \text{ in.} \times 0.0010 = 0.042 \text{ in.}$$

The increase in pressure at the edge is found by ratios, as before,

$$\frac{1}{1500} = \frac{1 + \Delta S}{1500 + \Delta p} = \frac{1.042}{1500 + \Delta p}$$

Hence

$$\Delta p = 63 \text{ psf}$$

The increase in pressure is thus found to be 4.2% above the average pressure of 1500 psf. Within the usual range of accuracy for soils, this is negligible. Consequently, the analysis of this second case is essentially correct as it stands. The first conclusion to be drawn from these calculations is that *this soil offers practically no resistance to rotations of a footing,* at least within the usual range of rotations that can be tolerated in the concrete column.

The small resisting moment incurred by the rotation (the small error in the foregoing calculations) can be found by statics; this moment is shown in Fig. 6-5 as the couple ΔP and its arm.

$$M_f = \Delta P\left(\frac{2B}{3}\right) = \frac{1}{2}(\Delta p)\frac{B}{2}B\frac{2B}{3} = \Delta p\frac{B^3}{6}$$

$$M_f = 63\left(\frac{7 \times 7 \times 7}{6}\right) = 3601 \text{ lb-ft}$$

This moment is seen to be less than 15% of the 25-kip-ft moment that would be incurred if the base were truly fixed. It is again a small amount within the usual range of accuracy for soils. The second conclusion to be drawn here is that *no significant fixity of the column base has been provided by the isolated footing.*

The foregoing discussions demonstrate that for an isolated footing founded on a soil having a low-to-medium strength:

1. The soil offers little resistance to rotations of the footing.
2. No significant fixity of a column base is provided.

The foregoing two conclusions are generalized in the next section and shown to be true for most of the smaller buildings and soil types encountered in routine practice. Since shallow foundations are generally suitable only for smaller buildings, the foregoing conclusions are thus true for most applications of shallow foundations.

GENERALIZATION OF EFFECTS OF ROTATIONS

The particular solution of the preceding section is repeated here for a general case. In developing this general case, it is assumed that the design conditions are at or near the most severe possible state. A sketch of the general column and footing is shown in Fig. 6-6. Formulas for rotations due to flexure are found in standard references [26].

The rotation at the base of the column in Fig. 6-6 is given by

$$\theta_f = \frac{M_c L}{6E_c I} \tag{6-1}$$

where M_c is the moment at the top of the column. It is assumed that the ratio M_c/I in Eq. (6-1) is at its highest, where by the flexure formula, $f_c = M_c c/I$ with $c = kd$,

$$\frac{M_c}{I} = \frac{f_c}{kd} \tag{6-2}$$

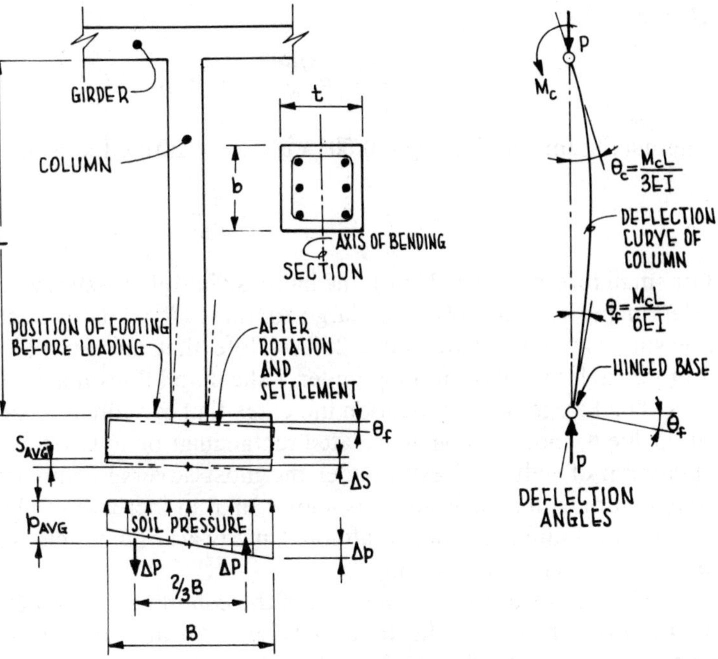

Figure 6-6 Typical column and footing.

For Eq. (6-2) it is assumed that f_c approaches $\frac{1}{2}f'_c$ and that the depth of compression kd approaches $t/3$ for a lightly loaded concrete column (see Fig. 8-6). To obtain the maximum value for the ratio, f'_c is assumed to be quite large, 5000 psi, and the size t to be quite small, 12 in. [The theory of bending in concrete is presented in detail in Chapter 8; at this point the reader is asked to accept the validity of Eq. (6-2).]

It is assumed that the maximum unsupported length of column, L, is 16 ft for first-floor columns; the modulus of elasticity of concrete, E_c, is taken at $57{,}000\sqrt{f'_c}$ (ACI 318-83). The foregoing values are substituted into Eq. (6-1). The solution for the rotation of the column section is then found to be

$$\theta_f = \left(\frac{M_c}{I}\right)\left(\frac{L}{6E_c}\right) = \left(\frac{f_c}{kd}\right)\left(\frac{L}{6E_c}\right) = \frac{\frac{1}{2} \times 5000 \times 16 \times 12}{\frac{1}{3} \times 12 \times 6 \times 57{,}000\sqrt{5000}} = 0.005 \text{ rad}$$

It is concluded that the highest reasonable value of rotation that can occur at the base of a concrete column is about 0.005 rad. (It should be recalled from Chapter 5 that the allowable "drift" angle for a shearwall and its footings is also 0.005 rad, which is consistent with the findings here.)

The soil pressure diagram of Fig. 6-6 reveals that for a general soil pressure p_{avg} at a settlement in the range of 1 in., the ratio of pressures is

$$\frac{p_{\text{avg}}}{1 \text{ in.}} = \frac{p_{\text{avg}} + \Delta p}{1 \text{ in.} + B\theta/2}$$

Hence

$$\frac{\Delta p}{p_{\text{avg}}} = \frac{B\theta}{2} \qquad B \text{ in inches} \tag{6-3}$$

The maximum rotation, $\theta_f = 0.005$ rad, is substituted into this equation, yielding

$$\frac{\Delta p}{p_{\text{avg}}} = \frac{B}{400} \tag{6-4}$$

For small footings ($B = 24$ in.), the increase in soil pressure given by the foregoing relationship is about 6%. For large footings ($B = 100$ in.), the increase in soil pressure is found to be about 25%. (Note that an extremely long rectangular footing such as a strip footing under a shearwall does not fit into this estimate.)

The foregoing investigation indicates that for concrete structures, soil pressures due to moments on an isolated rectangular or square footing may increase a maximum of only 25% even under the most adverse combinations of load. For steel columns, a similar analysis shows that an increase of about 35% above the average pressure could occur, although many arbitrary choices have to be made in arriving at such a conclusion.

The degree of fixity at the base of the column can now be found (for moment distribution or slope–deflection analyses). An approximate column moment is first found from the flexure formula, where the approximate section modulus is $S_c = bt^2/6$, with $f_c = \frac{1}{2}f'_c$ and $b = t$:

$$M_c = f_c S_c = \frac{\frac{1}{2}f'_c bt^2}{6} = \frac{f'_c t^3}{12} \tag{6-5}$$

The resisting moment of the footing is found by statics from the pressure diagram of Fig. 6-6:

$$M_r = \frac{1}{2}(\Delta p)\frac{1}{2}B^2\left(\frac{2B}{3}\right) = \Delta p\left(\frac{B^3}{6}\right) \tag{6-6}$$

The ratio M_r/M_c is then

$$\frac{M_r}{M_c} = 2\frac{\Delta p}{f'_c}\left(\frac{B}{t}\right)^3 \tag{6-7}$$

For Eq. (6-7), a nominal value of B/t is assumed to be 2.5, the highest soil pressure to be 5000 psf, and the lowest value for f'_c to be 3000 psi. Hence

$$\frac{M_r}{M_c} = 2\left(\frac{0.25 \times 5000/144}{3000}\right)(2.5)^3 = 0.09$$

The actual restraint of the footing is therefore less than 10% of the moment capacity of the column. (For those familiar with moment distribution, the maximum fixity that could exist is thus seen to be less than 20% of the base moment and could be considerably less.)

In summary, it is concluded that when a concrete column is rigidly attached to its spread footing, the footing can never rotate far enough to produce major changes in soil pressure. At most, the magnitude of the soil pressure at the edge of the footing cannot reasonably be more than 25% above the average pressure. As a consequence, the footing produces no reliable level of resistance to the elastic rotations of the column; thus there is no dependable degree of fixity produced at the base of the column. The foregoing conclusions form the basis of the design of all shallow footings herein.

ATTACHMENT OF COLUMNS TO FOOTINGS

In view of the foregoing conclusions, it is apparent that there is rarely any benefit to be obtained by fixing a column (or bearing wall) to its spread footing. Superficially, it would seem that the best solution is simply to provide a hinged base for the column, thereby eliminating any potential moment on the footing. There would then be no possibility that the potential 25% peak could ever occur.

A more detailed investigation follows, indicating that for steel columns, a hinged base for the column is in fact the simplest and cheapest solution in most circumstances. For concrete columns, it may be the simplest solution, but it is not usually the cheapest one.

Typical details for providing hinged bases and fixed bases are shown in Fig. 6-7. A fixed base for a steel column is seen to be considerably more elaborate than a hinged base and as a consequence is more costly. Except in unusual circumstances, there is therefore little justification to fix a steel column to an isolated footing; a hinged base is much preferred.

For concrete, the opposite is true. The rather elaborate hinged base shown in Fig. 6-7 is considerably more difficult and costly to build than the simple dowels of a fixed base. As a consequence, it is common practice to dowel the concrete column (or bearing wall) to its footing and to suffer the consequences of the more complicated design calculations.

It should be recognized that the complications in the design calculations arise only when the strength of the soil controls the allowable pressure. When settlement controls, there is no concern about the potential 25% peak pressure at the edge of the footing; the increase need not even be calculated. Only when strength controls is the increase in soil pressure a problem.

It should also be noted that the complications arise only for isolated spread footings or strip footings. Grade beams inherently provide a moment-resistant foundation along their length. Where loads are light enough to permit the use of a grade beam for the foundation, fixing the column bases to the grade beam does not introduce additional problems.

An alternative to grade beams is the narrow moment-resistant "tie beam" shown in Fig. 6-8. The tie beam spans between the columns and may actually rest on the top of the footing. The tie beam relieves the footing of any column moments; it also doubles as the cutoff wall for the floor slab and provides support for the wall above.

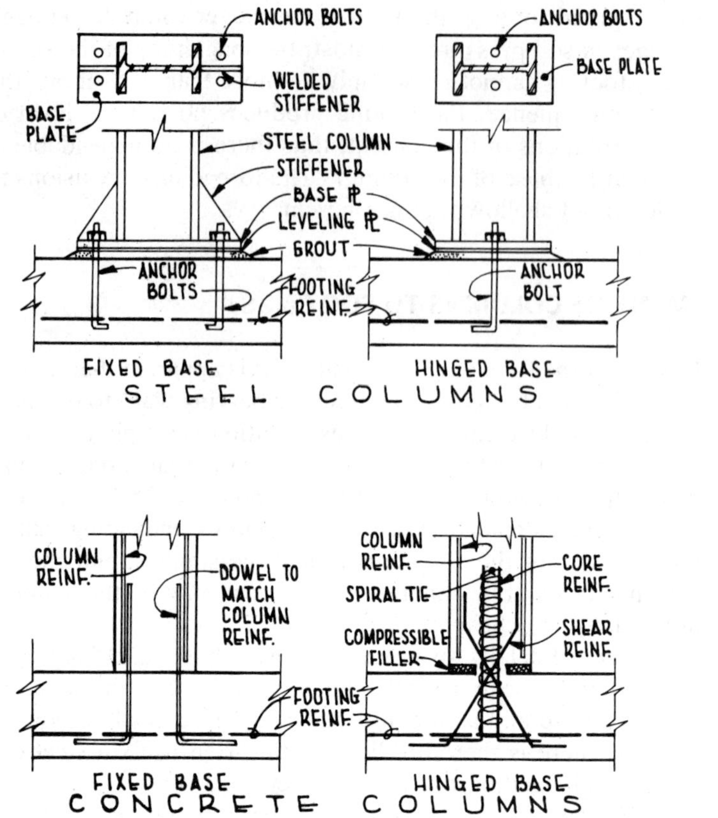

Figure 6-7 Typical column bases.

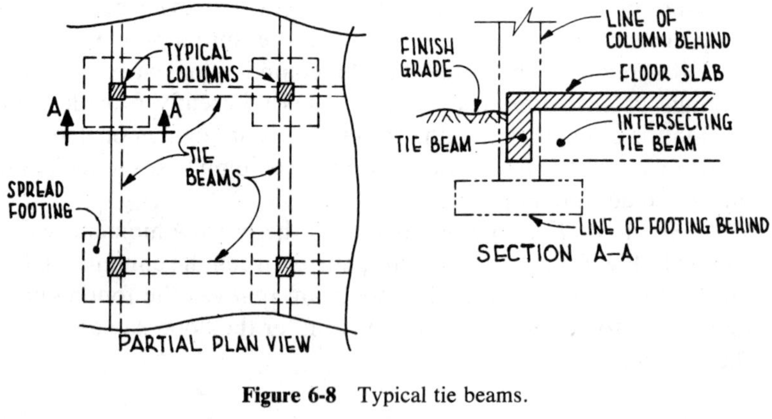

Figure 6-8 Typical tie beams.

The following items summarize the conclusions reached up to this point in the chapter:

1. When settlements control the allowable soil pressure, the strength of the soil may be assumed to be adequate to take any potential increases in pressure at the edges of the footing due to rotations. Spread footings may therefore be designed for the average soil pressure (vertical load ÷ contact area) regardless whether the columns (or bearing walls) are fixed or hinged to their footings.
2. When strength controls the allowable soil pressure and column bases are hinged, no peak pressure occurs in the soil; the soil pressure under the footing is uniformly distributed. The spread footings may therefore be designed for an average soil pressure (vertical load ÷ contact area) equal to this uniform pressure.
3. When strength controls the allowable soil pressure and column bases are fixed (very common in concrete buildings), a peak soil pressure will occur at the edge of the footing. The design should provide for this potential 25% increase above the average pressure.
4. When grade beams or tie beams are used at the foundation, the column moments are taken by these beams and the moments cannot cause any peaks in soil pressures. The grade beams or spread footings may therefore be designed for the average soil pressure (vertical load ÷ contact area).

The only exception in the cases above is seen to be in item 3, where strength controls the design and column bases are fixed to the footing. Except for this case, all footings could be designed using only the average soil pressure (vertical load ÷ contact area), ignoring moments and rotations of the footings. The incidence of item 3, however, is very common; it includes the footings that support shearwalls when the shearwall undergoes a drift of 0.005 rad.

Obviously, some simple method to transform item 3 is needed such that the same simple design procedure may be used in all cases. There are, of course, many such methods that could be developed. One such method is presented in the next section.

PEAK LOAD WHEN STRENGTH GOVERNS

It has been seen that fixing the base of a column to its spread footing can cause an increase in the soil pressure of uncertain magnitude at the edge of the footing. At its worst, however, such an increase will be less than about 25% above the average soil pressure for concrete columns, somewhat higher for steel columns. In view of this, it is desirable to develop a simplified method of design that uses only the average soil pressure; the uncertain peak pressure can then be accommodated without having to make a special analysis for it.

Such a method is readily possible using an established engineering approach. The applied vertical load can be artificially magnified by a suitable multiplier. The footing is then designed to limit the average soil pressure produced under this magnified load. Under the lesser actual load, the peak pressure will then always be less than the allowable pressure; its actual magnitude will no longer be of any concern. A comparison of the two cases is shown in Fig. 6-9.

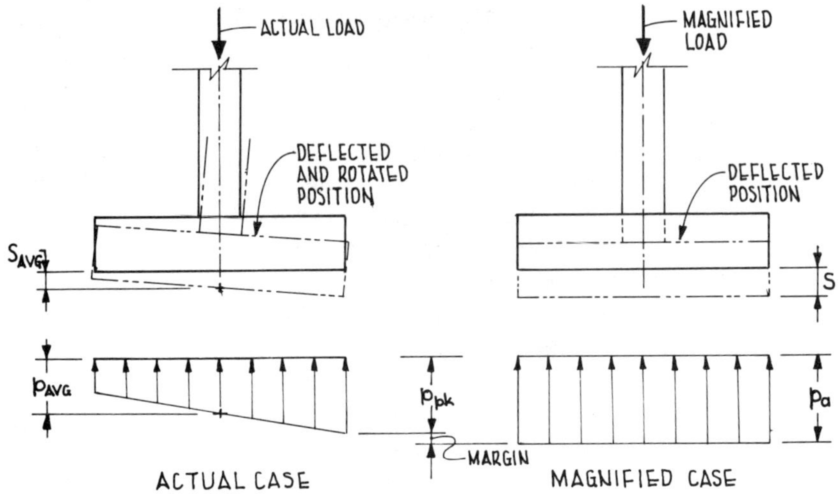

Figure 6-9 Actual load and magnified load.

The problem, of course, lies in selecting a suitable multiplier. It must produce an average pressure under the magnified load that is only slightly higher than the peak pressure under the actual load. Relative levels for these pressures are compared in the pressure diagram of Fig. 6-9.

A 25% increase would seem to produce the desired result except that it has been noted earlier that this value applies only to concrete buildings. For steel buildings, the peak pressure may be considerably higher, adding as much as another 10%. It is emphasized that both these values are based on the incidence of all the worst possible combinations occurring at the same time, a rather remote possibility.

In this book, an increase of 25% is adopted, to apply to all structures. Its justification comes only partly from the degree of conservatism that has been applied in determining the 25% increase in edge pressures. To this must be added the conservatism that is traditionally applied when selecting the allowable soil pressure. And to this must be added the conservatism that is applied in selecting the live loads that the structure itself must sustain. In the remote possibility that a steel building may actually experience all these maximum increases simultaneously, even then the resultant 10% overstress in the soil is still within the usual range of accuracy commonly accepted for soils.

The 125% magnifier is applied only where strength criteria controls the allowable soil pressure and where columns or bearing walls are rigidly fixed to their spread footings. The design procedure for that case then becomes quite simple:

1. The maximum vertical load that may occur at a footing is multiplied by a magnifier of 1.25,

$$P_{mag} = 1.25(\text{DL} + 100\%\text{LL})$$

2. The size of the contact area is computed for this magnified load, keeping the average soil pressure within the specified allowable soil pressure.

Note that lateral loads, wind, and earthquake do not enter into the design. Their major effect would be to produce column moments and moments do not affect the average pressure.

For the other case involving peak load, strength still controls the allowable soil pressure, but the columns are hinged to the footing. This case was presented at the beginning of the chapter and is shown in Fig. 6-1. No magnification of load is required where column bases are hinged:

1. The maximum vertical load is computed as

$$P_{max} = \text{DL} + 100\%\text{LL}$$

2. The size of the contact area is computed for this peak load, keeping the average soil pressure (a uniform pressure) within the specified allowable soil pressure.

The following examples will illustrate the procedure where the strength of the soil is limited.

EXAMPLE 6-1: Selection of Contact Area of a Spread Footing

Allowable soil pressure limited by strength.
Columns fixed to the spread footing.

Given:

Allowable $p_a = 4000$ psf
$P_{DL} = 130$ kips; $P_{LL} = 75$ kips
$M_{DL} = 350$ kip-ft, $M_{LL} = 500$ kip-ft
(moments occur at top of column)
$f'_c = 4000$ psi, $E_c = 3{,}600{,}000$ psi
$L = 13$ ft, $b = t = 28$ in.

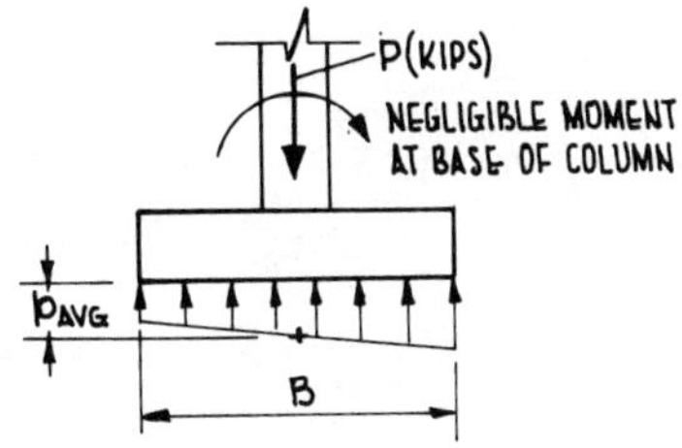

SOLUTION Compute magnified peak vertical load:

$$P_{mag} = (1.25)(P_{DL} + P_{LL}) = 1.25(130 + 75)$$
$$= 256 \text{ kips}$$

Compute the required contact area:

$$A = \frac{P_{mag}}{p_a} = \frac{256{,}000}{4000} = 64 \text{ ft}^2 \qquad \text{use 8 ft square}$$

Note that the rotations of the footing never entered the calculations.

Although not required, it is of interest to find the actual peak pressure in the foregoing example. Equation (6-1) is used to compute the rotation θ_f, where $I = bh^3/12$, and M_c is the moment at the top of the column,

$$\theta_f = \frac{M_c L}{6E_c I} = \frac{850{,}000 \times 12 \times 13 \times 12}{6 \times 3{,}600{,}000 \times 51{,}221} = 0.00144 \text{ rad}$$

This rotation is substituted into Eq. (6-4), with the result:

$$\frac{\Delta p}{p} = \frac{B\theta}{2} = \frac{96 \times 0.00144}{2} = 0.069$$

The peak pressure p_{pk} is thus 1.069 times the pressure at 1 in. settlement, taken here as p_{avg},

$$p_{pk} = \frac{P}{A}(1.069) = \frac{130 + 75}{8 \times 8}(1.069)$$
$$= 3424 \text{ psf} < 4000 \text{ psf allowed}$$

The peak pressure p_{pk} is seen to be 86% of the allowable value of 4000 psf, slightly conservative but acceptable.

EXAMPLE 6-2: Selection of Contact Area of a Strip Footing

Allowable soil pressure limited by strength.
Masonry bearing wall hinged to the footing.

Given:

Allowable $p_a = 2500$ psf
$P_{DL} = 6$ kips/ft; $P_{LL} = 2$ kips/ft
No computable moments

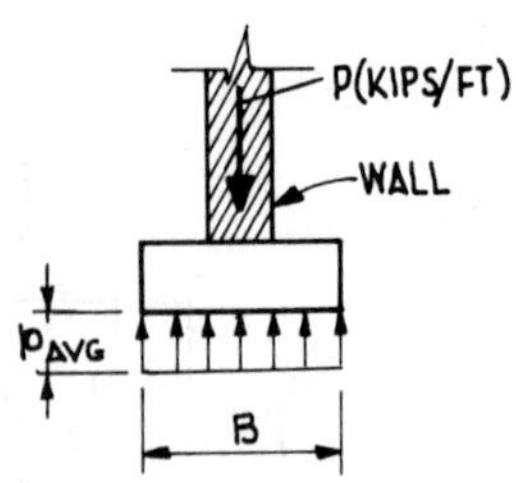

SOLUTION Compute the peak vertical load (no magnification):

$$P_{\max} = P_{\mathrm{DL}} + P_{\mathrm{LL}} = 6 + 2 = 8 \text{ kips/ft}$$

Compute the required contact area:

$$A = \frac{P_{\max}}{p_a} = \frac{8000}{2500} = 3.2 \text{ ft/ft}$$

Use strip footing 3 ft 6 in. wide.

It should be recognized that masonry walls would rarely have a moment at the base. Normally, the wall would be attached to a lateral support of some type at the top and to its strip foundation at the bottom. Lateral loads on the structure would then be taken by other shearwalls at right angles to this wall.

SUSTAINED LOAD WHEN SETTLEMENTS GOVERN

The simplified approach developed in the preceding section can also be adapted to the case where settlements must be limited. The sustained load that produces settlements in soils can be computed relatively easily. It is then used with the average soil pressure to compute the required contact area.

It was noted in earlier chapters that settlements in soil are time dependent. Settlements in clays are measured in terms of months or years. Settlements in sands occur faster, but they are still measured in terms of weeks or months.

Not all live loads are in place long enough to affect settlements. Certainly, loads such as furnishings, bookcases, carpets, cabinets, files, and so on, are in place long enough to produce settlements. Transient loads, including wind and earthquake, come and go with no effect on settlements; only the elastic deformations will occur under these short-term transient loads.

It is the usual practice to assume that the dead load plus some portion of the live load are in place long enough to produce settlements. For buildings serving routine architectural functions, the load causing settlements is commonly assumed to be the dead load plus about 50% of the live load. For buildings such as libraries, the live load could be as much as 75% and for auditoriums as little as 25%. The actual percentage of live load to be used is thus a matter of judgment. The usual limits are somewhere between 20 and 80%. For the sake of simplicity, all subsequent calculations will use dead load plus 50% live load.

The term "sustained load" is used herein to designate this long-term load. Other references sometimes call it the "reduced" load [4]. An example will illustrate the use of the sustained load in selecting the required size of the contact area.

EXAMPLE 6-3: Selection of Contact Area of a Spread Footing

Allowable soil pressure limited by settlement.
Column fixed to the spread footing.

Given:

Allowable $p_a = 3000$ psf
$P_{DL} = 65$ kips; $P_{LL} = 50$ kips
At top of column,
$M_{DL} = 50$ kip-ft/ft; $M_{LL} = 45$ kip-ft/ft

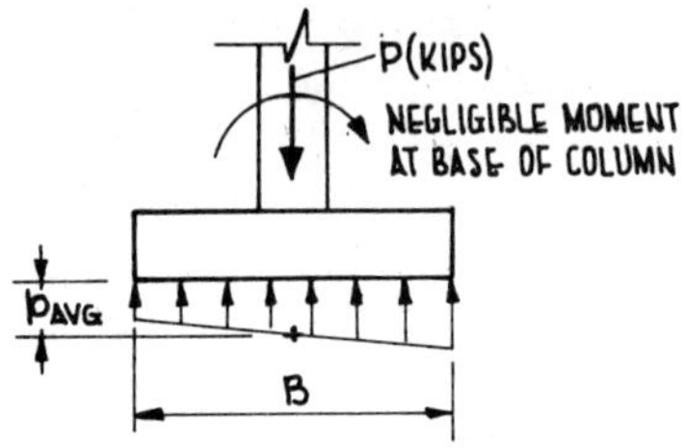

SOLUTION Compute sustained vertical load:

$$P_{SUS} = P_{DL} + 0.50P_{LL} = 65 + 25 = 90 \text{ kips}$$

Compute the required contact area:

$$A = \frac{P_{SUS}}{p_a} = \frac{90,000}{3000} = 30 \text{ ft}^2$$

Use footing 5 ft 6 in. square.

Again, it is pointed out that moments have no effect on the design. Also, the fact that the column is fixed to the footing has no effect where settlements control the allowable soil pressure.

MARGIN OF STRENGTH WHEN SETTLEMENTS GOVERN

The margin of strength to bearing failure has not yet been established for the case when the foundation is designed to limit settlements. Of particular concern is the margin of strength under those footings supporting shear panels. The soil pressure under such footings could be as much as doubled whenever lateral loads exist; the margin of strength must be verified under this potential overload.

In addition to the doubling of the load on these footings from above, the soil under these footings loses strength due to the added shear. This loss in strength is developed in detail in Chapter 3; the percent of loss is given in Table 3-2. The loss in soil strength, when coupled with a doubling of the pressure, produces such a sharp reduction in the margin of strength that a detailed examination is warranted.

For clay soils, the factor of safety to bearing failure is fixed at 3 when settlements govern the design; the margin of strength thus provided must accommodate any increases above the design load, DL + sustained LL. One such increase can happen when the full live load occurs, typically producing an increase of about 20%. In addition, if the footing were to be attached rigidly to the structure above, a further increase of up to 25% of the average pressure could occur. Further, for shear panels, the entire load could be doubled when the panel undergoes maximum rotation. The total pressure (with an initial factor of safety of 3 to bearing failure) would then be, for clays,

$$p_{max} = \frac{1}{3} q_0 \times \frac{6}{5} \times \frac{5}{4} \times 2 = 1.0 q_0 \tag{6-8}$$

where q_0 is the ultimate bearing strength. The value of q_0 must also be reduced, however, by the percent loss in strength shown in Table 3-2 to account for the added shear stress in the soil.

From Eq. (6-8) it is apparent that there would be no margin of strength left in a clay soil supporting a shear panel if the foundation were to be designed to meet settlement criteria. Such a load case cannot be permitted, of course. It can only be concluded that a foundation supporting a shear panel on a clay soil must always be designed for strength criteria rather than settlement criteria. The design load on each footing would then be the total load on the entire panel, 0.75(DL + LL) rather than the settlement load at one footing only, DL + sustained LL. The allowable bearing pressure would be the remaining strength after the losses of Table 3-2 have been deducted.

For sands, such a clean distinction cannot be drawn. As indicated in Fig. 3-16, the margin between the allowable capacity for settlements (the horizontal lines) and the allowable capacity for strength (the sloping lines) can vary considerably, being much higher for large values of B than for low values of B. For sands, there is thus a distinct gain in allowable soil strength as the footing size increases.

As a practical matter, the depth of burial of a small footing in a sandy soil will always be about 2 to 3 ft minimum. For a footing width of about 6 ft, the usual ratio of D_f/B in Fig. 3-16 is therefore about 0.5. From Fig. 3-16 it is seen that the margin of strength for a footing size of about 6 ft is quite comfortable; for low values of ϕ, the bearing capacity is roughly 2 to $2\frac{1}{2}$ times the settlement capacity.

When this nominal value of 0.5 for D_f/B is taken into Table 3-2, it is seen that for low values of ϕ, the maximum loss in strength in the soil due to the added shear stress is about 60%. The soil is therefore losing strength due to the added shear while it is gaining in strength due to footing size. The dividing line is estimated to be at about 6 ft; for a footing width of about 6 ft, the soil will gain about as much strength due to the larger footing size as it will lose due to the added shear stress. For footing sizes smaller than about 6 ft in width, there is a net reduction in strength; such footings should be designed for strength criteria. For footings larger than about 6 ft, there is a net increase in strength; these footings may be designed for settlement criteria. If there is any question about which should govern, the design should be made both ways and the most conservative result used as the final design.

The foregoing conclusions can easily be verified for a footing size of about 6 ft. The total factor of safety to bearing capacity failure for a footing this size is found from Fig. 3-16 for $D_f/B = 0.5$; an average factor of safety based on settlement capacity is seen to be around 5. With this factor of safety, a derivation similar to that of Eq. (6-8) reveals that, for sands,

$$p_{\max} = \frac{1}{5} q_0 \times \frac{6}{5} \times \frac{5}{4} \times 2 = 0.6 q_0 \tag{6-9}$$

where q_0 is again the ultimate bearing strength. It is recalled that an allowable overstress of 33% is permitted when lateral loads exist. When this overstress is considered, the maximum pressure under a footing 6 ft wide supporting a shear panel is seen to be pushing maximum but still acceptable; the footing might therefore be designed either for settlement criteria or for strength criteria.

STRIP FOOTINGS SUPPORTING SHEAR PANELS

It was observed earlier that long strip footings are not included in the foregoing approach to the design of spread footings. Such long strip footings are commonly used to support heavily loaded shear walls as indicated in Fig. 6-10. Although their design is not at all complex or difficult, it does constitute a special case insofar as this approach to footing design is concerned.

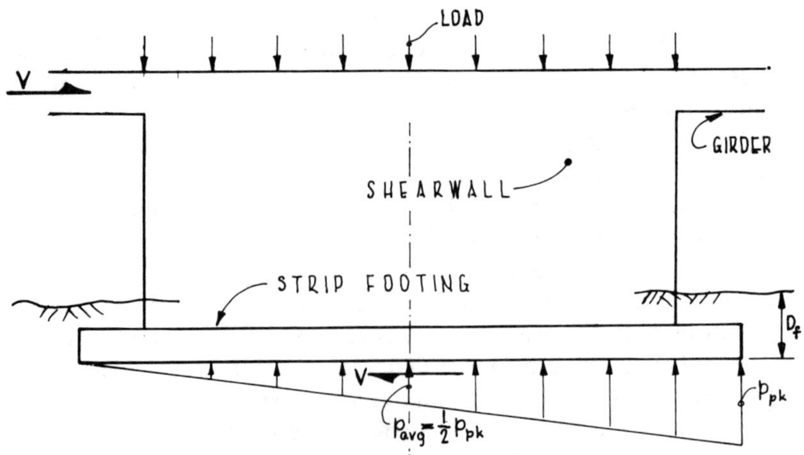

Figure 6-10 Typical strip footing at a shearwall.

Since a shearwall is always subject to shear, the design load will always be

$$P_W = 0.75(P_{DL} + P_{LL}) \tag{6-10}$$

At this design load, the wall undergoes a rotation, creating the triangular pressure diagram as shown. The peak pressure p_{pk} can occur at either end, its magnitude being twice the average pressure occurring at the center of the shearwall; its peak magnitude is of course limited to the maximum allowable bearing pressure for the soil. The required area of the strip footing is then very simply calculated as the load P_W as given by Eq. (6-10), divided by the average pressure p_{avg}, where p_{avg} is half the allowable bearing pressure.

The allowable bearing pressure must include the appropriate reductions due to the existence of the lateral loads as well as other routine corrections for soil weight, submergence, or footing shape. An example of this special case is included in the comprehensive examples in a subsequent section.

This response to load by a shearwall founded on a strip footing is far more complex than the simplistic approach presented here. The short-term elastic modulus, for example, is a very vague point in soil mechanics, hence the assumption of a triangular pressure diagram under the lateral loads may not be very accurate. For larger buildings, a pile foundation for such shearwalls may be well worth considering. For smaller buildings, the approach used here produces generally acceptable though conservative results.

SUMMARY OF DESIGN LOADS AND PRESSURES

A series of load combinations for footing design were introduced in Chapter 5 and a series of structural configurations have been introduced in this chapter. Considered together, these loads and configurations are used in making the final calculations of the contact area required to support a particular column or bearing wall. The primary considerations have been found to be:

1. Whether strength or settlements govern the magnitude of the allowable soil pressure
2. Whether lateral loads occur on the footing
3. Whether the footing supports a shear panel
4. Whether the footing is rigidly attached to the structure above

It is recognized that a profusion of load cases, soil pressures, and strength losses can thus occur within a single foundation system, even for a small building. There is no "blanket" rule; the foundation designer must recognize which loads and which pressures are to be used at a particular footing. For ease of comparison, however, the loads and pressures of the earlier discussions have been collected and summarized in Table 6-1.

In using Table 6-1, it must be remembered that in practice, whenever settlements govern the design, it often happens that the actual failure strength of the soil is never known. The entire design is simply based on some low value of bearing pressure that will limit settlements; the use of presumptive bearing pressures taken from the building code is a typical example of such circumstances. In such designs, the strength of the soil may have to be estimated if shear panels are to be used.

TABLE 6-1 SUMMARY OF DESIGN LOADS AND PRESSURES

When strength criteria limit the allowable bearing pressure:	When settlement criteria limit the allowable bearing pressure:
1. Wind and earthquake must be included in the design loading of all footings. The design load cases become DL + LL 0.75(DL + LL + WIND) 0.75(DL + LL + EQ)	1. Wind and earthquake loads do not affect settlements. The design load case is DL + sustained LL
2. Footing loads are to be multiplied by 1.25 if and only if the footing is rigidly attached to the structure above (applies also to footings supporting shear panels).	2. Fixing a footing rigidly to the structure does not affect settlements; no corrective measures are necessary for different types of attachment.
3. Frictional resistance to wind and earthquake must be verified. (Base shear must be computed, distributed, and checked to see that frictional resistance is adequate at all footings.)	3. Designing for frictional resistance is the same as in strength criteria.
4. Soil strength must be reduced by appropriate factors to account for footing shape, lateral loads, flooding, or other influences as appropriate. The "most probable" value of Q for use in Table 3-2 may be estimated as DL + 0.5LL.	4. Loss in soil rigidity is proportional to loss in strength. With the exception of corrections for transient lateral loads, the correction factors of the strength criteria apply as well to settlement criteria.
5. For footings supporting columns in a rigid frame, the design load is the vertical load only; the lateral load is used only in items 3 and 4 and all moments are accounted for in item 2.	5. For footings supporting columns in a rigid frame, the design is performed both for strength criteria and for settlement criteria; the more conservative of the two designs is used.
6. For foundations supporting shear panels[a] under lateral loading, the shear panels are assumed to rotate until soil pressure at one end is zero. The panel load then becomes $0.75P_W$, where P_W is the sum of all tributary dead and live loads at both ends of the panel and throughout its length. If the foundation is a strip footing, the soil pressure diagram becomes triangular, effectively doubling the average pressure due to gravity loads only. If the foundation is two spread footings, either of the two footings could be loaded by the entire panel load, effectively doubling the soil pressure under the footing. (The allowable 33% increase in soil pressure is already included in the $0.75P_W$.)	6. In clays, the design of footings supporting shear panels[a] is always governed by strength criteria, never by settlement criteria. In sands, the design of footings supporting shear panels is governed by strength criteria for footing widths B up to about 6 ft and by settlement criteria for widths B larger than about 6 ft; for footing widths near 6 ft, both sets of criteria should be checked and the more conservative results used.

[a] Shear panels may be either shearwalls or braced panels.

COMPREHENSIVE EXAMPLES

The use of the general criteria listed in Table 6-1 is illustrated in the following examples. For the sake of simplicity, the structural analysis for lateral loads is not included. Explanatory notes and comments are included within the examples.

EXAMPLE 6-4

A column in a concrete braced frame is subject to a dead load of 90 kips and a live load of 40 kips. The allowable soil pressure is limited by strength criteria to 3400 psf for strip footings. The underlying soil is a loose sand, SC, unit weight 116 pcf, subject to annual flooding. Depth of burial is set at about 2 ft, footings are to be square spread footings, and columns are to be fixed to the footings. Select the required contact area for the footing.

SOLUTION Since the design is governed by strength criteria, use the total load (no lateral loads occur on the columns in a braced frame except those columns supporting the braced panels):

$$\begin{aligned} P_W &= P_{DL} + P_{LL} = 90 + 40 \\ &= 130 \text{ kips} \end{aligned}$$

Magnify load to account for fixing the base of the column to the footing:

$$\begin{aligned} P_{mag} &= 1.25 P_W = 130 \times 1.25 \\ &= 163 \text{ kips} \end{aligned}$$

Correct the soil pressure to account for flooding:

$$\begin{aligned} P_a &= 3400 \times \tfrac{1}{2} \\ &= 1700 \text{ psf} \end{aligned}$$

Correct the soil pressure to account for square footing shape. From Table 3-1, the shape factor is 0.85:

$$\begin{aligned} P_a &= 1700 \times 0.85 \\ &= 1450 \text{ psf} \end{aligned}$$

Note that any further corrections to the allowable bearing pressure due to D_f/B or to a different soil weight would be applicable only if the allowable pressure had been taken from the graphs of Fig. 3-16. Since the allowable pressure in this example is taken from other sources (presumably the soils report), the corrections that would apply to Fig. 3-16 would not be appropriate here.

Compute the required area:

$$\text{area} = \frac{P_{mag}}{p_a} = \frac{163{,}000}{1450} = 112 \text{ ft}^2$$

Use a footing 10 ft 9 in. square.

EXAMPLE 6-5

A column in a concrete rigid frame carries a dead load of 60 kips, a live load of 46 kips, and a lateral wind load of 7 kips. The soil is classified as a loose sand, SM, unit weight 112 pcf. The allowable bearing pressure is limited by settlement criteria to 1900 psf for strip footings; ultimate strength is approximately 6600 psf. The locality is not subject to flooding. The column is to be doweled to its footing, producing a rigid connection. There is a moment at the top of the column of 98 kip-ft. The depth of burial of the footing is 2 ft; D_f/B is estimated at 0.25. Select a square spread footing for this column.

SOLUTION Since the column is subject to lateral load, the design should be done for settlement criteria, then checked for strength criteria (see Table 6-1).

Perform the design for settlement criteria. Compute sustained load.

$$P_{\text{SUS}} = \text{DL} + 0.50\text{LL} = 60 + 23$$
$$= 83 \text{ kips}$$

Check for friction at minimum vertical load:

$$\text{allowable } V = 0.3 \times P_{\text{DL}} = 0.3 \times 60 = 18$$
$$= 18 \text{ kips} > 7 \text{ kips (OK)}$$

Allowable bearing pressure is 1900 psf but must be corrected for footing shape; from Table 3-1 the shape correction factor is seen to be 0.85.

$$p_a = 1900 \times 0.85 = 1620 \text{ psf}$$

Compute the required area:

$$\text{area} = \frac{P_{\text{SUS}}}{p_a} = \frac{83{,}000}{1620} = 51.2 \text{ ft}^2 \qquad \text{(Case I)}$$

Perform the design for strength criteria, both when subject to a wind load and when no wind occurs. Without wind, the total load is

$$P_W = P_{\text{DL}} + P_{\text{LL}} = 60 + 46 = 106 \text{ kips}$$

This load is magnified by 1.25 for a fixed column base,

$$P_{\text{mag}} = 106 \times 1.25 = 133 \text{ kips}$$

Allowable pressure is 3300 psf with a factor of safety of 2, but must be corrected for rectangular shapes,

$$p_a = 3300 \times 0.85 = 2800 \text{ psf}$$

The required area is, with no wind load,

$$\text{area} = \frac{P_{\text{mag}}}{p_a} = \frac{133{,}000}{2800} = 47.5 \text{ ft}^2 \qquad \text{(Case II)}$$

With wind, the design load is

$$P_W = 0.75(P_{\text{PL}} + P_{\text{LL}} + P_{\text{WIND}}) = 0.75(60 + 46 + 0) = 80 \text{ kips}$$

For a fixed column base, multiply by 1.25,

$$P_{\text{mag}} = P_W \times 1.25 = 80 \times 1.25 = 100 \text{ kips}$$

The allowable bearing pressure must be corrected for footing shape and for the loss in strength due to lateral wind loads. The "most probable" value of Q for use in Table 3-2 is taken as DL + 0.5LL when maximum wind load occurs, hence $V/Q = 7/(60 + 23) = 0.1$. From Fig. 2-23, ϕ is found to be 30°. From Table 3-2, with D_f/B estimated at 0.25, the loss in strength is found to be 25%; hence

$$p_a = 3300 \times 0.85 \times (1 - 0.25) = 2100 \text{ psf}$$

The required area is, then,

$$\text{area} = \frac{p_W}{p_a} = \frac{100{,}000}{2100} = 47.5 \text{ ft}^2 \qquad \text{(Case III)}$$

The three results are compared, indicating that the load case for settlement criteria is more severe than that for strength criteria; the required area is 51.2 ft². Use a footing 7 ft 3 in. square, $A = 52.6$ ft², $D_f/B = 0.27$.

The next problem includes lateral loads due to earthquake. It should be noted that for the earthquake analyses of Chapter 5 in which the lateral load V was computed, only the dead load was used to calculate base shear. Even though live load would, in fact, exist, the live load would be movable and would not undergo the absolute acceleration of the building; with no acceleration, it would not contribute any force to the base shear. The live load still exists as a vertical load, however, and appears in the load Q when computing the ratio V/Q. The "most probable" value of Q that would occur at the time of maximum earthquake is therefore the same as it would be for maximum wind, that is, DL + 0.5LL.

EXAMPLE 6-6

A shear panel in a braced frame structure has columns at each end as shown. At the left column, $P_{DL} = 87$ kips, $P_{LL} = 62$ kips, $P_{EQ} = 32$ kips. At the right column, $P_{DL} = 106$ kips, $P_{LL} = 71$ kips, $P_{EQ} = 32$ kips.

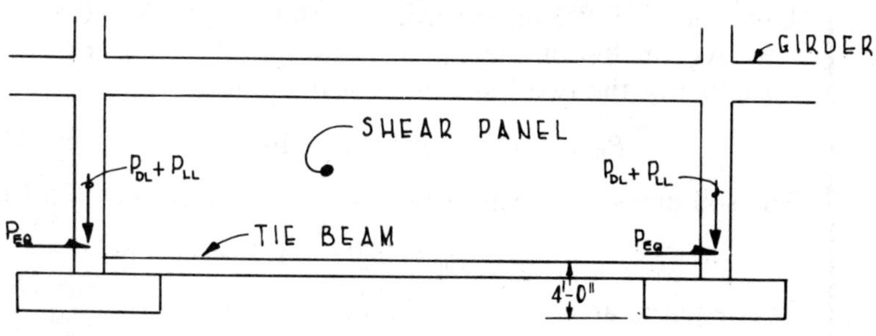

The underlying soil is a low-plasticity clay, classified as CL. Design is governed by settlements; allowable bearing pressure for settlements is given by the soils report as 2800 psf for strip footings (with a factor of safety of 3 to ultimate strength). Select the required contact areas for the footings.

SOLUTION From Table 6-1, it is seen that all footings carrying shearwalls must be designed for strength criteria; maximum loading must therefore be used. For earthquake loads, the total loads are

$$P_{LT} = P_{DL} + P_{LL} = 87 + 62 = 149 \text{ kips}$$

$$P_{RT} = P_{DL} + P_{LL} = 106 + 71 = 177 \text{ kips}$$

The wall may rotate until all loads are carried by one footing. For either of the two footings, therefore, including the 33% allowable increase in soil pressure,

$$P_W = 0.75(P_{DL} + P_{LL}) = 0.75(149 + 177)$$
$$= 245 \text{ kips}$$

Magnify P_W to account for fixity of the column base to the footings for strength criteria,

$$P_{mag} = 245 \times 1.25 = 306 \text{ kips}$$

Allowable soil pressure for settlements is 2800 psf, with a factor of safety of 3; the soil pressure must be corrected to reduce the factor of safety to 2 for strength criteria. It must also be corrected for footing shape and for loss in shear strength due to lateral loads. At DL + 0.5LL for the "most probable" value of Q in Table 3-2, it is found that

$$\frac{V}{Q} = \frac{32 + 32}{87 + 106 + 0.5(62 + 71)} = 0.25$$

From Table 3-2, percent loss = 29%.

$$p_a = 2800 \times \frac{3}{2} \times 1.2 \times (1 - 0.29) = 3580 \text{ psf}$$

With the entire load on either footing,

$$\text{area} = \frac{P_{mag}}{p_a} = \frac{306{,}000}{3580} = 85 \text{ ft}^2$$

Use footings 9 ft 3 in. square, $A = 86.0$ ft^2., $D_f/B = 0.5 \pm$ (O.K.)

As a matter of interest, a review for settlements can be readily performed for either of the two footings. For the right side,

$$P_{SUS} = P_{DL} + 0.5P_{LL} = 106 + 0.5 \times 71 = 142 \text{ kips}$$

The soil pressure at 1 in. settlement must be corrected for footing shape,

$$p_a = 2800 \times 1.2 = 3360 \text{ psf}$$

The required area at 1 in. of settlement is then

$$\text{area} = \frac{P_{SUS}}{p_a} = \frac{142{,}000}{3360} = 42.3 \text{ ft}^2$$

This area is only about half that required when earthquake loads occur, indicating that settlements under this rather large footing will be small compared to adjacent footings.

EXAMPLE 6-7

A shearwall 24 ft long in a diaphragm–shearwall structure system is subjected to a dead load of 8.4 kips per foot, a live load of 6.8 kips per foot, and a lateral wind load of 51 kips. It is to be founded on a strip footing 30 ft long as shown. The underlying soil is a medium sand, SM, having an SPT blow count of 16; unit weight is 119 pcf, and the locality is not subject to flooding. Strength of the soil governs the design. Select the proper width for the strip footing.

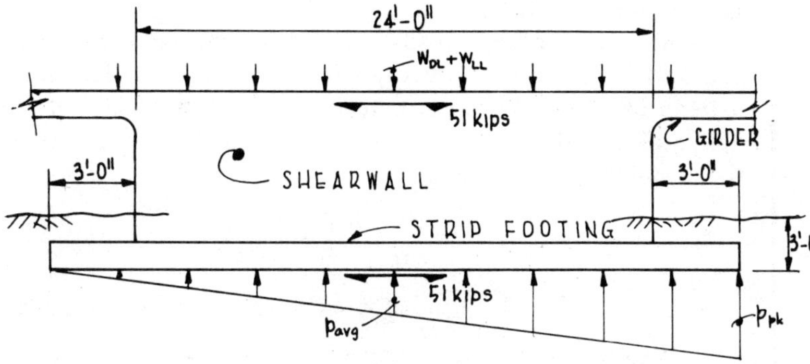

SOLUTION Design must conform to strength criteria as described in Table 6-1.

$$\text{Total dead load} = 24 \times 8.4 = 202 \text{ kips}$$

$$\text{Total live load} = 24 \times 6.8 = 163 \text{ kips}$$

$$P_W = 0.75(P_{\text{DL}} + P_{\text{LL}}) = 0.75(202 + 163) = 274 \text{ kips}$$

The allowable bearing pressure is selected from Fig. 3-16, using a value of ϕ of 32°, obtained from Fig. 3-17. B is estimated to be about 4 feet, hence $D_f/B = 0.75$. For these values, the bearing strength of the soil is projected to be 5500 psf. This value must be corrected, for an actual weight of 119 pcf,

$$p_a = 5500 \times \frac{119}{100} = 6550 \text{ psf}$$

The allowable pressure must also be corrected for loss of strength due to lateral loads. For use in Table 3-2, the "most probable" value of Q is found to be DL + $\frac{1}{2}$LL = 202 + 81 = 283 kips; hence $V/Q = 51/283 = 0.20$. With ϕ estimated from Fig. 2-23 to be 32° and $D_f/B = 0.75$, the average percent loss in p_a is found from Table 3-2 to be 37%; hence

$$p_a = (1 - 0.37) \times 6550 = 4130 \text{ psf}$$

The required width of the strip footing can now be calculated. It must be remembered, however, that the pressure diagram is triangular and that the pressure of 4130 psf just computed is the maximum pressure that may occur at either end. The average pressure at the center of the footing is $\frac{1}{2} \times 4130 = 2070$ psf. The area is then

$$\text{area} = \frac{P_W}{p_a} = \frac{274{,}000}{2070} = 132 \text{ ft}^2$$

For the 30-ft length, the width is

$$B = \frac{132}{30} = 4.4 \text{ ft}$$

Use a width of 4 ft 6 in.

As a matter of curiosity, the settlement under this footing can now be checked. The pressure corresponding to 1 in. of settlement is read from Fig. 3-16 as 3800 psf. When corrected for actual unit weight of 119 pcf, the allowable pressure at 1 in. of settlement is 4520 psf. At normal service conditions, the sustained pressure is

$$p_{\text{SUS}} = \frac{p_{\text{DL}} + 0.5p_{\text{LL}}}{B} = \frac{8.4 + 3.4}{4.5}$$

$$= 2620 \text{ psf} \ll 4520 \text{ psf at 1 in. of settlement}$$

As in Example 6-6, it is seen that settlement under this large footing will be much less than the allowable 1 in. of settlement commonly expected under such large footings.

SUMMARY

The following points summarize the findings and conclusions of this chapter:

- The allowable bearing pressure is the permissible *increase* in the vertical soil pressure that may be imposed within a given stratum.

- When the allowable soil pressure is limited by settlement considerations rather than by strength considerations, the allowable bearing pressure is the pressure that will produce a settlement of about 1 in. in a relatively large footing, commonly taken to be 8 ft square.

- Both the average pressure and average settlement occur at the centroid of the contact area; they are directly proportional.

- Moments on a concentrically loaded footing affect neither the average pressure nor the average settlement.

• The contact area is selected using only the vertical design load; moments are not needed.

• The vertical design load is magnified or reduced to suit the structural configuration; the required contact area of a footing is then computed simply by dividing the vertical design load by the allowable soil pressure.

• The vertical design loads for various conditions are:

1. Where the allowable soil pressure is limited by the strength of the soil,
   a. At hinged column bases:

$$P_{max} = \text{DL} + 100\%\text{LL}$$

   b. At fixed column bases:

$$P_{mag} = 1.25(\text{DL} + 100\%\text{LL})$$

2. Where the allowable soil pressure is limited by susceptibility to settlements, for either fixed or hinged column bases,

$$P_{SUS} = \text{DL} + \text{partial LL}$$

   where the partial live load is that part that is sustained long enough to affect settlements, commonly about 50%.

• Where a footing is rigidly fixed to an elastic column, the footing rotations that can occur are inherently limited to such small values that at most only a 25% increase in the soil bearing pressure may occur at the edge of the footing. Consequently, the footing cannot provide any reliable degree of fixity for the column.

• For the structural analysis of the building, a hinge should be assumed to form at the point where the centerline of the column or bearing wall intersects the founding line. The overall column length of the ground floor columns should thus be extended to include the footing thickness.

• At spread footings, dowels used to fix a concrete column to the spread footing need not be designed for more than a nominal 10% of the maximum capacity of the column in bending; the footing cannot rotate far enough to develop any more resisting moment than this.

• When lateral loads occur on shear panel systems, the soil pressure under the footings may be as much as doubled, while at the same time the strength of the soil below the footing drops sharply. In combination, these factors require that special attention be paid to the design of footings supporting shear panels; such a compounding of adverse effects does not occur at other types of footings.

• In clay soils, the design of footings supporting shearwalls is always governed by strength criteria. For sands, the design of footings supporting shearwalls can be expected to be governed by strength criteria for footing widths B up to about 6 ft and by settlement criteria for footing widths larger than about 6 ft.

REVIEW QUESTIONS

6-1. To what level of accuracy can the mechanical properties of soil be determined? Of concrete?

6-2. Why is it that the theoretical computation of a footing settlement may be grossly in error but the same general theory will provide a reasonably accurate comparison between the settlement at this footing and the settlement at another footing?

6-3. What is the effect of the overall settlement of a building on the stresses in the building structure?

6-4. What is the difference between settlement and differential settlement?

6-5. Why is it usually impractical to attempt a settlement analysis on a routine building supported by a shallow foundation?

6-6. Define the allowable bearing pressure on soil.

6-7. About how far below ground surface is the allowable bearing pressure given by the soils report assumed to be valid?

6-8. What is the generally accepted value for allowable settlement of a footing? At this level of total settlement what will likely be the maximum differential settlement between adjacent footings?

6-9. How much settlement is generally assumed when the design is governed by settlement criteria? By strength criteria?

6-10. In a spread footing, where does the average settlement occur?

6-11. What is the effect of column moments on average settlement?

6-12. In a typical soil at shallow depths, how much resistance to rotation of a shallow footing can be expected?

6-13. At a typical spread footing supporting a column, how much fixity can be assumed at the footing for the sake of the structural analysis of the building above?

6-14. About what maximum increase in soil pressure can be expected due to column moments on a footing? Why the limitation?

6-15. Under what structural configuration between a column and its footing can rotations of the footing be developed?

6-16. Why is it not practical to hinge all columns to their footings?

6-17. What configuration can be used at the foundation when it becomes necessary to fix the base of a column?

6-18. At a typical spread footing supporting a concrete column in a rigid frame, where does the point of rotation (the hinge) at the lower end of the column actually occur?

6-19. Where a column is rigidly fixed to its spread footing and strength criteria in the soil limits the allowable soil pressure, what design procedure is used to account for the peak soil pressure?

6-20. Where settlement criteria in the soil limits the allowable soil pressure, what design procedure is used to account for loads that are too transient to produce settlement?

6-21. Why are footings that support shearwalls such a point of concern in foundation design?

6-22. How much increase in soil pressure (maximum) can occur under the foundation of a shear panel under full wind or earthquake load?

6-23. How much loss in strength of the soil (maximum) can occur under the foundation of a shear panel due to the high lateral shears?

6-24. Compare the restoring moment offered by a shear panel supported by spread footings at each end to its resisting moment when supported by a continuous strip footing.

PROBLEMS

6-1. A spread footing 8 ft square carries a load of 256 kips, exerting a bearing pressure of 4000 psf on the supporting soil. The footing is forced to undergo a rotation of 0.20°. Determine the changes in soil pressure that occur due to this rotation, assuming a soil pressure of 4000 psf corresponds to a settlement of about 1 in.

6-2. A spread footing 8 ft square is used to support a concrete column 14 in. square and 14 ft long. The footing undergoes a rotation of 0.24° due to soil conditions. How much stress is induced back into the column? Assume that the column is fixed at its upper end and use the flexure equations of Fig. 6-3. Modulus of elasticity of the concrete is 3,600,000 psi, ultimate strength is 4000 psi, and allowable stress is 1800 psi.

Problems 6-3 through 6-14. Determine the required contact area for the footings under the indicated conditions.

Problem	Type of footing	Limiting criteria for p_a	p_a (psf)	Load P_{DL}	Load P_{LL}	Column or wall base
6-3	Spread	Strength	4000	115 k	121 k	Hinged
6-4	Strip	Strength	4000	5.8 klf	5.0 klf	Hinged
6-5	Spread	Settlement	2600	188 k	62 k	Fixed
6-6	Strip	Settlement	2600	8.1 klf	6.5 klf	Fixed
6-7	Spread	Strength	3200	154 k	131 k	Fixed
6-8	Strip	Strength	3200	9.2 klf	7.1 klf	Fixed
6-9	Spread	Settlement	1800	63 k	59 k	Hinged
6-10	Strip	Settlement	1800	5.8 klf	6.3 klf	Hinged
6-11	Spread	Strength	2200	72 k	41 k	Fixed
6-12	Strip	Strength	2200	3.9 klf	5.0 klf	Fixed
6-13	Spread	Settlement	2200	142 k	119 k	Hinged
6-14	Strip	Settlement	2200	5.3 klf	4.8 klf	Fixed

6-15. A concrete column in a braced frame carries a dead load of 116 kips and a live load of 92 kips. The allowable soil pressure is limited by settlement criteria; the supporting soil is a silty sand, $\phi = 31°$, $\gamma = 117$ pcf. Depth of burial is set at 4 ft, footings are to be fixed to the columns. The area is subject to annual flooding. Select the required contact size for the footing.

6-16. A column in a rigid frame is subject to a dead load of 141 kips, a live load of 109 kips, and a lateral earthquake load of 12 kips. The supporting soil is a clay, CL, having a unit weight of 112 pcf. The allowable soil pressure is limited by settlement criteria to 2200 psf for strip footings. The locality is subject to annual flooding. The depth of burial is set at 3 ft 6 in., and footings are to be fixed to the columns. Select the required contact size for the footing.

6-17. A shearwall 24 ft long carries a uniform dead load of 7.5 kips/ft, a live load of 4.2 kips/ft, and a lateral load of 68 kips. The underlying soil is a loose sand. The wall is rigidly fixed to its footing. Design is limited by settlements. The area is not subject to flooding. Select a strip footing for the shearwall, using a depth of burial of 3 ft.

6-18. A concrete column in a rigid frame carries a dead load of 94 kips, a live load of 106 kips, and a lateral load of 15 kips. The column is rigidly fixed to its footing. The underlying soil is a loose sand, $\gamma = 119$ pcf. Settlement criteria governs the foundation design. The area is not subject to flooding. Depth of burial has been set at 4 ft. Select the required size for a square spread footing.

6-19. A shear panel in a braced frame structure has columns at both ends, each rigidly fixed to its square spread footing. At the left side, $P_{\text{DL}} = 52$ kips, $P_{\text{LL}} = 61$ kips, $P_{\text{EQ}} = 21$ kips. At the right side, $P_{\text{DL}} = 109$ kips, $P_{\text{LL}} = 118$ kips, $P_{\text{EQ}} = 21$ kips. The supporting soil is a medium sand, $\phi = 33°$, $\gamma = 123$ pcf. Settlement criteria governs the design. The area is subject to flooding. Depth of founding has been set at 4 ft. Select the required footing sizes for the columns at the ends at the braced panel.

6-20. A column in a braced frame is supported by a clay soil, CL, having an unconfined compressive strength of 2500 psf and a unit weight of 106 pcf. Dead load on the column is 88 kips and live load is 69 kips. The column is rigidly fixed to its footing. Depth of burial is 3 ft. Settlements govern the design. The area is subject to flooding. Select the required size for a square spread footing.

6-21. Determine the footing size for Problem 6-15 if the soil is a clay, CL, $\gamma = 114$ pcf, $q_u = 1900$ psf.

6-22. Determine the footing size for Problem 6-16 if the soil is a sand, SC, $\gamma = 116$ pcf, $\phi = 31°$.

6-23. Determine the footing size for Problem 6-17 if the soil is a medium clay with strength governing the design.

6-24. Determine the footing size for Problem 6-18 if the soil is a soft clay, $c = 0.5$ ksf, $\gamma = 106$ pcf.

6-25. Determine the footing size for Problem 6-19 if the soil is a stiff clay, $\gamma = 117$ pcf, with strength governing the design.

7

Induced Interactions

The interaction between a single isolated footing and its supporting soil was shown in Chapter 6 to be a direct action–reaction process. Its immediate effects were discussed in detail and a general method of analysis was developed. Footings rarely occur singly, however. Foundations are composed of groups of footings, sometimes in close proximity to each other, sometimes having large footings adjacent to small footings, sometimes having spread footings adjacent to strip footings. In addition to the direct interaction between a single footing and its supporting soil, there can also be a secondary or induced interaction between any one footing and the footings adjacent to it.

Not only is there a possible interaction from footing to footing, there may also be an interaction from one soil stratum to an underlying weaker (or more compressible) soil stratum. This stratum-to-stratum interaction can obviously contribute directly to settlements. It can also induce further interactions or secondary effects from footing to footing.

The propagation of effects from footing to footing or from stratum to stratum occurs due to the existence of the bulb of pressure that extends outward and downward from the contact area under each footing. While dispersion of pressure occurs rapidly and pressures decrease sharply in only a few feet, the pressure can still cause undesirable effects on footings nearby. This chapter is devoted to a brief study of this dispersion of pressure under a footing and the effects such a dispersion can have on neighboring footings.

It is again noted that in the approach used in this book, the actual magnitude of settlements at a particular footing is rarely of concern. The overriding concern is the differential settlements of one footing compared to the settlements of other footings in the vicinity. This comparison of relative settlements rather than prediction of absolute values is utilized repeatedly in the following discussions.

CALCULATION OF PRESSURE DISPERSION

The exact distribution of pressure beneath a footing (the Boussinesq pressure distribution) was presented in Chapter 3. The distinctive "bulbs" of pressure contours are shown again in Fig. 7-1 for immediate reference. The full smooth bulbs shown in Fig. 7-1 can occur only if the soil stratum is homogeneous and continuous; if a change in soil characteristics occurs, some distortion of the bulbs must also occur.

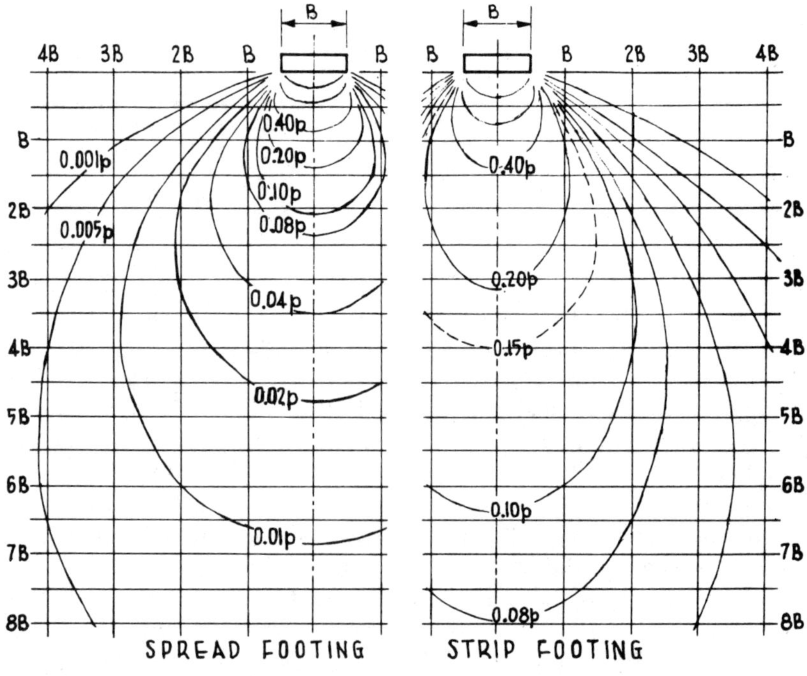

Figure 7-1 Boussinesq pressure bulbs.

It was also noted in Chapter 3 that in a group of spread footings having roughly equal contact pressures, the spread footing creating the largest pressure bulb will settle most and the one creating the smallest pressure bulb will settle least. As a rule, the largest pressure bulb will be the one corresponding to the largest total load (and footing size) and the smallest pressure bulb will be the one corresponding to the smallest total load (and footing size); settlements will therefore be proportional to total load, not just to contact pressures.

To simplify calculations, the Boussinesq pressure bulbs are often approximated by the pyramid or wedge shapes shown in Fig. 7-2. These approximate shapes, however, are still called the pressure "bulbs" in all subsequent discussions. The existence of these pressure bulbs must be borne in mind at every stage of the design of foundations. For the approximate bulbs of Fig. 7-2, side slopes are taken at 2 vertical to 1 horizontal for all ordinary soils. Pressure on any horizontal plane is assumed to be constant. The influence of pressures beyond the sloped sides and bottom surfaces indicated in Fig. 7-2 is presumed to be of little consequence and can be ignored.

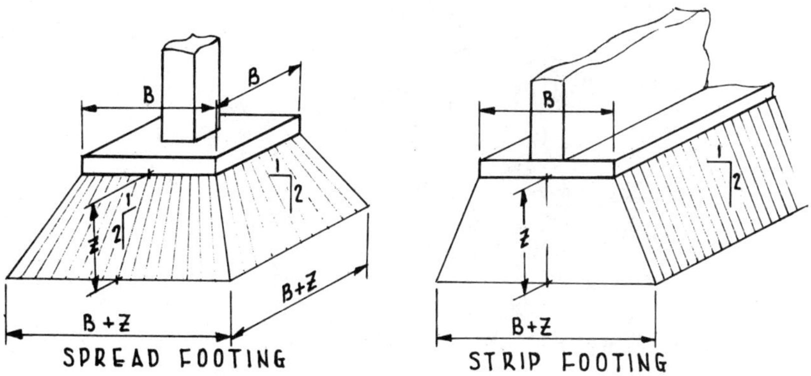

Figure 7-2 Approximate pressure bulbs.

All calculations for settlements may be based on this "average" pressure at each level below the founding line. Over the years, the use of this average pressure in calculating average settlements has been found to produce acceptable results. It is, however, only an approximation of the Boussinesq pressure bulb.

Where the pressure bulb extends into a second soil stratum having different characteristics from the first, any distortion of the approximate pressure bulb is considered to be small enough that it can be ignored. The dispersion angle remains 2 vertical to 1 horizontal for both soils. Graphic solutions for zones of influence are, of course, accurate enough for final design under these assumptions.

The pressure under an isolated spread footing is dispersed in two directions and dissipates rapidly, as indicated in Fig. 7-1. At a depth of $2B$ below the founding line it is already down to 10% of the contact pressure. In comparison, the pressure under a strip footing or grade beam disperses in only one direction and can penetrate to a much greater depth, up to $4B$ below the founding line.

Wherever the pressure bulbs overlap, a buildup of pressure occurs, with a commensurate buildup of settlements within the overlapping zones. Such overlaps can occur when footings are placed too close together, when large loads and their large footings are placed adjacent to small loads and their small footings, or where spread footings and strip footings are intermixed in the same foundation system. A sketch of such overlapping zones of influence is shown in Fig. 7-3.

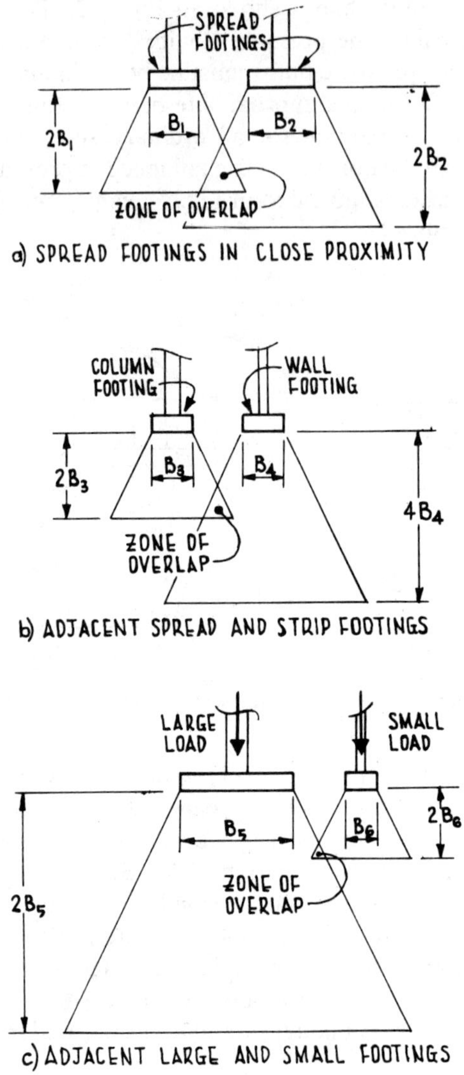

Figure 7-3 Overlapping pressure bulbs.

In general, precise computations involving such fine points as overlapping pressure bulbs are subject to question. The soil properties that must be used in such calculations are rarely consistent enough or accurate enough to permit such fine tuning of the design. It is far preferable to eliminate the overlapping pressure zones shown in Fig. 7-3 than to try to deal with them; rude but positive measures to circumvent these and other such induced interactions are discussed in the following sections.

EFFECTS AT UNDERLYING SOIL STRATA

Stratification of a supporting soil is of course a common foundation condition. Even where the supporting stratum is deep enough to contain the entire pressure bulb, cuts, or fills to level the building site may in effect produce stratification. Tilted or uneven stratification as shown in Fig. 7-4 is so common that it should be expected wherever stratification is present.

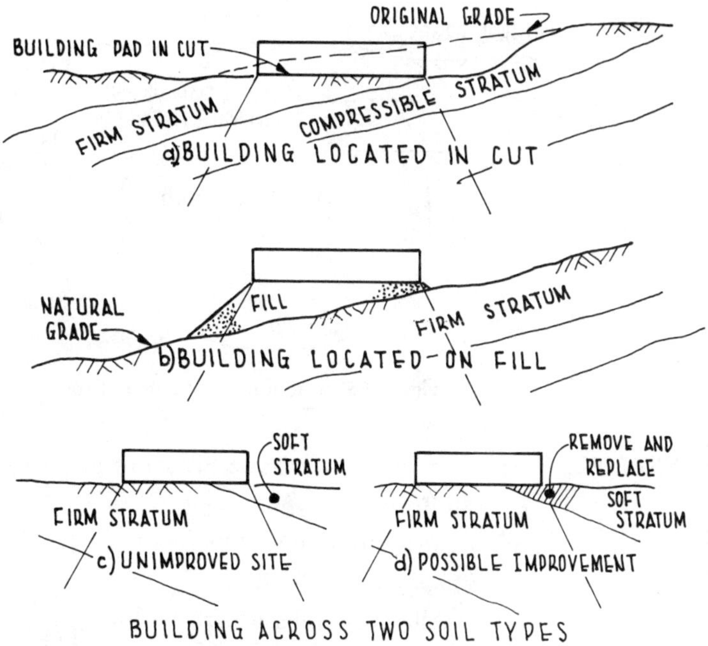

Figure 7-4 Typical soil stratifications.

Where a stratum of compressible soil of variable thickness supports the pressure bulbs as shown in Fig. 7-4a and b, the foundation will be subject to nonuniform settlement. The entire building will tilt very slightly as shown, with a maximum settlement of 1 in. at the footing that settles most. The settlements of the buildings of Figures 7-4a and b will still fall within the specified limits and the foundation condition is in fact acceptable. Whether the design should include measures to equalize the settlements and reduce the slight tilt is left to the individual designer; the tilt is likely to be undetectable, it does no particular harm to the building and all settlements remain within the specified 1-in. limits.

The stratification of Fig. 7-4c, however, is a different matter. In that case the settlement will not produce a linear tilt but a distinct increase in settlement at the right end of the building. Cases such as this are beyond the scope of an elementary text, but a common remedy for such a problem is to remove the offending wedge of material by excavation as shown in Fig. 7-4d, and to replace it with a uniform thickness of select backfill.

If the stratification happens to be reasonably uniform as shown in Fig. 7-5, it can be expected that settlements at equal soil pressures will also be reasonably uniform. The soils report in such cases should specify the allowable soil pressure at the top of each stratum, where the underlying strata are weaker or more compressible than the one above. The foundation design must include a check to see that none of these specified allowable pressures is exceeded.

An example will illustrate the computation of pressures at various levels under a spread footing.

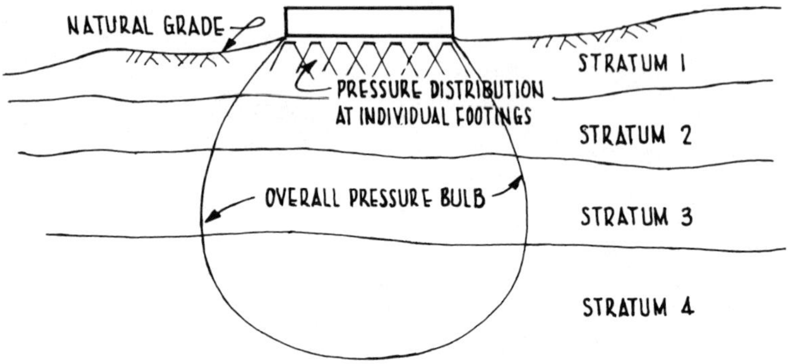

Figure 7-5 Uniform stratification of soils.

EXAMPLE 7-1: Computation of Soil Pressure at Various Depths

Uniformly stratified soil.
Approximate solution.

Given:

Allowable pressures:

Stratum 1: p_a = 4000 psf
Stratum 2: p_a = 2000 psf
Stratum 3: p_a = 1500 psf
Stratum 4: p_a = 2000 psf

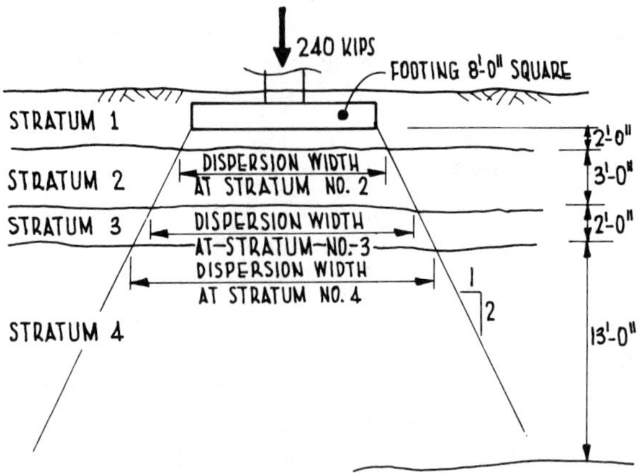

SOLUTION Calculate the dispersion widths.

Top of stratum 2: 8 ft + 2 × 2/2 = 10 ft
Top of stratum 3: 10 ft + 2 × 3/2 = 13 ft
Top of stratum 4: 13 ft + 2 × 2/2 = 15 ft

Calculate the contact pressures.

p_1 = 240 keys/8 × 8 = 3750 psf
At top of stratum 2: p_2 = 240 keys/10 × 10 = 2400 psf
At top of stratum 3: p_3 = 240 keys/13 × 13 = 1420 psf
At top of stratum 4: p_4 = 240 keys/15 × 15 = 1067 psf

These results are compared with the allowable soil pressures given above. It is seen that the allowable pressure on the soil in stratum 2 is exceeded; hence the given footing design is not acceptable.

Although corrective measures were not specifically a part of this example, such measures might be considered. One such solution would be to increase the footing size to 9 ft square, thereby reducing all the soil pressures and thereby bringing the pressure on stratum 2 within its allowable value. Another solution would be to raise the founding line 1 ft, thereby increasing the dispersion width at the top of stratum 2 and reducing the pressure enough to bring it within its allowable value. Verification of these measures is included as one of the outside problems at the end of this chapter.

RELATIVE SETTLEMENTS BETWEEN FOOTINGS

It has been observed earlier that in the design of a group of footings, the maximum estimated settlement of the larger footings will usually be limited to 1 in. The difference in settlements between any two adjacent footings, large or small, will then be somewhat less than 1 in., presumably not more than $\frac{3}{4}$ in. In general, properties of soils are not dependable enough to justify any higher level of confidence than this. Thus it is not known exactly what the settlement of any footing will be at any one time, only that the settlements overall will be kept within certain bounds.

Such a blanket approach is easy to defend, in view of the hundreds of load conditions that might occur on a foundation. The foundations, in turn, must be supported by a soil whose properties can change after each rainstorm. A blanket solution is one of the few practical approaches that can be applied under such a multiplicity of possible conditions.

Even within this approach, however, there has long been a trend in the industry to try to equalize the settlements under all footings as much as possible. Such problems in "proportioning for equal settlements" have even appeared occasionally on licensing examinations for civil engineers. Some federal agencies

[35] require that buildings built under their auspices include such provisions in the foundation design. A brief discussion of such proportioning is therefore appropriate even in an elementary textbook such as this one.

It was shown in Chapter 3 that foundation settlements S for spread footings are proportional both to the soil pressure p and the footing width B:

$$S = K_0 p B \tag{7-1}$$

where the coefficient of proportionality K_0 is a property of the particular soil mass, varying throughout the depth of the stratum.

The primary objection to the use of Eq. (7-1) to calculate the exact magnitude of a footing settlement was that the evaluation of the coefficient K_0 became impractical or impossible. However, although it may not be practical to find the exact magnitude of the settlement of a spread footing, a comparison between two adjacent spread footings within the same soil mass can eliminate the unknown K_0,

$$\frac{S_1}{S_2} = \frac{p_1 B_1}{p_2 B_2} \tag{7-2}$$

where S_1 and S_2 = settlements of footings 1 and 2
p_1 and p_2 = sustained soil pressures under footings 1 and 2
B_1 and B_2 = widths of footings 1 and 2

Equation (7-2) is derived simply by writing Eq. (7-1) for footing 1, then for footing 2, then dividing the first equation by the second. Equation (7-2) is also valid when comparing two strip footings or two grade beams, but not for comparing mixed types of footings.

Like Eq. (7-1), the accuracy of Eq. (7-2) is dependent on the relative sizes of the two footings being compared. The accuracy is best where the footings are of comparatively equal sizes. For footings of different sizes, the accuracy decreases as the difference in sizes increases.

Even so, the use of Eq. (7-2) allows an improvement in control over differential settlements to be exercised. As noted earlier, the total settlement of a structure is of little importance; the structure is sensitive only to the difference in settlements between adjacent footings, not to the total settlement. Equation (7-2) allows the computation of such relative settlements to be made more readily.

In a group of square spread footings, for example, one of the footings can be chosen as a reference footing. The settlements of all other footings in the group can then be compared to the reference footing. The settlement of each footing is then expressed as a percentage of the settlement of the reference footing.

The following example illustrates such a comparison. Note that no attempt is made in this example to produce equal settlements at each footing. The settlement of the largest footing is selected arbitrarily as the reference settlement against which all other settlements are compared.

EXAMPLE 7-2: Computation of Reference Settlements

Soil pressure limited by strength to 3000 psf. Vertical design loads as shown.

Given:

Footing B2 is selected as the reference footing for this case since its load (and size) is largest.

Compare settlements at all other footings to that of footing B2.

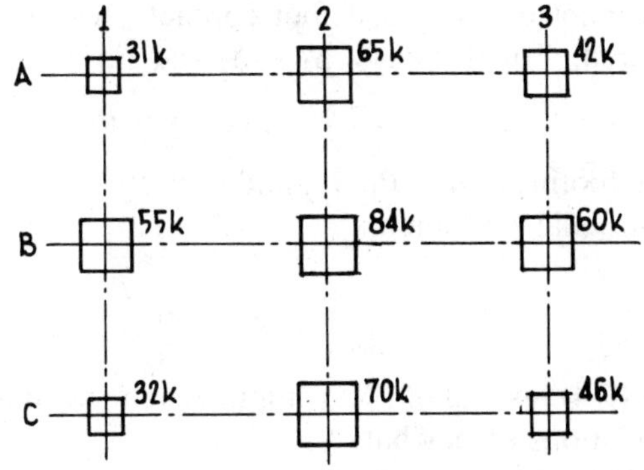

SOLUTION In the following tabulation, the required contact area is computed by dividing the load by the allowable soil pressure. The dimension B is computed as the square root of the contact area. The pressure p is computed as P/B^2.

Footing	Load (ksf)	Width, B	Pressure (ksf)	Relative settlement
A1	31	3 ft 6 in.	2.53	$0.58\,S_{ref}$
A2	65	5 ft 0 in.	2.60	$0.85\,S_{ref}$
A3	42	4 ft 0 in.	2.63	$0.69\,S_{ref}$
B1	55	4 ft 6 in.	2.72	$0.80\,S_{ref}$
B2	84	5 ft 6 in.	2.78	$1.00\,S_{ref}$
B3	60	4 ft 6 in.	2.96	$0.87\,S_{ref}$
C1	32	3 ft 6 in.	2.61	$0.60\,S_{ref}$
C2	70	5 ft 0 in.	2.80	$0.92\,S_{ref}$
C3	46	4 ft 0 in.	2.88	$0.75\,S_{ref}$

Settlements are computed from Eq. (7-2). For example, the settlement of footing A1 is found by

$$S_{A1} = S_{B2}\frac{p_{A1} \times B_{A1}}{p_{B2} \times B_{B2}} = 1.00\,S_{ref}\frac{2.53 \times 3.5}{2.78 \times 5.5} = 0.58\,S_{ref}$$

All others are computed similarly and listed in the tabulation above.

In Example 7-2, the variation in the soil pressure occurs as a result of rounding off the dimensions to the nearest 6 in., a common practice. There is seen to be considerable variation in settlements, the least being at footing A1, which settles only 58% as much as footing B2. Even so, if the settlement of footing B2 is known to be less than 1 in., all settlements will be well within the $\frac{3}{4}$-in. differential settlement that is usually accepted.

Alternatively, the footings of Example 7-1 can also be proportioned for equal settlements. Since the soil pressure is limited by strength criteria, the value of 3000 psf cannot be exceeded, but contact pressures may be made considerably less if desired. For this case, $S_1 = S_2$ and Eq. (7-2) degenerates to

$$p_1B_1 = p_2B_2 \tag{7-3}$$

For square footings, $p_1 = P_1/B_1^2$, and $p_2 = P_2/B_2^2$, where P_1 and P_2 are total loads. Hence, for equal settlements,

$$B_1 = B_2\frac{P_1}{P_2} \tag{7-4}$$

Using Eq. (7-4) instead of (7-3) eliminates the need to compute pressures, simplifying calculations somewhat.

The footings of Example 7-2 can now be recomputed to produce equal settlements. At the limiting pressure, the smallest footing will settle least and no other footing can be allowed to settle more than this amount. The settlement of the smallest footing is therefore the reference settlement.

EXAMPLE 7-3: Computation of Footing Sizes for Equal Settlements

Soil pressure limited by strength to 3000 psf. Vertical design loads as shown.

Given:

Footing A1 is chosen as the reference footing since its load, size, and settlement are smallest.

Compare sizes and pressures of all footings to those of footing A1 when all settlements are kept roughly equal.

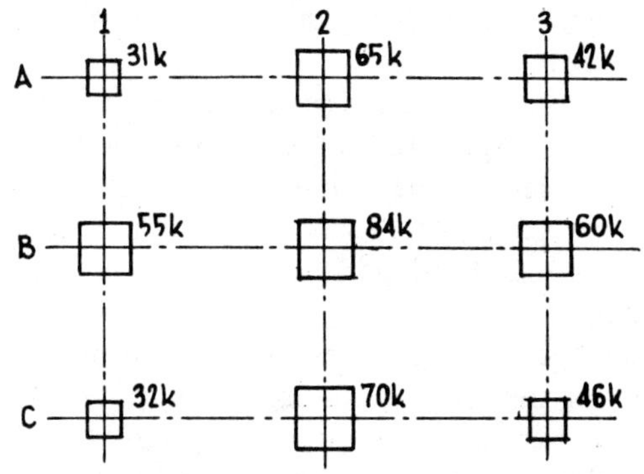

SOLUTION

Footing	Load (kips)	Width, B	Pressure (ksf)	Relative settlement
A1	31	3 ft 6 in.	2.53	$1.00S_{ref}$
A2	65	7 ft 6 in.	1.16	$0.98S_{ref}$
A3	42	5 ft 0 in.	1.68	$0.97S_{ref}$
B1	55	6 ft 6 in.	1.30	$0.95S_{ref}$
B2	84	9 ft 6 in.	0.93	$1.00S_{ref}$
B3	60	7 ft 0 in.	1.22	$0.96S_{ref}$
C1	32	4 ft 0 in.	2.00	$0.90S_{ref}$
C2	70	8 ft 0 in.	1.09	$0.98S_{ref}$
C3	46	5 ft 6 in.	1.52	$0.94S_{ref}$

The dimension B is computed from Eq. (7-4), using footing A1 as the reference footing. For example, the size of footing A2 is computed as

$$B_{A2} = B_{A1}\frac{P_{A2}}{P_{A1}} = 3.5\frac{65}{31} = 7.34 \text{ ft} \qquad \text{say, 7 ft 6 in.}$$

The widths of other footings in the table are calculated the same way. Pressures are simply P/B^2. Relative settlements in the table are computed from Eq. (7-2) as a final check. The variations in the final settlements are due to rounding the width B to the next highest 6 in.

The increase in some of the footing sizes between Examples 7-2 and 7-3 is striking. To produce equal settlements, the larger footing sizes had to be almost doubled, from 5 ft 6 in. for footing B2 in Example 7-1 to 9 ft 6 in. in Example 7-3. Usable pressures were of course reduced accordingly. It can be seen immediately that proportioning for equal settlements does involve additional cost.

The additional cost, however, is primarily that of bulk concrete. Whether the extra cost is warranted can only be decided by the designer, in consideration of the particular structure and the particular site and the particular client at the particular time. Many designers, however, routinely design all foundations for equal settlements.

In Example 7-3, the soil pressure was limited by strength criteria. For settlements to be kept equal, the settlements of the larger footings had to be reduced to that of the smaller footings, since the smaller footings would settle least under the allowable contact pressure. The reduction was accomplished by increasing the sizes of the larger footings, thereby reducing their contact pressures and their settlements.

If, however, the soil pressure were to be limited by settlement criteria, the settlements of the smaller footings would have to increase to match the settlements of the larger footings, since in this case the larger footings would settle most under the allowable contact pressure. Such an increase in settlements is accomplished by reducing the sizes of the smaller footings, thereby increasing their contact pressures and their settlements. The following example illustrates this procedure.

EXAMPLE 7-4: Computation of Footing Sizes for Equal Settlements

Soil pressure limited by settlements to 3000 psf.
Vertical design loads as shown.

Given:

Footing B2 is chosen as the reference footing since its load, size, and settlement are largest.

Compare sizes and pressures of all footings to those of footing B2 when all settlements are kept roughly equal.

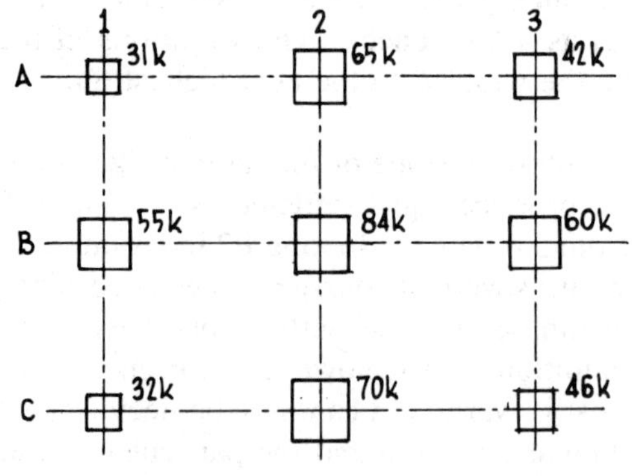

SOLUTION

Footing	Load (kips)	Width, B	Pressure (ksf)	Relative settlement
A1	31	2 ft 6 in.	4.96	$0.81\,S_{ref}$
A2	65	4 ft 6 in.	3.21	$0.94\,S_{ref}$
A3	42	3 ft 0 in.	4.67	$0.92\,S_{ref}$
B1	55	4 ft 0 in.	3.44	$0.90\,S_{ref}$
B2	84	5 ft 6 in.	2.78	$1.00\,S_{ref}$
B3	60	4 ft 0 in.	3.75	$0.98\,S_{ref}$
C1	32	2 ft 6 in.	5.12	$0.84\,S_{ref}$
C2	70	5 ft 0 in.	2.80	$0.92\,S_{ref}$
C3	46	3 ft 0 in.	5.11	$1.00\,S_{ref}$

As before, the dimension B for each footing is computed using Eq. (7-4), but in this case using footing B2 as the reference footing. For footing A1, for example,

$$B_{A1} = B_{B2}\frac{P_{A1}}{P_{B2}} = 5.5\frac{31}{84} = 2.03 \text{ ft} \qquad \text{say, 2 ft 6 in.}$$

Pressures are computed simply as P/B^2. Relative settlements are computed using Eq. (7-2) but are needed only as a final check. As before, the variation in settlements is due to rounding the dimension B to the next higher 6 in.

When the results of Example 7-4 are compared to Examples 7-2 and 7-3, it is seen that the footing sizes have been kept to the smaller values, but that the contact pressures are alarmingly high. Even though an increase in contact pressure under a small footing would not usually be a problem where settlement criteria controls the design, a contact pressure 70% greater than the limiting pressure (such as footing C3) is cause for concern. Whether the strength of the soil would be adequate for such an increase would normally be a topic for the soils report; the soils engineer would, of course, have to be advised of such potential problems before beginning investigations.

The results of Example 7-4 reaffirm the need for preparing a reasonably accurate foundation design before the soils investigation begins. The soils report can then be directed toward specific problem areas rather than toward general practices. The savings in cost, for example, between the foundations of Examples 7-4 and 7-3 could be quite significant and well worth the extra effort involved in preparing a preliminary design.

It should be apparent that a median case might be struck between the extremes of Examples 7-3 and 7-4. Where soil pressure at the larger footing is limited by settlement criteria, a nominal increase in contact pressure of 33% or even 50% might be allowed for the smaller footings. The procedure of Example 7-3 using this higher allowable contact pressure would then afford a solution that would fall somewhere between the two extremes. Such a compromise is indeed feasible; its application is included as Problem 7-8.

EFFECTS OF CLOSE PROXIMITY

When two footings are placed close together, their pressure bulbs may overlap, as shown in Fig. 7-6. It should be apparent that the pressure at any level in the overlap is simply the sum of the two contributing pressures. The settlements in this zone will of course be commensurately higher than those at the same level at either side.

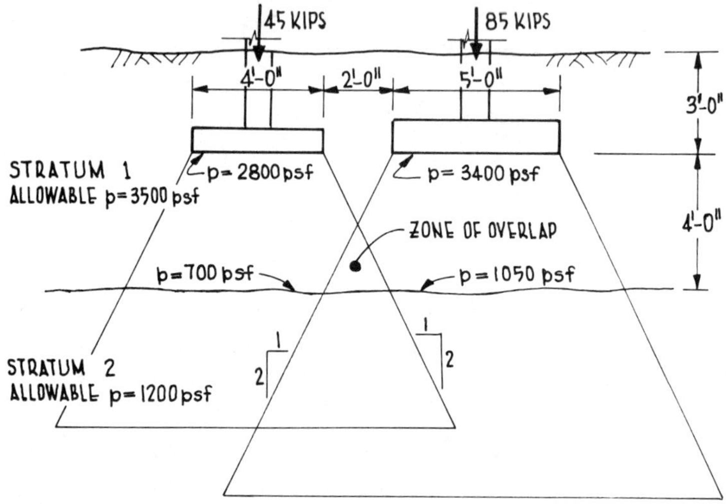

Figure 7-6 Zone of overlap.

It is a simple matter to compute the total pressure in the overlap zone. In Fig. 7-6, the pressure at the top of stratum 2 is the sum of the two overlapping pressures, 1050 + 700 = 1750 psf. Since the allowable pressure on the lower stratum is only 1200 psf, a serious overstress would seem to occur in this example.

The accuracy of such calculations is highly questionable. The pressure bulb itself is an approximation, made under the assumption that pressures are constant at any level within the bulb and that the dispersion angle of the bulb is 2 vertical to 1 horizontal. It was shown in Chapter 3, however, that the pressures are not uniform and that the angle is in fact variable. While the average pressure at any level may be a reasonably accurate approximation, computations of the precise pressure at a particular point are not so believable.

When pressure bulbs overlap as shown in Fig. 7-6, the settlements can produce a "tilt" of the individual footings toward each other. An exaggerated case is shown in Fig. 7-7. The exact amount of tilt would, of course, depend on the magnitude of pressures and the compressibility of the soil.

It was shown in Chapter 6 that even large moments on a rigid elastic column

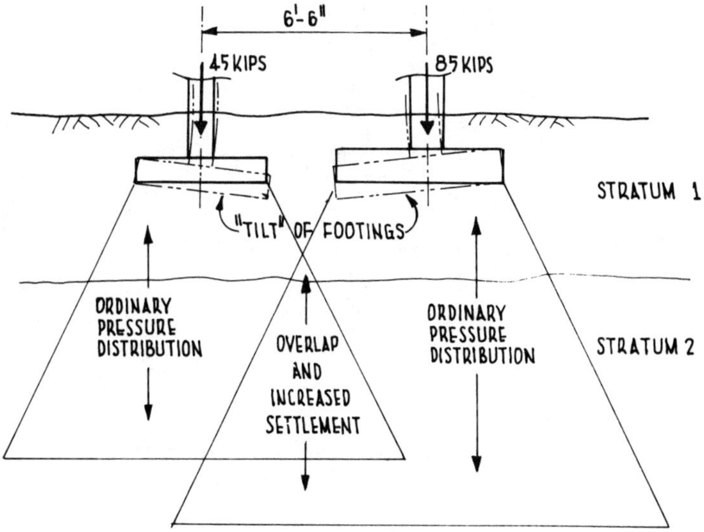

Figure 7-7 Rotations due to overlap.

do not produce enough rotations at the footings to cause significant variations in soil pressure. That characteristic, however, is a two-edged sword. The converse conclusion must also be drawn: *when a column is fixed to its footing, even small rotations of the footing can induce significant moments back into the rigid elastic column*. The tilt of the footing shown in Fig. 7-7 could therefore easily induce destructive moments back into the column.

As noted earlier, it may be possible to make calculations and come up with some numbers to represent the rotation of the footings of Fig. 7-7. Such numbers are rarely believable. It is far more prudent to eliminate the problem than to be forced to rely on such computations.

One way to eliminate the problem would be to combine the two footings of Fig. 7-7 into a single footing, thereby eliminating the problem of rotations of the individual footings. Such a combined footing is shown in Fig. 7-8. The combined footing has a constant width as shown.

The selection of the dimensions of the combined footing shown in Fig. 7-8 is somewhat arbitrary but entirely straightforward. The dimensions must be set to produce a uniform soil pressure under the combined footing. A uniform soil pressure will be produced if the resultant of loads passes through the centroid of the contact area as shown.

The magnitude of the resultant of the two loads is obviously the sum of the two forces, $R = 45 + 85 = 130$ kips. The location of the resultant is found by simple statics. Moments are summed about point B to find the distance $\bar{x}$,

$$\sum M_B = 0 = 45 \times 6.5 - 130\bar{x}$$

$$\bar{x} = 2.25 \text{ ft} = 2 \text{ ft } 3 \text{ in.}$$

The distance $\bar{x} = 2$ ft 3 in. is shown in Fig. 7-8.

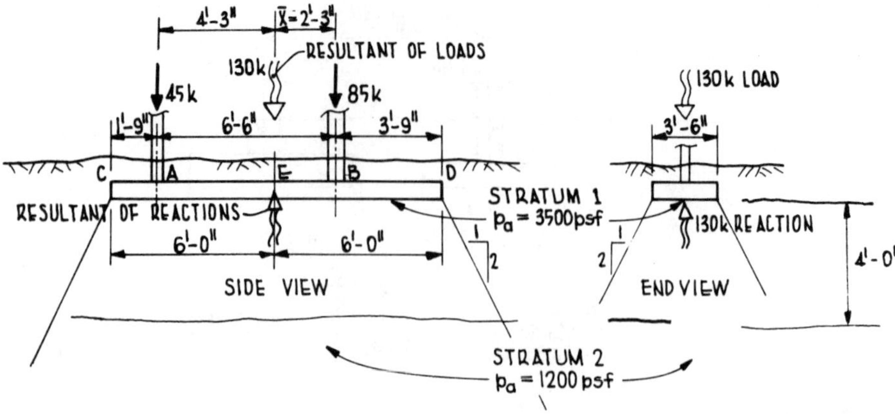

Figure 7-8 Combined footing.

The point E shown in Fig. 7-8 is thus the center of the rectangular footings. It remains only to establish the distances CA and BD to define completely the required overall length. The distance CA is set at 1.75 ft to provide a reasonable amount of clearance around the column at point A, producing the final half-length of 6 ft as shown; the total length is then 12 ft.

The width of the combined footing is now set to suit the requirements for contact pressure. For this case, there are two requirements to be met: The contact pressure may not exceed 3500 psf at the founding line; the average pressure in the pressure bulb may not exceed 1200 psf at the top of stratum 2. In the calculations, the width is set to suit the first requirement and is then checked to see that the second requirement is met.

The sum of vertical forces yields a solution for the width w, where the allowable soil pressure is 3500 psf,

$$\sum F_v = 0 = R - pLw = 130{,}000 - 3500 \times 12w$$

$$w = 3.09 \text{ ft} \qquad \text{say, 3 ft 6 in.}$$

The pressure at the top of stratum 2 is found similarly:

$$\sum F_v = 0 = R - pLw = 130{,}000 - p(12 + 4)(3.5 + 4)$$

$$p = 1083 < 1200 \text{ psf} \quad \text{(OK)}$$

The pressure is seen to be less than the allowable pressure of 1200 psf; the dimensions of 3 ft 6 in. × 12 ft for the combined footing are therefore acceptable.

It should be noted that the use of a combined footing in this case has circumvented any need for sophisticated calculations. The pressure under the combined footing is uniform and the danger of erratic settlements has been eliminated. Where column spacing is close, the use of a combined footing offers an easy and economical means to avoid footing-to-footing interactions.

EFFECTS OF UNEQUAL LOADS

When two adjacent footings carry loads having highly different magnitudes, the larger load may cause undesirable settlements under the adjacent smaller load. Such a case is shown in Fig. 7-9. Although the interaction may be more severe where footings are spaced closely together, the effects can also occur even at normal column spacings of 20 ft or so.

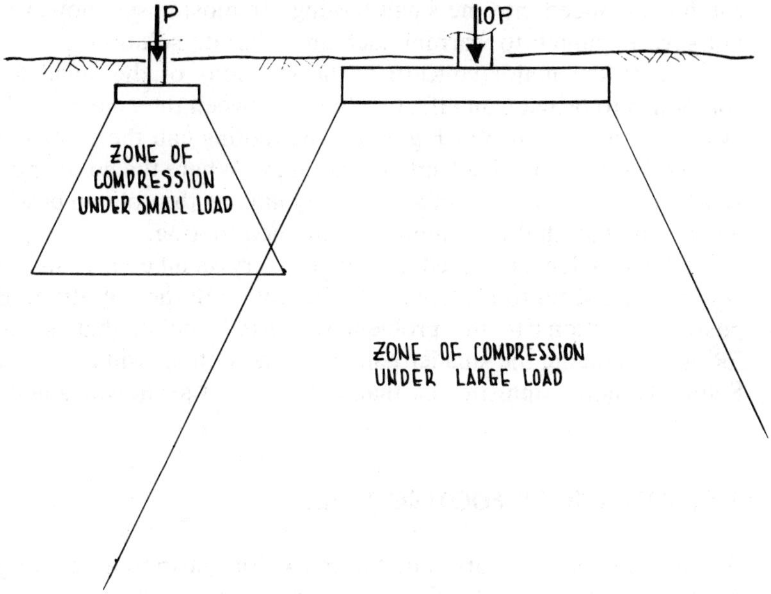

Figure 7-9 Influence of unequal loads.

The occurrence of a very lightly loaded column being placed adjacent to a very heavily loaded column is not a rare feature in building design. A covered colonnade adjacent to a multistory building produces such a circumstance, as does a covered loading area adjacent to an industrial building. Wherever such a feature occurs, it should be recognized in the foundation design as a special condition.

It should be obvious from Fig. 7-9 that the existence of the small footing will have little or no effect on the large footing. The small pressure bulb may or may not intersect the large one, but either way, its pressures are so small compared to the pressures in the large bulb that its effects on the large bulb can be ignored.

The pressure bulb of the large footing, however, can extend entirely underneath the small footing as indicated in Fig. 7-9. Any settlements produced by the large bulb will produce some degree of settlement in the small footing above. Depending on relative sizes and locations, the small footing will be displaced downward and might also be rotated somewhat.

Since the maximum settlement of the large footing is itself limited to 1 in., any settlements it may induce into the small footing will certainly be less than that. The vertical displacements by themselves will therefore be within the usual design limitations and need not be considered further. The potential rotations of the small footing, however, merit some attention.

In the preceding section, it was found that when two footings are in very close proximity, a combined footing can eliminate the problem of induced rotations. A combined footing might also be used here to eliminate the problem of the rotations induced into the small footing. In most cases, however, the problem is not severe enough to warrant such an elaborate solution.

Since the major concern is the rotations of the small footings, a simple solution is to provide an effective hinge between the column and the footing such as one of those shown in Fig. 6-7. The footing can then rotate without inducing destructive moments back into the column. Where the number of such columns is small, the additional cost of deliberately hinging the column bases will probably be much less than that of building a combined footing.

The problems incurred in placing a very small column load near a very large column load seem to receive a lot of attention in the literature. Either of the two possible solutions to the problem suggested above, that is, using a combined footing or hinging the smaller column to its footing, will circumvent the problem. Solutions more sophisticated than these do not seem warranted.

EFFECTS OF INTERMIXED FOOTING TYPES

At equal contact pressures and footing widths, strip footings will settle considerably more than spread footings; the difference is estimated herein to be roughly twice as much. The disparity occurs since the pressure under a spread footing is dispersed in two directions; its pressure bulb therefore diminishes quite rapidly. In contrast, the pressure under a strip footing can disperse in only one direction; pressures therefore disperse more slowly and the pressure bulb goes deeper. A sketch showing the two pressure bulbs is shown in Fig. 7-10.

The disparity does not produce as serious a problem as it may appear at first glance. Analogous to the derivation of Eq. (7-2), the comparative settlement analyses of Chapter 3 permit a rough comparison of the two settlements:

$$\frac{S_{sp}}{S_{st}} = \frac{p_{sp}B_{sp}}{2p_{st}B_{st}} \tag{7-5}$$

where S_{sp}, S_{st} = settlements of the spread footing and strip footing, respectively
p_{sp}, p_{st} = contact pressures of the spread footing, respectively
B_{sp}, B_{st} = least widths of the spread footing and strip footing, respectively

At equal contact pressures and for equal widths, the strip footing is seen to settle about twice as much as the spread footing, as stated earlier. The widths, however, are rarely comparable. In almost all cases one might encounter in practice, the width of the strip footing, B_{st}, will be considerably less than the width of

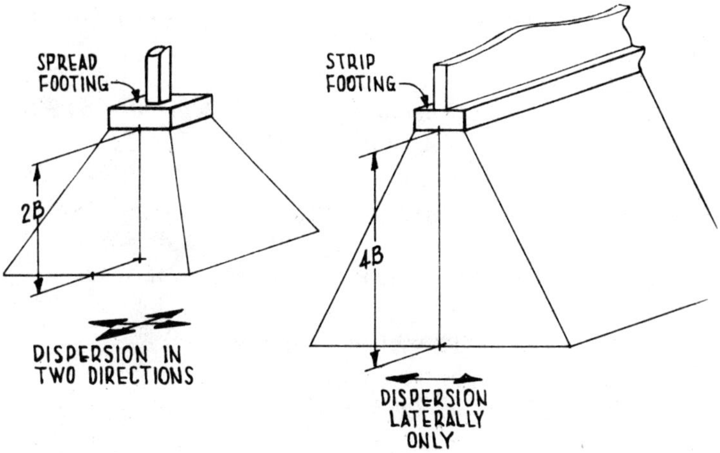

Figure 7-10 Contrasting pressure bulbs.

the spread footing, B_{sp}, which will bring the two settlements into closer agreement with each other.

An example will illustrate the comparison of the two settlements.

EXAMPLE 7-5: Comparison of Settlements

Strip footing intermixed with spread footings.
Vertical design loads as shown.

Given:

Allowable soil pressure, p_a is 3500 psf.

Compare settlement of the largest footing, B1, to the settlement of the bearing wall.

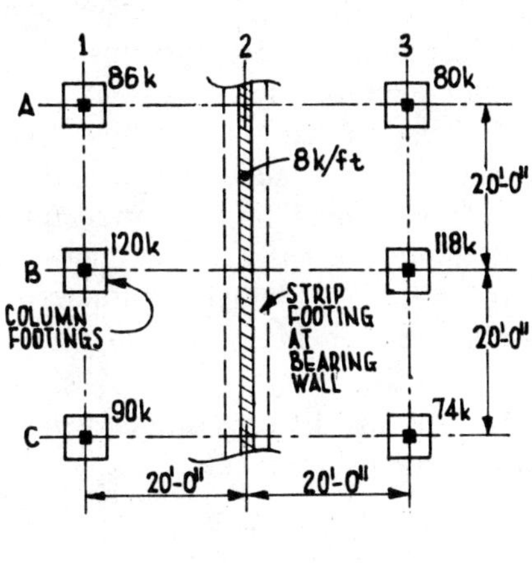

SOLUTION The footing sizes are selected to suit the given vertical design loads.

Spread footing: Area $= P_{sp}/p_a = 120/3500 = 34.3\ \text{ft}^2$
Use $B_{sp} = 6$ ft square
Actual $p_{sp} = P_{sp}/B_{sp} = 120/36 = 3330$ psf

Strip footing: Area $= P_{st}/p_a = 8/3500 = 2.28\ \text{ft}^2$
Use $B_{st} = 2$ ft 6 in. wide
Actual $p_{st} = P_{st}/B_{st} = 3200$ psf

Settlements are compared using Eq. (7-5)

$$\frac{S_{sp}}{S_{st}} = \frac{p_{sp}B_{sp}}{2p_{st}B_{st}}$$

$$S_{sp} = S_{st}\frac{3330 \times 6}{2 \times 3200 \times 2.5} = 1.25\ S_{st}$$

The settlement of the column footings in this case is seen to be slightly higher than that of the wall footing. If the settlement of the column footing is limited to 1 in., the differential is less than 0.25 in., well within the usual allowable range. Note that the loads in this example are comparable, considering the additional weight of the wall. The load on the wall over a 20-ft length is seen to be 160 kips, compared to the column load of 120 kips which supports the same length of structure.

For an alternative design, however, the uncertainties that arise in intermixing the footing types could be avoided simply by supporting the bearing wall on spread footings at each end rather than on a continuous strip footing. (Such a configuration is shown in Fig. 5-13.) All footings in the foundation would then be of the same type and the problem of different settlement characteristics would not arise.

The foregoing discussions involved strip footings. It should be apparent, however, that the supporting soil cannot distinguish whether the structure above the founding line is a bearing wall on a strip footing or a series of columns on a grade beam. The settlement characteristics of grade beams would therefore be identical to those of strip footings.

With this observation that grade beams and strip footings will have the same settlement characteristics, a second alternative design becomes immediately apparent. The columns could be supported on a grade beam and the bearing wall on a strip footing; the earlier differences between the two types of foundations and their different settlement characteristics would again be eliminated. The comparison of settlements would in this case revert to that derived as Eq. (7-2).

Example 7-5 will now be reexamined using this second alternative concept for control of differential settlements.

EXAMPLE 7-6: Comparison of Settlements

Strip footing intermixed with grade beams.
Vertical design loads as shown.

Given:

Allowable soil pressure is 3500 psf.
Compare the settlements of the grade beam to those of the strip footing

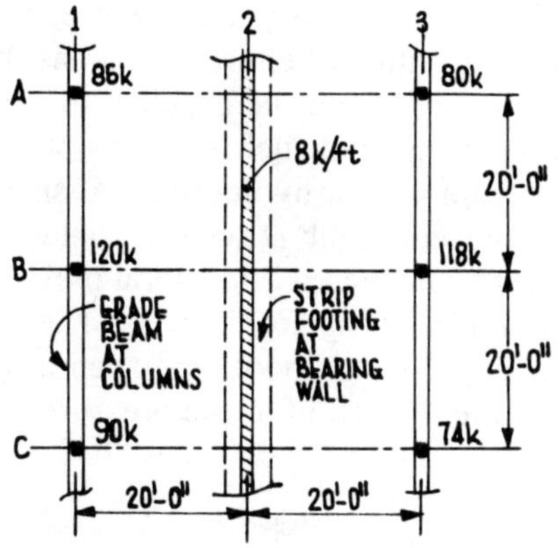

SOLUTION The footing sizes are selected to suit the given vertical design loads.

Grade beam: Area $= P_{gb}/p_a = 120/3500 = 34.3 \text{ ft}^2$
For 20-ft length, $B_{gb} = 34.3/20 = 1.71$, say 1 ft 9 in.
Actual $p_{gb} = P_{gb}/20 \times B_{gb} = 120/35 = 3430$ psf

Strip footing: Area $= P_{st}/p_a = 8/3500 = 2.28 \text{ ft}^2$
Use $B_{st} = 2$ ft 6 in. wide
Actual $p_{st} = P_{st}/20B_{st} = 3200$ psf

Settlements are compared using Eq. (7-2):

$$\frac{S_{st}}{S_{gb}} = \frac{p_{st}B_{st}}{p_{gb}B_{gb}}$$

$$S_{st} = S_{gb}\frac{3200 \times 2.5}{3430 \times 1.75} = 1.33\,S_{gb}$$

The settlement of the wall in this case is seen to be slightly greater than that of the columns, just opposite to the results of Example 7-6. Both cases, however, are well within the allowable limits for differential settlements.

It is concluded that intermixing spread footings and strip footings does not usually introduce serious problems, even though the different footing types have different settlement characteristics. Even so, the intermixing of footing types should not be taken for granted; where footing types are to be intermixed, a review of their settlements is always warranted.

EFFECTS OF ADJACENT EXCAVATIONS

To this point, it has been tacitly assumed that there were no interruptions or voids in the soil in the near vicinity of a footing. Such interruptions do happen, however, and their existence can require that particular attention be paid to the affected footings. As with earlier problem areas, however, it will again be seen that it is far easier to circumvent the problem than it is to deal with it.

A common example of such an interruption is a basement that intercepts the pressure bulb from a nearby footing. A sketch of such a case is shown in Fig. 7-11. The pressure bulb is seen to contact the basement wall well within its zone of influence. The increase in lateral pressure on the basement wall will have a definite effect on the design of the wall. The design of retaining walls, however, is beyond the scope of this book and such a design is not presented here. The concern here is the distortion of the pressure bulb and its effects on the footing.

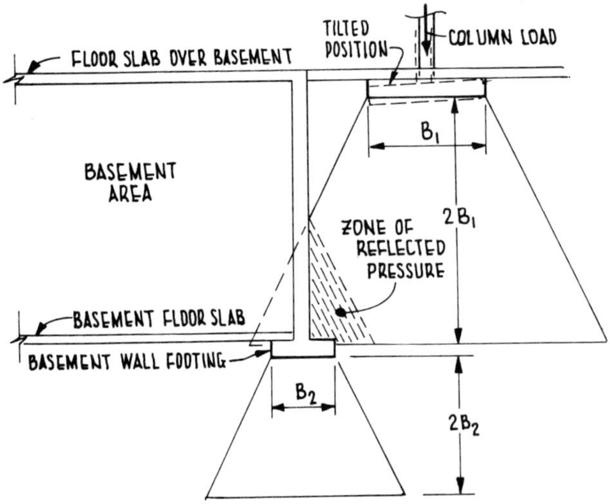

Figure 7-11 Pressure bulb interrupted by basement wall.

From the point where the pressure bulb contacts the basement wall, the pressures are reflected back into the bulb, creating a confused and overlapping pressure buildup in the bulb. The exact pattern of the pressure buildup need not be determined, however. It is enough to recognize that the buildup exists and that it interferes with the normal settlement of the footing.

Since the pressure buildup will occur next to the basement wall, it is presumed that the settlements will increase slightly on that side and the footing will tilt an undetermined amount as shown by the dashed line. The computation of the amount of settlement and the amount of rotation would be very complex calculations. The calculations would depend heavily on having accurate and representative values of the soil properties.

Such calculations are not necessary. It is enough to know that the total settlement will be less than 1 in. and that some rotations may occur. Since settlements are within the allowable amount, they are of no further concern. Any detrimental effects due to rotations of the footing can be avoided simply by assuring that the structural design includes an effective hinge between the column and the footing.

In this case and other cases where discontinuities in the soil stratum produce distortions in the pressure bulb, it may be possible to eliminate the problem in this way. Where settlements are known to be less than 1 in. but where footing rotations pose a potential hazard to the structure above, an effective hinge at the footing may provide an acceptable means to eliminate the hazard.

SUMMARY

The following points summarize the findings and conclusions of this chapter.

- For computing the average increase in soil pressure at some level below a footing, the side slopes of the pressure bulb may be taken at 2 vertical to 1 horizontal; that portion of the bulb outside these side slopes has such little influence it can be disregarded.

- The influence of the pressure bulb may be assumed to extend below the founding line to a depth of $2B$ for spread footings and to a depth of $4B$ for strip footings and grade beams, where B is the least width of the footing; that portion of the bulb deeper than these distances can be disregarded in settlement calculations.

- At equal contact pressures and with equal least widths B, a strip footing or a grade beam will settle considerably more than a spread footing; for the sake of comparisons, the difference has been estimated herein to be up to twice as much.

- Where the settlements of the larger footings in a group of footings is limited to 1 in., there is no particular need to proportion all the footings in the group for equal settlement.

- When it is elected to proportion a group of footings for equal settlements and the allowable pressure is limited by the strength of the soil, the settlement of the larger footings must be reduced to match that of the smallest settlement in the group; the smallest settlement is usually the one corresponding to the smallest load. Conversely, where the allowable soil pressure is limited by considerations of settlement rather than of strength, the settlement of the smaller footings must be increased to match that of the largest settlement in the group; the largest settlement is about 1 in. and is usually the one corresponding to the largest load.

- When it is elected to proportion a group of footings for equal settlements and the soil pressure is limited by settlements rather than by strength, it may be justifiable to increase the usable soil pressure under the smaller footings by 30 to 50% to produce a more efficient design.

• Settlements are proportional to actual loads and contact pressures; it must be recognized that any reduction in allowable soil pressure to compensate for short-term lateral loads will also reduce the final settlement.

• It was shown earlier that the existence of very large moments on a rigid elastic column can produce only small rotations at its footing. Conversely, however, inducing small rotations of a column footing can produce very large moments in the column above; such induced rotations of the footing are to be minimized in the foundation design.

• Where two adjacent footings are spaced so closely together that their pressure bulbs overlap and interactive footing rotations become a problem, a single combined footing supporting both loads will, if feasible, eliminate the problem.

• Where two adjacent footings support loads of vastly different magnitudes such that rotations induced into the smaller footing could cause a problem in the structure above, an effective hinged connection to the smaller footing will eliminate the potential problem.

• Where columns are intermixed with bearing walls in the design of a building and the different settlement characteristics between spread footings and strip footings become a consideration, the following alternatives arise:

1. Provide an effective hinge at each spread footing that occurs adjacent to a strip footing.
2. Do not intermix footing types (e.g., provide grade beams rather than spread footings to support the columns and use strip footings to support the walls); all footings are then of the same type and differences in settlement characteristics do not occur.
3. As an alternative means to avoid intermixing footing types, use spread footings throughout, both for the walls and for the columns; spread footings supporting the walls should be on the same module as the spread footings supporting the columns.

• Where a footing is placed in such close proximity to a discontinuity in the soil stratum that its pressure bulb is intercepted (such as at a basement wall), an effective hinged connection at the footing will avoid potential problems with induced footing rotations.

REVIEW QUESTIONS

7-1. How are footing-to-footing interactions generated?

7-2. What is the approximate side slopes of the approximate pressure bulb insofar as settlement calculations are concerned?

7-3. How deep is the approximate pressure bulb assumed to extend under a rectangular spread footing?

7-4. How deep is the approximate pressure bulb assumed to extend under a strip footing or a grade beam?

7-5. What is the assumed pressure at the bottom of the approximate pressure bulb for a rectangular footing? For a strip footing or grade beam?

7-6. What happens to the theoretical dispersion of pressure when the approximate pressure bulb extends through two or even three strata of different types of soil?

7-7. When pressure bulbs from two adjacent footings overlap, what is the primary effect on the footings themselves?

7-8. When stratified layers of soil are far out of level, what particular hazard might occur in the soil strata under a building?

7-9. In a diaphragm-and-shearwall structure, what problems might be encountered in proportioning for equal settlements?

7-10. How does comparison of settlements rather than computation of absolute settlements provide more believable results?

7-11. When proportioning for equal settlements, how is it possible to increase the allowable soil pressure under small footings by as much as 50% with no adverse effects?

7-12. Why would it be necessary to advise those doing the soils investigation that the foundation design will include proportioning for equal settlements?

7-13. Why is a slight "tilt" of a footing due to interacting pressure bulbs such a potential hazard to the structure above?

7-14. What is the usual remedy to offset "tilt" in a spread footing?

7-15. Why are calculations concerning footing rotations or tilting so subject to doubt?

7-16. What is a combined footing? What is its primary advantage? How is it different from a grade beam?

7-17. As a very general rule, how much more settlement (maximum) can be expected under a strip footing than a square footing having the same width and contact pressure in the same soil mass?

7-18. What is the effect on the footing when its pressure bulb is interrupted by an excavation or by a basement wall?

7-19. What is the point of caution in intermixing footing types, such as spread footings and grade beams?

7-20. What is the potential danger to the structure in placing a small lightly loaded footing adjacent to a large heavily loaded footing?

PROBLEMS

7-1. A square spread footing is to be placed on a stratified clay mass as shown in the sketch. Determine the increase in pressure at the top and at midheight of each stratum of clay within the pressure bulb.

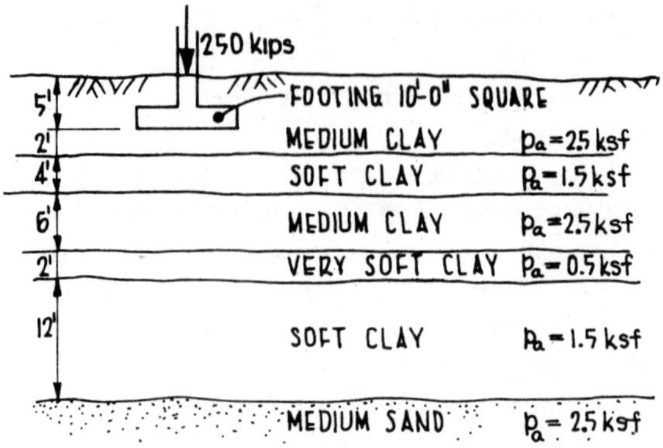

7-2. Determine the soil pressures in Problem 7-1 if the depth of founding were to be raised 18 in.

7-3. A strip footing is to be placed on a stratified sand mass as shown in the sketch. Determine the increase in pressure at the top of each stratum and at midheight.

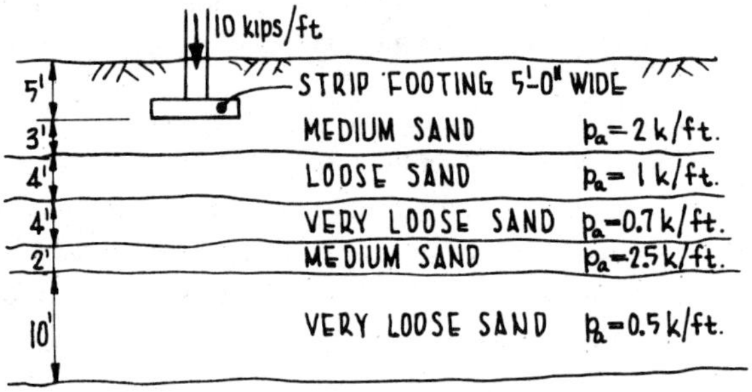

7-4. Determine the soil pressures in Problem 7-3 if the depth of founding were to be raised 3 ft 6 in.

7-5. For Example 7-1, verify that increasing the size of the footing to 9 ft square will reduce all pressures under the footing to acceptable levels. Verify also that raising the depth of founding 1 ft will reduce all pressures to allowable values.

7-6. The column loads for a small building are shown in the sketch. The allowable soil pressure is 3000 psf, based on the strength of the underlying sandy soil. Select the size for each footing, then compare the resulting settlements between footings for the sizes just selected.

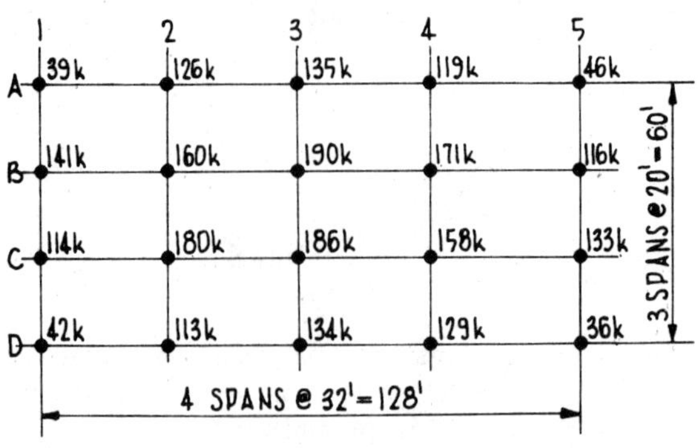

7-7. For the building loads of Problem 7-6, proportion the footing sizes to produce equal settlements, but without exceeding the allowable soil pressure of 3000 psf.

7-8. For the building loads of Problem 7-6, allow as much as a 50% increase in bearing pressure under the smaller loads (loads less than 50 kips) and as much as a 25% increase under the medium loads (loads between 50 and 145 kips). Compare the resulting settlements between footings for these sizes.

7-9. A column load of 120 kips occurs 12 ft away from a column load of 36 kips. Select the proportions for a combined footing to carry the two columns. Maximum allowable soil pressure is limited by settlements to 2500 psf.

7-10. A column in a downtown building has a design load of 54 kips occurring 1 ft from the property line. The nearest adjacent interior column is 16 ft away, carrying a design load of 161 kips. Allowable soil pressure is limited by settlements to 2000 psf. Select the proportions for a combined footing to carry these columns.

7-11. Six columns in a state capitol building are equally spaced in a circle 36 ft in diameter to carry the heavy concrete dome over the entry area. The design load on each column is 164 kips. Select the proportions for a hexagonal combined footing to carry these six column loads. Soil is a medium sand; pressure is limited by stringent settlement criteria to 1600 psf.

7-12. Two bearing walls are 8 ft apart, forming a long corridor. One wall carries a design load of 18 kips/ft, the other a design load of 7 kips/ft. Soil pressure is limited by settlement criteria to 1500 psf. Select the proportions for a combined strip footing to carry the two walls and find the required locations of the centerlines of the two walls on the strip footing.

7-13. A long warehouse building 48 ft wide has bearing walls around its perimeter, carrying 24 kips/ft. A line of columns runs the full length of the building along its centerline, spaced at 16 ft on center and carrying a load of 476 kips each. Allowable soil pressure is 4000 psf, based on settlement criteria. Select a width for the strip footings under the walls and select the size of the spread footings under the columns such that all settlements will be roughly equal.

7-14. For the warehouse of Problem 7-13, use a grade beam to carry the line of columns and select footing sizes such that settlements of the line of columns will be roughly equal to the settlements of the perimeter walls.

7-15. For the warehouse of Problem 13, use square spread footings placed at 16 ft intervals along the wall line to support the wall, assuming that the wall is to be carried on a tie beam spanning from footing to footing. Select the footing sizes both for the wall and the columns such that settlements will be roughly equal.

8

Portland Cement Concrete Members

Regardless of whether a modern building is built of concrete, steel, masonry, or timber, its foundations will almost certainly be built of portland cement concrete. Once built and covered, however, the foundation will rarely ever be accessible or even visible for routine inspection; its reliability must therefore be absolutely assured at the time it is built. Although portland cement concrete has a history of reliability in such circumstances, the action of soil, water, dissolved chemicals, and stress are ever present; such effects must be reconsidered in the design of each new foundation.

Portland cement concrete is made using a coarse aggregate of stone, a fine aggregate of sand, and a cementing paste of portland cement and water. The end result is a human-made stone that can be poured into molds while plastic and allowed to harden into its final configuration. Portland cement does not need air to harden or "set"; it will harden as well under water as exposed to air.

Portland cement concrete is quite weak in tension. In most structural applications, its tensile strength is assumed to be zero and steel is provided wherever the structure will experience tension. The result is concrete reinforced for tension or, more simply, reinforced concrete.

Portland cement itself is susceptible to attack by sulfates when the sulfates are in solution and in direct contact with the concrete. Sulfate solutions commonly appear in porewater and in groundwater wherever sulfur compounds such as calcium sulfate (gypsum) occur. Special blends of portland cement are made that are resistant in limited degrees to sulfate attack.

Portland cement provides a tough effective chemical protection for its steel reinforcement as long as the alkalinity of the cement is above a certain "threshhold" level. Under some forms of chemical attack, the level of alkaline protection deteriorates, allowing the onset of corrosion of the reinforcement deep inside the concrete. The steel then corrodes, forming corrosion materials that have a much higher volume than the original steel. The concrete, being weak in tension, is unable to contain such an increase in volume and spalls off, exposing the reinforcement thereafter to unprotected direct attack.

The most common source of such an attack on reinforcement is from chlorides. Typically, the chlorides may come from sodium chloride (salt) in seawater or from deicing salts (calcium chloride) used to clear ice from streets and structures in the winter. Wherever chlorides are present, a distinct hazard to the reinforcement exists, although chlorides seem to have little deleterious effect on the portland cement concrete itself.

These two forms of attack, sulfate and chloride, are two of the most common hazards to reinforced concrete foundations. Unfortunately, the chemistry that makes portland cement resistant to sulfate attack also serves to reduce the natural threshold protection that it affords its reinforcement. Groundwater or soil moisture that contains both gypsum and salt can therefore be particularly devastating to a foundation; potential solutions to this double-barreled attack are discussed later in this chapter.

PORTLAND CEMENT

Portland cement is a hydraulic cement, developed in 1824 by an English bricklayer named Joseph Aspdin. The name "portland" was applied because the color of the finished concrete resembled that of a natural stone then being quarried on the Isle of Portland, near Dorset, England. Portland cement is made from almost universally available materials, clay and limestone, with a small amount of gypsum added to control setting time.

Five standard types of portland cement are manufactured:

Type I: regular portland cement
Type II: increased resistance to sulfate attack
Type III: high early strength
Type IV: lowered heat of hydration
Type V: high resistance to sulfate attack

Types I, II, and IV in the foregoing list have already been mentioned. Type III cement achieves its specified strength in 7 days rather than in the 28 days required by other types, but it does so at a penalty in increased cost and at a higher heat of hydration. All five types develop heat as a by-product of hydration; type IV cement produces considerably less heat of hydration than the other types.

Heat of hydration is not usually a problem in small castings where the heat can be readily dissipated. For extremely large castings, however, such as heavy equipment foundations, the trapped heat can pose serious problems. Any casting in which heat must travel more than about 2 ft to the nearest cooling surface can produce problems with heat dissipation.

STRENGTH

As with all other structural materials, the strength of the concrete forms the basis of all structural calculations. The design strength is set at a specified minimum value; the design calculations are then based on a concrete having at least that much strength. When the concrete is being cast, samples of the concrete are taken; they are subsequently tested at 28 days to assure that the specified strength was in fact attained.

The strength of concrete is the maximum or ultimate compressive stress attained when a sample is tested to failure at 28 days. A typical strength–time curve for concrete is shown in Fig. 8-1. It should be noted that concrete typically attains roughly 60% of its 28-day strength at about 7 days, a useful bit of information to know when forms are to be stripped or precast elements are to be lifted.

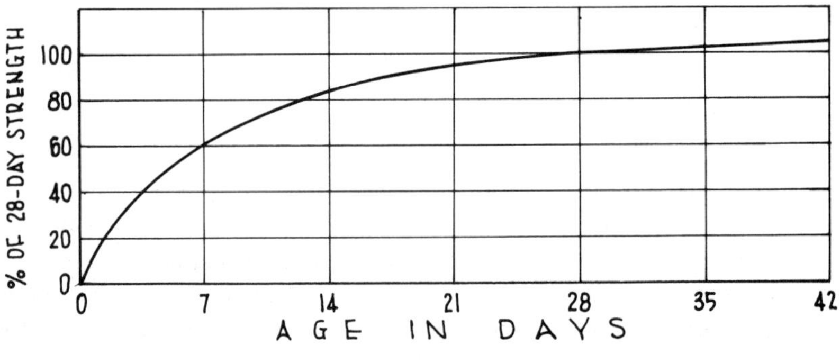

Figure 8-1 Strength gain in concrete.

For foundations, it has been customary in the past to use lower strengths of concrete since the roughly formed castings may be rather wasteful of concrete. More recent experience indicates that low-strength concretes (2000 psi or 2500 psi) can be rather short-lived. Concrete strengths of 4000 psi and even more are now being specified for foundations, particularly where there is danger of sulfate or chloride attack.

The strength of concrete is dependent on the number of pounds of water used per pound of cement; this ratio is called the "water–cement ratio" of the mix. An excess of water will not only reduce strength, it will also reduce hardness, durability, resistance to chemical attack, resistance to freeze–thaw, and all other desirable properties of the concrete. Thus the construction worker who adds water to the mix to make it flow more readily has, by this single act, significantly reduced all the desirable characteristics of the finished concrete.

In introductory comments such as these, it is difficult to overstate the importance of the water–cement ratio. As implied in Fig. 8-2, there is a minimum water–cement ratio required for hydration of all the cement molecules. Any water in the mix in excess of that amount will reduce the final strength of the concrete.

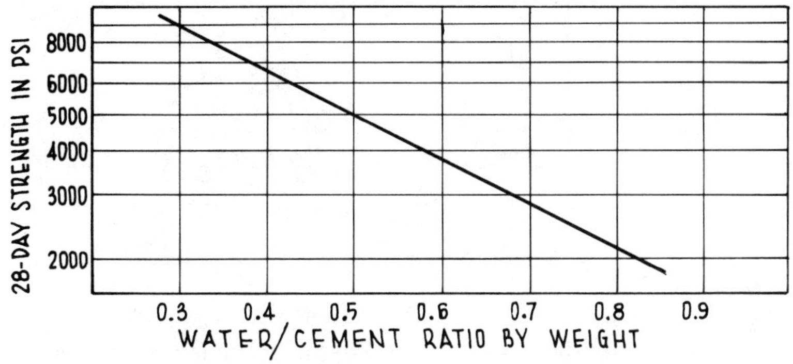

Figure 8-2 Concrete strength.

Almost all concrete used in today's industry will have water–cement ratios between 0.40 and 0.80. At higher water–cement ratios workability is improved but shrinkage becomes a problem. At lower water–cement ratios the shrinkage is less serious but the mix becomes more expensive and harder to work. In this book, only the more common water–cement ratios (between 0.4 and 0.8) are considered.

STRESS–STRAIN RELATIONSHIPS

An exaggerated stress–strain curve for concrete is shown in Fig. 8-3. An approximation of the initial slope of this curve, shown as line *OA*, is defined as the modulus of elasticity of the material; its value is the stress divided by the strain at any point. The modulus of elasticity of concrete in pounds per square inch (the slope of line *OA*) can be computed empirically by the relationship

$$E_c = 57{,}000\sqrt{f'_c} \tag{8-1}$$

where f'_c is the ultimate strength of the concrete in pounds per square inch.

Since the actual stress–strain curve of Fig. 8-3 has no well-defined yield point, an idealized curve *OAB* is substituted. This idealized curve is used in all subsequent derivations. The position of the line *AB* has been found to provide the best correlation to comprehensive test data when taken at $0.85f'_c$. It should be

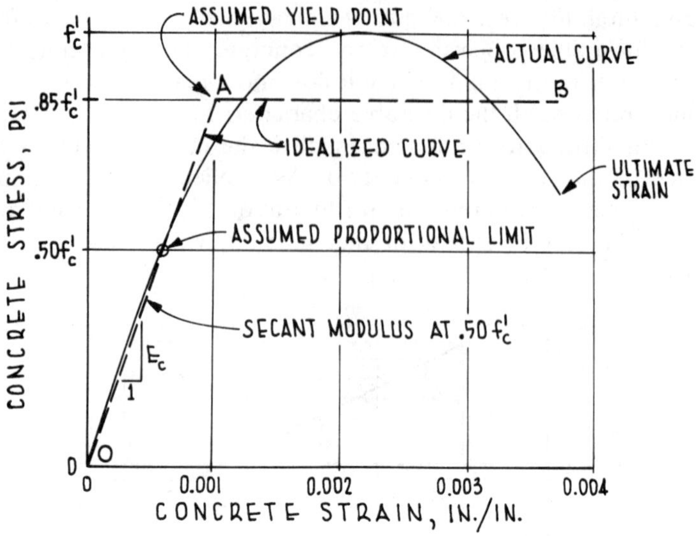

Figure 8-3 Idealized stress–strain curve for concrete.

recognized that the curve of Fig. 8-3 had to be exaggerated to show the relationships between the two curves; the actual differences are not so extreme.

Despite the apparent simplicity of the concept, the idealized curve *OAB* in Fig. 8-3 is a very important feature of the stress analysis. It will be referenced repeatedly in the derivations of the flexure formula in subsequent sections. The idealization (or approximation) of the actual curve into the straight-line segments shown as *OAB* will serve to simplify the flexural analysis considerably with little loss in accuracy.

REINFORCEMENT

Reinforcement is manufactured under several international specifications for a variety of yield stresses, or grades. In this book, consideration is limited to reinforcement grades 40, 50 and 60. For these grades, the yield stress is 40,000, 50,000, and 60,000 psi, respectively. For all grades of steel, the modulus of elasticity, E_s, is 29,000,000 psi.

Reinforcement is rolled in round bars, having deformed surfaces designed to improve the connection to the adjacent concrete (see Fig. 8-4). In the United States, the identifying number of a reinforcing bar, through a No. 8 bar, indicates its diameter in eighths of an inch; thus a No. 5 bar has a nominal diameter of $\frac{5}{8}$ in. Bar sizes Nos. 9, 10, and 11 are also round, but their diameters are set to provide the same areas as older square bars: 1, $1\frac{1}{8}$, and $1\frac{1}{4}$ in., respectively. A No. 9 bar has a diameter 1.128 in., a No. 10 bar has a diameter 1.27 in., and a No. 11 bar has a diameter 1.41 in. Bar sizes larger than No. 11 are not included in this book. Table A-2 of the Appendix lists the sizes and cross-sectional properties of rein-

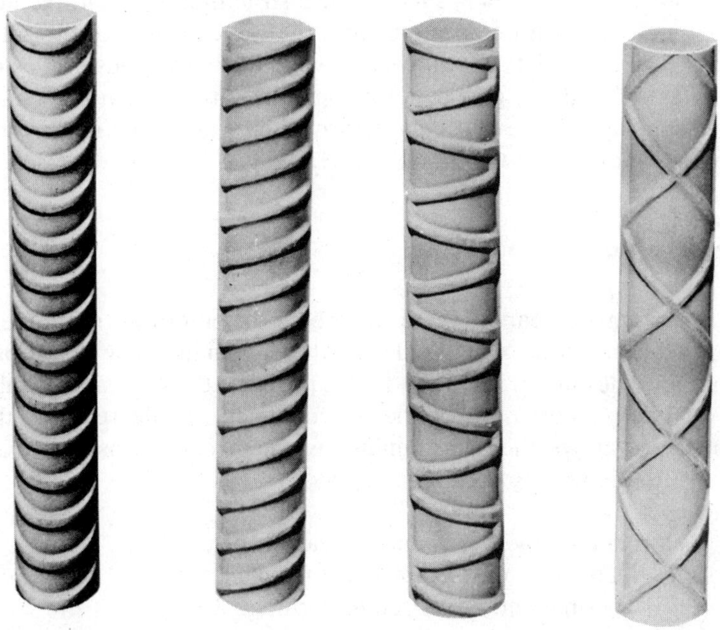

Figure 8-4 Deformed reinforcing bars. (Courtesy of the Concrete Reinforcing Steel Institute.)

forcing bars. The tables are limited to the most commonly used sizes, Nos. 3 through 11. Bar sizes less than No. 3 are not used structurally, and bar sizes greater than No. 11 are used primarily in heavy construction.

Stress–strain curves for steel are treated in detail in elementary strength of materials. For its applications in reinforced concrete, an idealized curve similar to that of concrete (Fig. 8-3) is required. Such idealized curves for steel are shown in Fig. 8-5, with the complete stress–strain curve shown alongside for

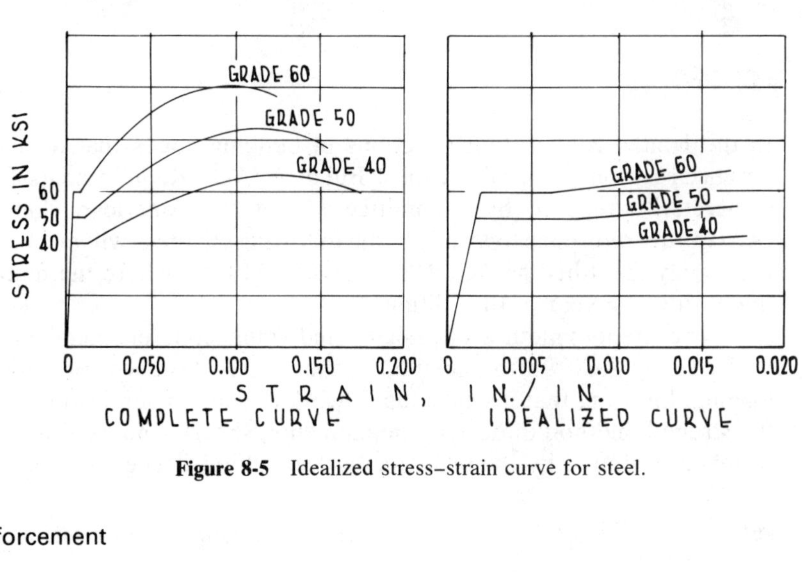

Figure 8-5 Idealized stress–strain curve for steel.

reference. Since steel is such an ideal structural material, the idealized curves of Fig. 8-5 are but little different from the same portion of the actual curve.

Table A-1 lists the allowable stress for the various grades of reinforcement. Tables A-2 and A-3 give the cross-sectional areas of various combinations of bars and spacings. Note that only deformed bars are listed in the tables; smooth bars are not included in this book.

MINIMUM COVER

The amount of clear concrete cover over the outermost reinforcement is governed by the design code, discussed at further length in the next section. It is this cover that provides fire protection. It also prevents the intrusion of salts or oxygen that would cause corrosion of the reinforcement. Code requires that the cover be increased above the code minimums where conditions are severe. For foundations, the following minimums apply.

Minimum cover for concrete cast in place:
- Concrete cast against earth and permanently exposed to earth — 3 in.
- Concrete cast against forms but subsequently exposed to salt water or earth or weather:
  - (a) Bars larger than No. 5 — 2 in.
  - (b) Bars No. 5 and smaller — 1 in.

The foregoing minimum values for cover are not overly generous. Where chlorides or salt in the groundwater (or soil porewater) are known to be present, cover should be increased at least 50% over the given values. Where both sulfates and chlorides exist, an unreinforced foundation might well be considered, but if reinforcement is found to be essential, the foregoing values for concrete cover should be doubled.

DESIGN CODES

In the United States, those sections of building codes that deal with concrete design are usually based on the "Building Code Requirements for Reinforced Concrete," prepared by Committee 318 of the American Concrete Institute (ACI). Any reference to a code in the following chapters will mean this code. It is commonly identified as ACI 318-77 or ACI 318-83, where the number following the dash is the year of the edition.

The strength method of design of individual concrete members is the method preferred by ACI. In this method, a member is designed for a high "ultimate" loading, knowing that it will actually work at some much lower service level. In the strength method of design, the moments, shears, and axial forces acting on a member are found by the usual analytical methods. They are then multiplied by appropriate "load factors" and combined together to produce peak "ultimate" loads. The size of the member and its reinforcement are then proportioned to sustain these ultimate loads at the ultimate strength of the section.

At its ultimate load, a member is assumed to have reached its ultimate strength, or highest computed strength. All materials are allowed to be in full yield. In establishing the ultimate load, the multipliers, or load factors, are carefully set such that under day-to-day service loads the stresses in the member will be down at some acceptable level. ACI does not require any further check on these day-to-day service stresses, nor is there any limit placed on the magnitude of these service stresses. Nonetheless, it is prudent practice to keep the service stresses in concrete down to about half the ultimate strength of the concrete wherever possible.

In using the strength method, the moments, shears, and axial forces acting on the member are computed by the usual elastic methods that would be used for any other structural material. It is only when selecting the size of the concrete member and its reinforcement that the ultimate loads are used. The strength method is therefore seen to be simply a means of sizing and reinforcing a member, not a general method of structural analysis and design.

LOAD FACTORS AND STRENGTH REDUCTION FACTORS

In the strength method for the design of concrete, two factors are used to project the loads from their service levels up to their ultimate levels. The first of these is the "load factors," already mentioned. The second is the "strength reduction factor," which recognizes that construction tolerances in concrete can in fact accumulate, causing an overall reduction in the strength of a member.

The load factor for dead load is 1.4, for live load is 1.7, and for wind or earthquake is 1.7. These factors are prescribed by code and are not subject to individual interpretation. The lower factor of 1.4 that is applied to dead load is considered to be adequate, since dead load can be determined much more accurately than the other loads.

When combined with the load combinations of Eqs. (5-4a) and (5-4b), the total factored moment M_u for flexural design becomes

$$M_u = 1.4M_{\mathrm{DL}} + 1.7M_{\mathrm{LL}} \tag{8-2a}$$

$$M_u = 0.75[1.4M_{\mathrm{DL}} + 1.7M_{\mathrm{LL}} + 1.7M_{\mathrm{WIND}} \text{ (or } M_{\mathrm{EQ}})] \tag{8-2b}$$

where the subscripts DL, LL, WIND, and EQ refer to dead load, live load, wind, and earthquake, respectively.

For minimum loads or stress reversals, the factored load combination of Eq. (5-4c) is given separately by code:

$$M_u = 0.9M_{\mathrm{DL}} + 1.3M_{\mathrm{WIND}} \text{ (or } 1.4M_{\mathrm{EQ}}) \tag{8-2c}$$

The same factors apply, of course, for factored beam shear V_u:

$$V_u = 1.4V_{\mathrm{DL}} + 1.7V_{\mathrm{LL}} \tag{8-2d}$$

$$V_u = 0.75[1.4V_{\text{DL}} + 1.7V_{\text{LL}} + 1.7V_{\text{WIND}} \text{ (or } V_{\text{EQ}})] \tag{8-2e}$$

$$V_u = 0.9V + 1.3V_{\text{WIND}} \text{ (or } 1.4V_{\text{EQ}}) \tag{8-2f}$$

It is useful at times to have some idea of a single overall "average" load factor in the design rather than the multiplicity of factors listed in Eqs. (8-2). Taking the dead load in a concrete structure to be about 60% of the total working load and live load to be the remaining 40%, Eq. (8-2a) yields the simplified equation $M_u = 1.52M_w$, where M_w is the total working load. Equation (8-2d) will, of course, yield the same overall factor of 1.52 for shear. Although only a general guide, this overall factor of 1.52 will be useful in checking calculations.

In addition to the foregoing load factors, code requires that the computed ultimate strength of concrete members be reduced somewhat to account for any accidental buildup in tolerances. The amount of the reduction varies, depending on the type of loading; a higher reduction is required for columns than for beams. The following strength reduction factors, ϕ, apply to the indicated type of loading at ultimate capacity:

Beam flexure	$\phi = 0.90$
Axial tension, with or without flexure	$\phi = 0.90$
Axial compression, with or without flexure, square or rectangular tied members	$\phi = 0.70$
Shear or torsion	$\phi = 0.85$
Bearing	$\phi = 0.70$

The final nominal ultimate loads to be carried by a structural member are then computed as

$$M_n = \frac{M_u}{\phi} \quad \text{and} \quad V_n = \frac{V_u}{\phi} \tag{8-3a,b}$$

where M_u and V_u are the maximum values found by Eqs. (8-2). These values of M_n and V_n computed by Eqs. (8-3) are the final values to be used in the design of the concrete members. They are the nominal ultimate loads to be carried by the concrete section. The following examples will illustrate the computation of M_n and V_n for some typical cases.

EXAMPLE 8-1: Computation of Nominal Ultimate Loads

Concrete member in flexure.
No wind or earthquake loads occur.

Given:

$M_{\text{DL}} = 60$ kip-ft, $V_{\text{DL}} = 38$ kips
$M_{\text{LL}} = 44$ kip-ft, $V_{\text{LL}} = 29$ kips

Find:

Nominal ultimate loads M_n and V_n on the member

SOLUTION Since there are no lateral loads, there is only one load case, given by Eqs. (8-2a) and (8-2c):

$$M_n = \frac{1.4M_{\mathrm{DL}} + 1.7M_{\mathrm{LL}}}{\phi}$$

$$V_n = \frac{1.4V_{\mathrm{DL}} + 1.7V_{\mathrm{LL}}}{\phi}$$

The appropriate loads are substituted, with $\phi = 0.90$ for moment and $\phi = 0.85$ for shear, yielding

$$M_n = \frac{1.4 \times 60 + 1.7 \times 44}{0.90} = 176 \text{ kip-ft}$$

$$V_n = \frac{1.4 \times 38 + 1.7 \times 29}{0.85} = 58 \text{ kips}$$

These computed values of M_n and V_n are the values to be used for design of the member.

EXAMPLE 8-2: Computation of Nominal Ultimate Load

Concrete member in flexure.
Wind and earthquake loads included.

Given:

M_{DL}	= 86 kip-ft,	V_{DL}	= 40 kips
M_{LL}	= 77 kip-ft,	V_{LL}	= 27 kips
M_{WIND}	= 36 kip-ft,	V_{WIND}	= 12 kips
M_{EQ}	= 48 kip-ft,	V_{EQ}	= 18 kips

Find:

Nominal ultimate loads M_n and V_n, and minimum loads M_n and V_n

SOLUTION The load combinations of Eqs. (8-2) will apply. For maximum values, the earthquake values are seen to be greater than those due to wind and therefore will govern the maximum values of M_n and V_n. For minimum values of M_n and V_n, however, wind will govern. For nominal ultimate moment, see Eqs. (8-2a) through (8-2c):

$$M_n = \frac{1.4M_{\mathrm{DL}} + 1.7M_{\mathrm{LL}}}{\phi}$$

$$M_n = \frac{0.75(1.4M_{\mathrm{DL}} + 1.7M_{\mathrm{LL}} + 1.9M_{\mathrm{EQ}})}{\phi}$$

$$M_n = \frac{0.9M_{\mathrm{DL}} + 1.3M_{\mathrm{WIND}}}{\phi}$$

For nominal ultimate shear, see Eqs. (8-2d) through (8-2f):

$$V_n = \frac{1.4 M_{DL} + 1.7 V_{LL}}{\phi}$$

$$V_n = \frac{0.75(1.4 V_{DL} + 1.7 V_{LL} + 1.9 V_{EQ})}{\phi}$$

$$V_n = \frac{0.9 M_{DL} + 1.3 M_{WIND}}{\phi}$$

The appropriate values of moment and shear are substituted, with $\phi = 0.90$ for flexure and $\phi = 0.85$ for shear.

$$M_n = \frac{1.4 \times 86 + 1.7 \times 77}{0.9} = 279 \text{ kip-ft}$$

$$M_n = \frac{0.75(1.4 \times 86 + 1.7 \times 77 + 1.9 \times 48)}{0.9} = 285 \text{ kip-ft}$$

$$M_n = \frac{0.9 \times 86 + 1.3 \times 36}{0.9} = 138 \text{ kip-ft}$$

$$V_n = \frac{1.4 \times 40 + 1.7 \times 27}{0.85} = 120 \text{ kips}$$

$$V_n = \frac{0.75(1.4 \times 40 + 1.7 \times 27 + 1.9 \times 18)}{0.85} = 120 \text{ kips}$$

$$V_n = \frac{0.9 \times 40 + 1.3 \times 12}{0.85} = 61 \text{ kips}$$

By inspection of the foregoing results the design values for M_n and V_n are selected:

At maximum levels: $M_n = 285$ kip-ft; $V_n = 120$ kips
At minimum levels: $M_n = 138$ kip-ft; $V_n = 61$ kips

USE OF SECTION MODULUS

The conventional approach to the design of structural sections is to use the flexure formula to compute the required section modulus. The flexure formula is, of course, well known from strength of materials. For stresses at the compressive side of the beam, the flexure formula becomes

$$f_c = \frac{M_w}{S_c} \tag{8-4a}$$

where M_w = working moment on the section
f_c = compressive stress
S_c = section modulus to the compressive side of the section

Rewritten in its design form, the flexure formula is

$$\frac{M_w}{f_c} = S_c \tag{8-4b}$$

In routine use, the values of M_w and f_c would be known; the required magnitude of the section modulus can be readily determined from Eq. (8-4b). A member having at least this magnitude of S_c will be adequate to carry the moment M_w at the stress f_c.

For design in steel, the section modulus of a steel section is rarely computed; it is almost always found by looking it up in the steel manual [36]. For design in timber, the section modulus may be computed ($S_c = 0.1667\,bh^2$) if one happens to remember the actual finish dimensions b and h for timber, but it, too, is usually looked up in a section modulus table for standard timber sizes. For design of concrete sections, the procedure is no different; the section modulus is looked up in a standard table and is used in exactly the same way.

In the following sections, the section modulus of a concrete section is derived, first at ultimate loads and later at service levels. The section can be designed, therefore, for either condition. For both cases, the flexure formula forms the basis of the computations.

FLEXURE AT ULTIMATE LOAD

The preceding sections give the nominal ultimate load that is to be sustained by a concrete section when it is loaded to its ultimate capacity. The ultimate capacity of a concrete section is the highest possible resistance to load that the section can ever offer, with both concrete and steel fully yielded. The computation of this ultimate capacity, or ultimate strength, is developed in the following discussions.

For computing ultimate strength, it is required by code that the maximum strain in the concrete be taken at 0.003 in./in. It is also required by code that strains shall be assumed to vary linearly across the section. These requirements are developed from extensive experimental data on concrete sections at failure and form the basis of the strength method of design. A sketch illustrating these requirements is presented in Fig. 8-6.

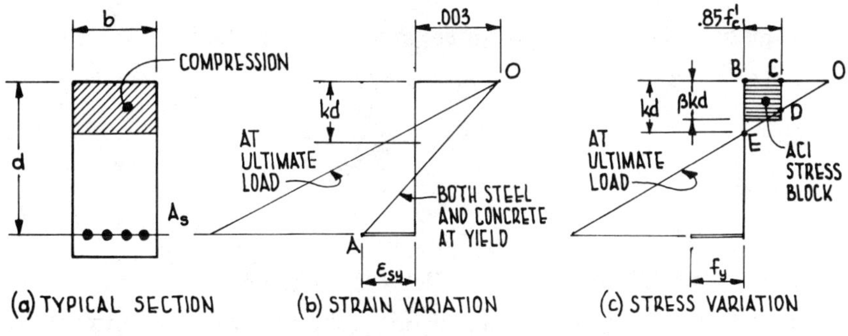

Figure 8-6 Ultimate conditions.

It bears repeating at this point that concrete has no strength in tension. In the flexural analysis, therefore, wherever the concrete enters tension it is assumed to crack; the stress immediately falls to zero. Such a condition is shown in the stress diagram of Fig. 8-6c.

It is important to recognize the stress–strain relationships at ultimate load that are indicated in Fig. 8-6. As the section is loaded, it first rotates elastically, with all stresses in the elastic range. As loads increase, the stresses in both the concrete and the steel enter yield, as shown by line *OA* in Fig. 8-6b. As rotations continue beyond the yield point of the materials, the stress cannot increase further even though strains continue to increase; the final result is the trapezoidal stress block *BCDE* for the concrete, with the stress in the steel at its yield value f_y.

The yield stress in concrete is taken at $0.85f'_c$, as shown earlier in the idealized curve *OAB* of Fig. 8-3. The variation in stress across the compressive area would actually be the trapezoidal stress block *BCDE*, but it does not matter, since code also specifies an empirical stress block to be used in computations. The ACI stress block is shown hatched in Fig. 8-6, having a height βkd; the value of β is specified by code:

$$\beta = 0.85 - 0.05(f'_c - 4000) \qquad f'_c \text{ in psi} \tag{8-5}$$

In addition, β cannot be less than 0.65 or more than 0.85.

A very useful parameter in reinforced concrete is the steel ratio ρ, where

$$\rho = \frac{A_s}{bd} \tag{8-6}$$

and all symbols are those shown in Fig. 8-6. From Fig. 8-6, the sum of horizontal forces yields an expression for k, where, with $A_s = \rho bd$,

$$\sum H = 0 = 0.85f'_c b\beta kd - f_y \rho bd \tag{8-7a}$$

Hence

$$k = \frac{f_y \rho}{0.85f'_c \beta} \tag{8-7b}$$

The sum of moments about the tensile steel yields

$$M_n = 0.85f'_c b\beta kd(d - \tfrac{1}{2}\beta kd) \tag{8-8}$$

Equation (8-7b) is substituted and the result solved for the plastic section modulus Z_c,

$$\frac{M_n}{0.85f'_c} = Z_c \tag{8-9a}$$

where

$$Z_c = \frac{f_y \rho}{0.85f'_c}\left(1 - \frac{\frac{1}{2}f_y \rho}{0.85f'_c}\right) bd^2 \tag{8-9b}$$

Equation (8-9a) is directly analogous to the elastic flexure formula, given earlier as Eq. (8-4b), and can be used in exactly the same way. At ultimate load, Eq. (8-9b) is used to obtain the required plastic section modulus Z_c. Z_c is a property of the section and may be computed and tabulated for immediate reference.

Such a tabulation for Z_c given by Eq. (8-9b) is given in Table A-4 for various values of f'_c, f_y, and ρ. The table includes the two most common grades of steel, 40 and 60. Values of steel ratio ρ shown in the table include all the values that are allowed by code.

The minimum value of steel ratio ρ that may be used in flexural members is prescribed by code:

$$\rho_{\min} = \frac{200}{f_y} \tag{8-10}$$

Table A-4 contains no entries below this value.

The maximum value of steel ratio ρ that may be used is also prescribed by code. It is 75% of the "balanced" steel ratio, where the balanced steel ratio is defined by code as

$$\rho_b = \frac{0.85 f'_c}{f_y} \beta \frac{87{,}000}{87{,}000 + f_y} \qquad f_y \text{ in psi} \tag{8-11}$$

Table A-4 contains no entries above 75% of this value.

In addition to the minimum A_s given by Eq. (8-10) for flexural reinforcement, code requires that concrete be reinforced to control any temperature effects or shrinkage that might occur. This requirement does apply to slabs or beams that are subject to flexure in one direction only; temperature steel must be placed in the other direction even though there is no computed stress. The minimum area of steel A_s for such temperature steel is $0.0020bt$ when grade 40 or grade 50 steel is used and $0.0018bt$ when grade 60 steel is used, where t is the overall thickness of the member.

When a beam is designed using the strength method, there is no requirement to check the beam for its day-to-day service stresses. For those who care to keep track of these stresses, however, an approximate value of service stress is included in the tables. The column headed "svc f_c" lists an approximate level of service stresses to be expected, listed first in newtons/mm^2, then in pounds/in.^{2}.

This rough check is readily derived. It was noted earlier that an average overall load factor for concrete members is about 1.52. With a value of ϕ of 0.9 for flexure,

$$M_n = \frac{1.52 M_w}{0.9} = 1.7 M_w \text{ (average)} \tag{8-12}$$

The values of M_n and M_w are found from Eqs. (8-4a) and (8-9a) and substituted, with the final result

$$f_c = \frac{1}{2} f'_c \frac{Z_c}{S_c} \tag{8-13}$$

This value of f_c is the value tabulated in Table A-4 under the column heading "svc f_c." The value of S_c used in Eq. (8-13) is the elastic section modulus; the derivation of S_c is presented in a later section with the working stress analysis.

It is good practice to keep this "svc f_c" down to about half the ultimate strength of concrete when designing flexural members, even though the code does not stipulate any such limits on service stresses. Although only a rough estimate, the tabulated value of "svc f_c" is a guide in keeping day-to-day service stresses down to prudent levels.

The following examples will illustrate the use of the tables. For the sake of simplicity, the examples are limited to flexural members subject only to dead and live loads; lateral loads are omitted. The load factors are then 1.4 for dead loads and 1.7 for live loads. The strength reduction factor for flexure is 0.9.

When the dimensions b and d are selected for a concrete member, the dimension b should be about $0.6d$ for an economical section. This ratio is not at all critical, and may vary widely where other considerations affect the size of b or of d. However, using $b = 0.6d$ does afford a good starting point for selecting the size of a member when there are no limitations on dimensions.

EXAMPLE 8-3: Strength Method of Design of Concrete Members

Design of a rectangular section in flexure.
No limitations on dimensions.

Given:

$M_{DL} = 36$ kip-ft
$M_{LL} = 42$ kip-ft
Load factors 1.4 for DL
and 1.7 for LL; $\phi = 0.9$
$f'_c = 3000$ psi
Grade 60 steel
Minimum steel cover 1 in.

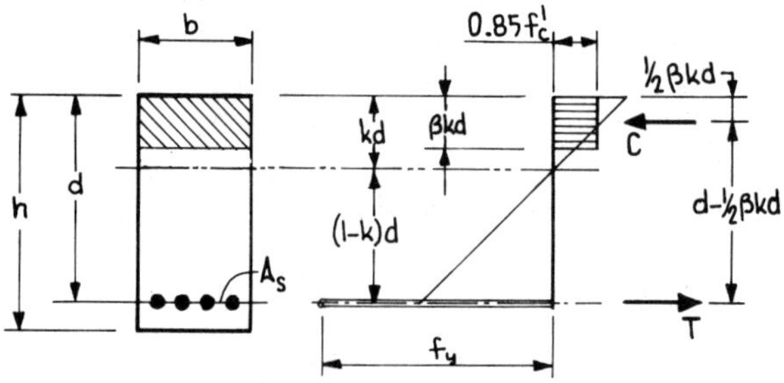

SOLUTION Compute the nominal ultimate moment M_n:

$$M_n = \frac{M_u}{\phi} = \frac{1.4M_{DL} + 1.7M_{LL}}{\phi} = \frac{1.4 \times 36 + 1.7 \times 42}{0.9}$$

$$= 122 \text{ kip-ft}$$

Enter Table A-4 and keep "svc f_c" less than half the ultimate strength of concrete. Select

$$\rho = 0.006 \quad \text{and} \quad Z_c = 0.131\,bd^2$$

Apply the flexure formula, assuming that $b = 0.6d$,

$$\frac{M_n}{0.85f'_c} = Z_c \qquad \frac{122 \times 10^3 \times 12}{0.85 \times 3000} = 0.131bd^2 = 0.131 \times 0.6d \times d^2$$

Solve for d:

$$d = 18.4 \text{ in.}, \ b = 11.0 \text{ in.}$$

Compute the required area of steel:

$$A_s = \rho bd = 0.006 \times 11 \times 18.4 = 1.21 \text{ in.}^2$$

From Table A-3, select A_s = 4 No. 5 bars, $A_s = 1.23$ in.^{2}. Dimensions are rounded, usually to the nearest inch. Use $d = 19$ in., $b = 11$ in., A_s = 4 No. 5 bars, $h = 21$ in. Note that the overall dimension h does not enter directly into the calculations but is set at some even value once d is known, keeping the required concrete cover over the steel at least 1 in.

It should be emphasized that in selecting section constants from the tables of the Appendix, all sections listed in the tables are usable and are within all requirements of ACI. It is simply good practice to keep the service stresses at about half the ultimate strength. Nonetheless, the choice of a section from the table having a steel ratio of 1.6% and its accompanying service stress of 2600 psi would have been acceptable under the code.

It should be recognized that the procedure used in the foregoing example is directly parallel to that used for design of steel or timber members. The concept of using the flexure formula to find a section modulus is retained, even though the section is at ultimate load. The same general procedure still applies when the depth of the section is limited, as illustrated in the next example.

EXAMPLE 8-4: Strength Method of Design of Concrete Members

Design of a rectangular section in flexure.
Depth of section limited.

Given:

M_{DL} = 37 kip-ft
M_{LL} = 25 kip-ft
Load factors 1.4 for DL
and 1.7 for LL; $\phi = 0.9$
Grade 40 steel
$f'_c = 4000$ psi
d limited to 10 in.

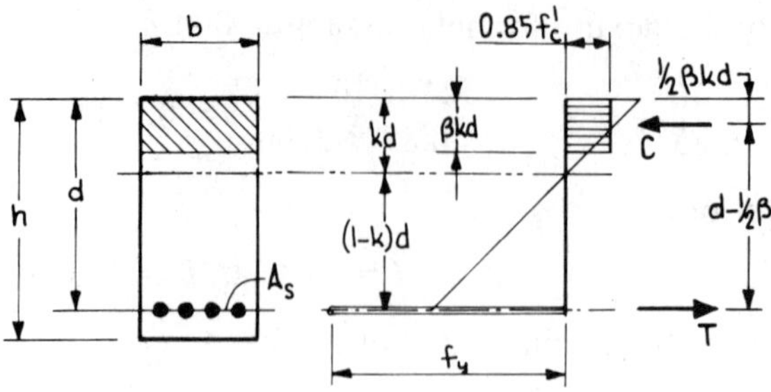

SOLUTION Compute the nominal ultimate moment M_n:

$$M_n = \frac{M_u}{\phi} = \frac{1.4M_{\text{DL}} + 1.7M_{\text{LL}}}{\phi} = \frac{1.4 \times 37 + 1.7 \times 25}{0.9}$$

$$= 105 \text{ kip-ft}$$

Enter Table A-4 and keep the service stress in concrete about half the ultimate strength. Select

$$\rho = 0.016 \quad \text{and} \quad Z_c = 0.171bd^2$$

Apply the flexure formula, using $d = 10$ in.:

$$\frac{M_n}{0.85f'_c} = Z_c \qquad \frac{105 \times 10^3 \times 12}{0.85 \times 4000} = 0.171bd^2 = 0.171 \times b \times 10 \times 10$$

Solve for b, $b = 22$ in.

Compute the required area of reinforcement:

$$A_s = \rho bd = 0.016 \times 22 \times 10 = 3.52 \text{ in.}^2$$

From Table A-3, select A_s = 4 No. 9 bars. Use $b = 22$ in., $d = 10$ in., A_s = 4 No. 9 bars.

In foundations, the width b of a grade beam is often fixed to match the columns above. The next example illustrates the procedure when the width b is thus limited.

EXAMPLE 8-5: Strength Method of Design of Concrete Members

Design of a rectangular section in flexure.
Width of section limited.

Given:

$M_{\text{DL}} = 84$ kip-ft
$M_{\text{LL}} = 66$ kip-ft
Load factors 1.4 for DL and 1.7 for LL; $\phi = 0.9$
Grade 60 steel
$f'_c = 3000$ psi
b limited to 12 in.

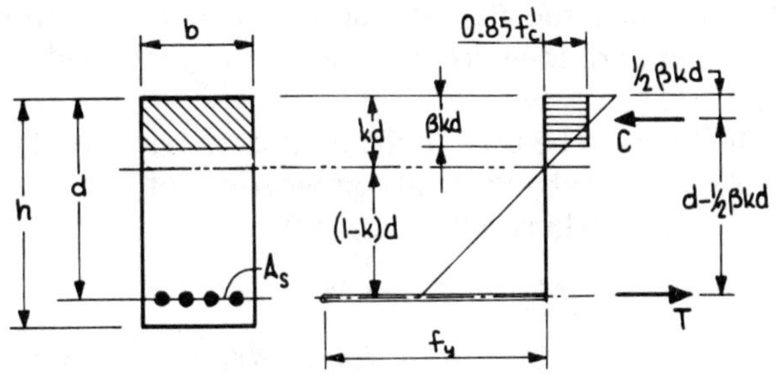

SOLUTION Compute the nominal ultimate moment M_n:

$$M_n = \frac{1.4 \times M_{\rm DL} + 1.7 M_{\rm LL}}{0.9} = \frac{1.4 \times 84 + 1.7 \times 66}{0.9}$$

$$= 255 \text{ kip-ft}$$

Enter Table A-4 and keep the service stress in concrete to about half the ultimate strength. Select

$$\rho = 0.0060 \quad \text{and} \quad Z_c = 0.131\,bd^2$$

Apply the flexure formula, using $b = 12$ in.:

$$\frac{M_n}{0.85f'_c} = Z_c \qquad \frac{255 \times 10^3 \times 12}{0.85 \times 3000} = 0.131 \times 12 \times d^2$$

Solve for d, $d = 28$ in.

Compute the required area of reinforcement:

$$A_s = \rho bd = 0.0060 \times 12 \times 28 = 2 \text{ in.}^2$$

From Table A-3, select 2 No. 9 bars. Use $b = 12$ in., $d = 28$ in., $A_s = 2$ No. 9 bars.

The use of narrow, deep beams as grade beams, such as the beam of Example 8-5, is quite common in foundation design.

FLEXURE AT WORKING LEVELS*

Design of structural concrete members using working stresses (rather than ultimate load) is also permitted by code. Called the "alternate design method," the working stress method is frequently used in foundation design. Its primary advantage is that it provides better crack control; its primary disadvantage is that it requires more steel.

In the working stress method, a maximum allowable stress is established well within the elastic range of the material. The applied loads are combined, as

* This section may be omitted in a condensed design course).

usual, to produce the highest working load the structure should ever experience. The members are then sized to carry this peak load at this maximum allowable working stress.

The loads to be used in the elastic analysis are the combined loads at working levels, as given by the load combinations of Eqs. (5-4). The design moment on the section at working levels M_w is then

$$M_w = M_{DL} + M_{LL} \tag{8-14a}$$

$$M_w = 0.75[M_{DL} + M_{LL} + M_{WIND} \text{ (or } M_{EQ})] \tag{8-14b}$$

$$M_w = M_{DL} + M_{WIND} \text{ (or } M_{EQ}) \tag{8-14c}$$

where the subscripts are those defined with Eqs. (8-2). Shears on the section at working levels are computed similarly:

$$V_w = V_{DL} + V_{LL} \tag{8-14d}$$

$$V_w = 0.75[V_{DL} + V_{LL} + V_{WIND} \text{ (or } V_{EQ})] \tag{8-14e}$$

$$V_w = V_{DL} + V_{WIND} \text{ (or } V_{EQ}) \tag{8-14f}$$

The section modulus of concrete members loaded in the elastic range can be readily derived. At working levels in concrete it is assumed, as in other materials, that strains vary linearly across the section. It is also assumed that concrete has no strength in tension; wherever concrete goes into tension, it is assumed to crack and its stress remains zero. A typical concrete section is shown in Fig. 8-7, with the stress and strain diagrams shown alongside.

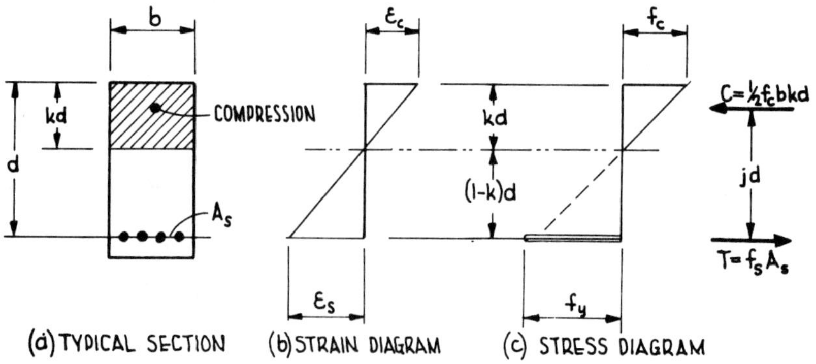

Figure 8-7 Elastic conditions.

By ratios from the strain diagram of Fig. 8-7,

$$\frac{\varepsilon_s}{\varepsilon_c} = \frac{1-k}{k} \tag{8-15}$$

By definition, $\varepsilon_s = f_s/E_s$ and $\varepsilon_c = f_c/E_c$. These values are substituted into Eq. (8-15) and the results solved for f_s, hence

$$f_s = \frac{nf_c(1 - k)}{k} \tag{8-16}$$

where $n = E_s/E_c$. Refer again to Fig. 8-7. Forces are summed horizontally, yielding

$$\sum H = 0 = \tfrac{1}{2}f_c bkd - f_s \rho bd \tag{8-17}$$

where the notation for stress ratio $\rho = A_s/bd$ is again used to denote steel areas. Equation (8-16) is substituted and the result solved for k,

$$k = \sqrt{(n\rho)^2 + 2n\rho} - n\rho \tag{8-18}$$

Moments are now summed about the neutral axis, yielding

$$M_w = \tfrac{1}{2}f_c bkd \left(\frac{2kd}{3}\right) + f_s A_s(d - kd) \tag{8-19}$$

Equation (8-16) is substituted and the result solved for the section modulus S_c,

$$\frac{M_w}{f_c} = S_c \tag{8-20a}$$

where

$$S_c = \left[\frac{k^2}{3} + n\rho\,\frac{(1 - k)^2}{k}\right] bd^2 \tag{8-20b}$$

and k is given by Eq. (8-18). The value of the section modulus given by Eq. (8-20b) is readily tabulated for various values of n, ρ, and f'_c. Such a table is presented in the Appendix (Table A-4) for concrete strengths ranging from 2000 to 5000 psi. The elastic section properties in the tables are those listed under the column heading "At Service Levels."

The moment of inertia I_0 included in Table A-4 is found from the definition of the section modulus,

$$S_c = \frac{I_0}{c} \qquad \text{hence} \quad I_0 = S_c kd \tag{8-21}$$

The ratio f_s/f_c is found from Eq. (8-16),

$$\frac{f_s}{f_c} = \frac{n(1 - k)}{k} \tag{8-22}$$

The allowable stresses in the concrete and in the steel are prescribed by code. A list of these stresses is presented in Table A-1. For working stress design, the load combinations of Eqs. (5-4) still apply, but the ultimate load factors and strength reduction factors do not. The usual procedure in using Table A-4 is to enter the table with this computed value of the stress ratio, f_s/f_c. All the section constants are then immediately available, including the steel ratio ρ.

Some examples will illustrate the working stress method for the design of reinforced concrete members.

EXAMPLE 8-6: Working Stress Method of Design of Concrete Members

Design of a rectangular section in flexure.
No limitations in dimensions, no lateral loads.

Given:
$M_{DL} = 36$ kip-ft
$M_{LL} = 42$ kip-ft
Grade 60 steel
$f'_c = 3000$ psi

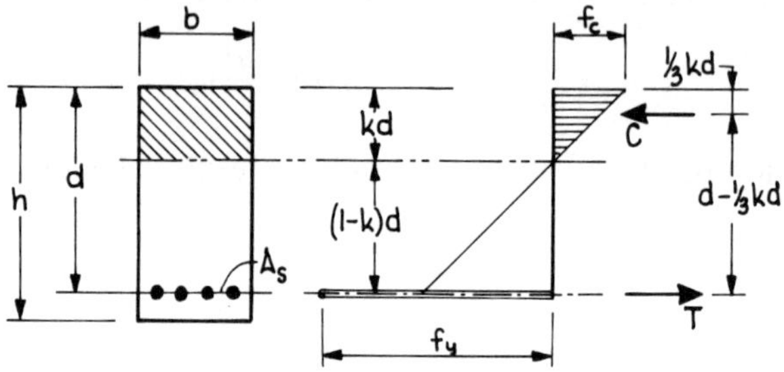

SOLUTION From Table A-1, $f_c = 0.45f'_c = 1350$ psi, $f_s = 24{,}000$ psi. Compute $f_s/f_c = 24{,}000/1350 = 17.78$. Enter Table A-4; select $\rho = 0.010$ and $S_c = 0.154bd^2$. Compute the working load:

$$M_w = M_{DL} + M_{LL} = 36 + 42 = 78 \text{ kip-ft}$$

Compute the elastic section modulus, assuming that $b = 0.6d$;

$$\frac{M_w}{f_c} = S_c \qquad \frac{78 \times 10^3 \times 12}{1350} = 693 \text{ in.}^3 = 0.154 \times 0.6d \times d^2$$

Solve for d; $d = 20$ in., $b = 12$ in.

Compute the required steel area:

$$A_s = \rho bd = 0.010 \times 20 \times 12 = 2.40 \text{ in.}^2$$

From Table A-3, select 4 No. 7 bars. Use $b = 12$ in., $d = 20$ in., $A_s = 4$ No. 7 bars.

Note that this problem is exactly that of Example 8-3, but that the end results are significantly different. It is important to recognize that the strength method and the working stress method are *not* equivalent (i.e., the two methods do not produce the same end result). The next example illustrates the procedure when the depth of the section has to be restricted for some reason. Note that it is the same as Example 8-4, but again, the end result is different from that of Example 8-4.

EXAMPLE 8-7: Working Stress Method of Design of Concrete Members

Design of a rectangular section in flexure. Depth of section limited.

Given:
M_{DL} = 37 kip-ft
M_{LL} = 25 kip-ft
Grade 40 steel
f'_c = 4000 psi
d limited to 10 in.

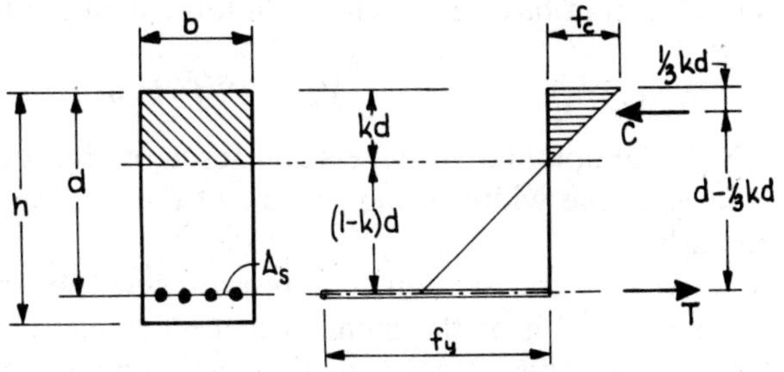

SOLUTION From Table A-1, $f_c = 0.45 f'_c = 1800$ psi, $f_s = 20{,}000$ psi. Compute $f_s/f_c = 20{,}000/1800 = 11.1$. Enter Table A-4; select $\rho = 0.0190$ and $S_c = 0.181\,bd^2$. Compute the working load:

$$M_w = M_{DL} + M_{LL} = 37 + 25 = 62 \text{ kip-ft}$$

Compute the elastic section modulus, assuming that $d = 10$ *in.*:

$$\frac{M_w}{f_c} = S_c \qquad \frac{62 \times 10^3 \times 12}{1800} = 0.181 \times b \times 10 \times 10$$

Solve for b, $b = 23$ in. (rounded).

Compute the required steel area:

$$A_s = \rho bd = 0.019 \times 23 \times 10 = 4.37 \text{ in.}^2$$

From Table A-3, select 6 No. 8 bars. Use $d = 10$ in., $b = 23$ in., $A_s = 6$ No. 8 bars.

BEAM SHEAR AT ULTIMATE LOAD

Shear in foundations does not usually require extensive computations, primarily because foundations are rarely reinforced for shear. Foundations are simply made large enough to carry the shearing forces within the capacity of concrete acting alone, and the complications introduced by shear reinforcement then do not occur. Since formwork and falsework are minimal in foundations, the cost of the increased size is one simply of bulk concrete; it is usually much cheaper to make a foundation slightly thicker than it is to add reinforcement for shear.

The procedures for designing concrete beams for shear are simple and well developed and their effectiveness has been well proven over the years. Since there are hundreds of combinations of loads that could vary the shear patterns, the design for shear has evolved into a semiempirical "blanket" method which assures that the member will be capable of carrying the extremes in shear. Although the extremes may not actually occur in all beams, code requires that all beams be capable of sustaining them.

At ultimate load, the maximum amount of shear force that can be applied to a concrete section having no shear reinforcement is prescribed by code:

$$V_n = 2\sqrt{f'_c}\, b_w d \tag{8-23}$$

where V_n is the nominal ultimate shear force on the section as computed by Eqs. (8-2) and b_w is the width of the "web" of the section; f'_c and d are the same as defined earlier.

The web of the section, b_w, is prescribed as the width of a rectangular section or the width of the stem of a tee-shaped section or the diameter of a circular section. These typical sections are shown in Fig. 8-8.

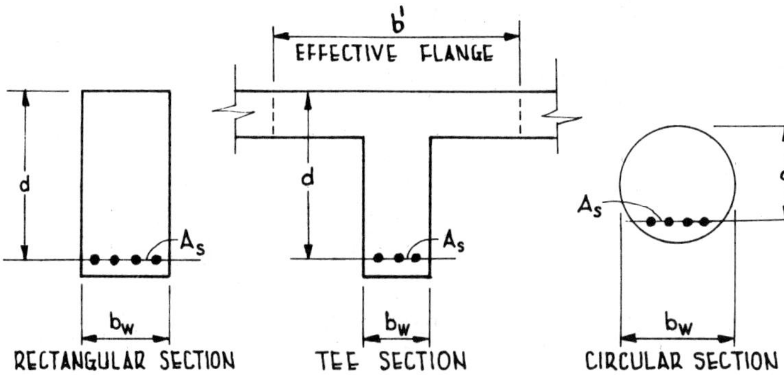

Figure 8-8 Concrete widths in shear.

The terms used in Eq. (8-23) are usually interpreted as an "average" ultimate shear stress of $2\sqrt{f'_c}$ acting across the somewhat arbitrary web area given by $b_w d$. Although this is a convenient interpretation and this "average" ultimate shear stress is a useful parameter, the code makes no such claim. The equation is simply a highly simplified but proven means to compute the maximum shear force that can be taken by a concrete section; it applies equally well to rectangular beams and slabs as well as tee shapes and circular shapes.

Patterns of shear stress become diffused near a concentrated load such as a support. Recognizing this phenomenon, code defines the critical section for shear computations as the section taken at a distance d from the face of the support, where d is the effective depth of the section. The shearing force permitted on this critical section is, of course, the combined and factored force V_n given by Eq. (8-3b); this point is illustrated in one of the following examples.

EXAMPLE 8-8: Strength Method of Design of Concrete Members

Review of beam shear.
Size of section selected by flexural design methods.

Given:
Section as shown, already designed for flexural considerations.
$V_{DL} = 8$ kips; $V_{LL} = 6$ kips
$f'_c = 3000$ psi

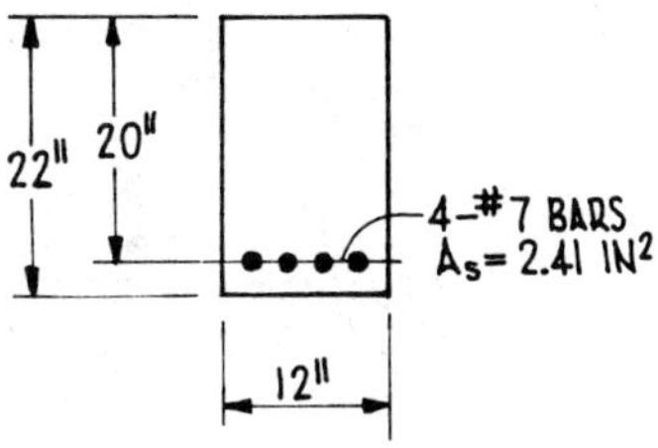

SOLUTION To determine whether this section is adequate in shear, the nominal ultimate shear force that the section can sustain is computed:

$$V_n = v_c b_w d = 2\sqrt{f'_c}\, b_w d = 2\sqrt{3000} \times 12 \times 20$$
$$= 26{,}400 \text{ lb}$$

This value is compared to the nominal ultimate value of shear to which the section is subjected:

$$V_n = \frac{V_u}{\phi} = \frac{1.4 V_{DL} + 1.7 V_{LL}}{\phi}$$
$$= \frac{1.4 \times 8 + 1.7 \times 6}{0.85} = 25 < 26.4 \text{ kips} \quad \text{(OK)}$$

Comparing the values, the section is seen to be barely adequate in shear. Had the section not been adequate, it would be necessary to increase the size of the section until it becomes adequate, then select new flexural reinforcement to suit these larger dimensions. This case occurs in the examples of Chapter 9.

EXAMPLE 8-9: Strength Method of Design of Concrete Members

Review of shear capacity of a beam.
Flexural design already completed.

Determine whether the section shown in the following sketch is adequate for shear. Reinforcement for flexure has already been selected as indicated. Shear and moment diagrams include dead load and are drawn for the nominal ultimate values of V_n and M_n; they also include the strength reduction factor ϕ. Use grade 60 steel, $f'_c = 3000$ psi.

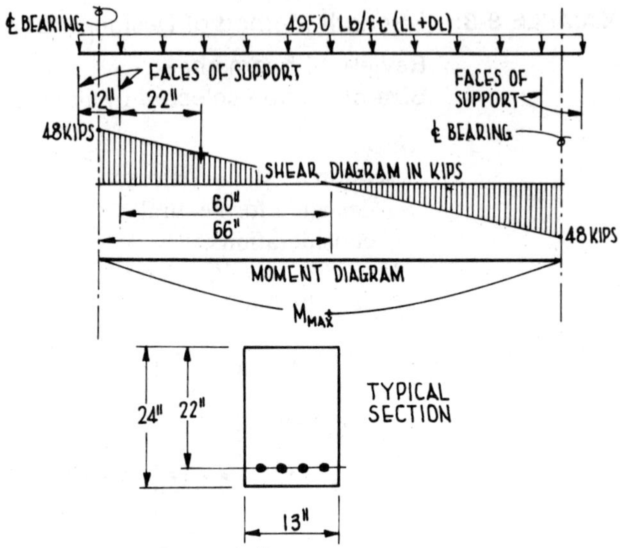

SOLUTION Compute the critical value of shear V_{cr} at a distance d from face of support, by ratio:

$$V_{cr} = \frac{60 - 22}{66} 48 = 28 \text{ kips} \quad (\text{includes } \phi)$$

Compute the shear force that the section can sustain:

$$V_c = 2\sqrt{f'_c}\, b_w d = 2\sqrt{3000} \times 13 \times 22$$
$$= 31.3 \text{ kips}$$

When V_c is compared to V_{cr}, the section is seen to be adequate in shear.

It should be noted that in neither of the two foregoing examples concerning shear did the flexural reinforcement enter the calculations. This blanket method is based very simply on an average shear stress across the section, regardless of what is happening in the section due to flexure. The only unusual point is that the critical shear force is taken at a distance d from face of support, which reduces the design value of the shear force considerably.

BEAM SHEAR AT WORKING LEVELS

Where no shear reinforcement is required, the review of sections for shear stress in the working stress method is directly parallel to that used at ultimate loads. The only differences are that the shearing force on the section is the unfactored shear given by Eq. (8-13) at working levels and that the allowable shear stress is distinctly smaller. At working levels,

$$V_w = v_w b_w d \tag{8-24}$$

where $V_w = 1.1\sqrt{f'_c}$ and all other symbols are those defined earlier. The general procedure for applying Eq. (8-24) is of course identical to that just demonstrated for applying Eq. (8-23).

BOND AND ANCHORAGE

The adhesion of concrete to its reinforcement is called *bond*. It is one of the more uncertain properties of concrete and can vary widely with little apparent cause. To reduce the dependence on theory when this uncertain property is being used, requirements for embedment of reinforcement have been established by extensive testing over a wide variety of conditions. Code requirements are based on these test results rather than on theoretical considerations.

The term *development length* is used frequently in discussing bond and anchorage. The development length of a reinforcing bar is the length that must be embedded in concrete to develop the full strength of the bar. Supposedly, a bar embedded in concrete to a depth equal to its development length will itself fail just before it is pulled out of the concrete.

The working stress method does not have a separate means for determination of development lengths. Embedment must develop ultimate strength without regard to working stress. The code therefore has only one set of requirements for both the strength method and the working stress method.

For deformed bars not larger than No. 11 and having less than 12 in. of fresh concrete below them, the development length in tension is computed by

$$L_d \geqq 0.04 A_b \frac{f_y}{\sqrt{f'_c}} \quad \begin{array}{l}\text{but not less than } 0.0004 d_b f_y \\ \text{and not less than 12 in.}\end{array} \tag{8-25}$$

where L_d = development length, in.
A_b = cross-sectional area of the bar, in.^{2}
d_b = diameter of the bar, in.

and all other symbols are those defined previously.

Horizontal bars having more than 12 in. of fresh concrete cast below them can lose a considerable amount of bond strength due to collections of air, water, and laitance at their underside. Code requires 40% more development length for such bars, called *top bars*. Development lengths for top bars in tension are therefore 1.4 times the length computed by Eq. (8-25).

Where development lengths are restricted, the end of a tension bar may be bent into a standard configuration known to improve its anchorage to the concrete. Called *hooks*, the two standard configurations prescribed by code are shown in Fig. 8-9.

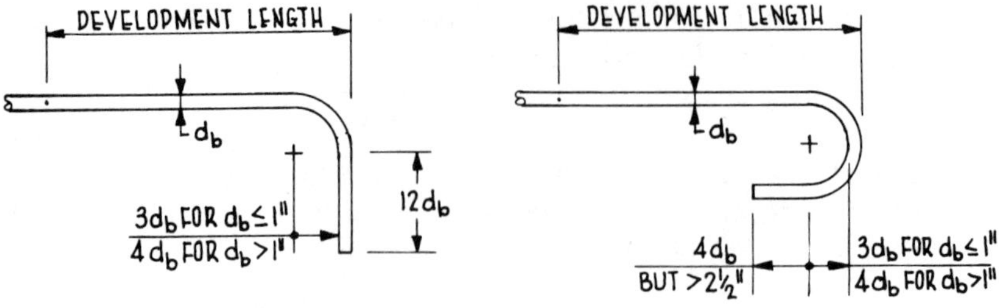

Figure 8-9 Standard tension hooks.

The development lengths for hooked bars in tension are computed by

$$L_d \geqq 0.02 d_b \frac{f_y}{\sqrt{f'_c}} \quad \text{but not less than } 8d_b \text{ and not less than 6 in.} \tag{8-26}$$

Hooks are not effective in compression.

The development lengths given by Eq. (8-25) for straight bars in tension and by Eq. (8-26) for hooked bars in tension are tabulated in Tables A-5 and A-6. It should be noted that the development length is given to the back of the hook; clear concrete cover is then computed from this point, simplifying the calculations somewhat.

For members in compression, such as columns, the development length of the reinforcement is found from similar empirical equations prescribed by code. There is no category of top bars for development lengths in compression. For bar sizes Nos. 3 through 11, development length is given by

$$L_d \geqq 0.02 d_b \frac{f_y}{\sqrt{f'_c}} \quad \text{but not less than } 0.0003 d_b f_y \text{ and not less than 8 in.} \tag{8-27}$$

Development lengths for deformed bars in compression are computed from Eq. (8-27) and are included in Table A-5.

Reinforcing bars may be spliced within a concrete member simply by lapping them a specified length. The efficiency of a lap splice is 100%; full strength is transferred. Code does not permit bars larger than No. 11 to be lap spliced except in exceptional circumstances.

Code separates lap splices in tension into three classes, depending on the amount of excess steel at a particular splice location and the percentage of bars being spliced at the location, whether 50%, 75%, or 100%. Table 8-1 defines the three classes of splices. It should be apparent from the requirements of Table 8-1

TABLE 8-1 TENSION LAP SPLICES

$\frac{A_s \text{ provided}}{A_s \text{ required}}$	Percent A_s spliced at the location		
	50%	75%	100%
≥2.0	Class A	Class A	Class B
<2.0	Class B	Class C	Class C

that class A splices rarely occur. Almost all splices fall within class B or class C, and even class B splices usually occur as a result of deliberately arranging to splice 50% of the bars at one location and the remainder at another location. In practice, class C splices are by far the most common splices. The lap lengths for the three classes are based on the development lengths L_d given by Eq. (8-25):

Class A splice: Lap length $= 1.0L_d$
Class B splice: Lap length $= 1.3L_d$
Class C splice: Lap length $= 1.7L_d$

but in no case may the lap length be less than 12 in. As before, these splice lengths have been computed and tabulated for ready reference in Table A-5. It should be noted that the factors 1.0, 1.3, and 1.7 are applied before the minimum criteria applicable to Eq. (8-25) and are applied and that the effects of being top bars apply equally to splices.

Reinforcement may also be spliced in a compression member; one very common place where such splices may occur in foundations is in splicing the dowels from a grade beam to the column reinforcement. Required lap length for splicing in compression does not include a category of top bars, nor are there any class A, B, or C conditions. The compressive splice length is prescribed by code:

$$L_d \geqq 0.02 d_b \frac{f_y}{\sqrt{f'_c}} \quad \text{but not less than } 0.0005 d_b f_y \text{ and not less than 12 in.} \tag{8-28}$$

Compressive splice lengths have been computed from Eq. (8-28) and tabulated in Table A-5.

Some examples will illustrate the use of the tables of development lengths.

EXAMPLE 8-10: Strength Method or Working Stress Method

Determination of anchorage for reinforcement. Select a workable anchorage.

Given:
Rectangular beam as shown
Grade 40 steel
$f'_c = 3000$ psi

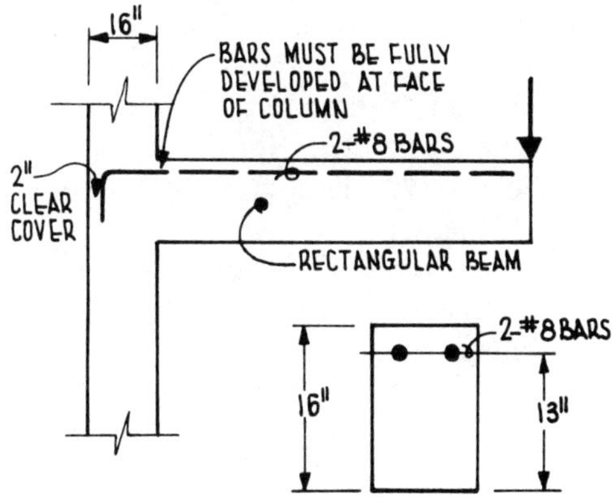

SOLUTION The required development length for a straight No. 6 bar is given in Table A-5. Since the bars shown in the sketch have more than 12 in. of fresh concrete below them, they are classed as top bars, and L_d = 18 in. in tension. There is not enough room at the support to provide this development length. If a hook is used, the development length from Table A-6 is 11 in., which can be provided. Use hooked bars, at least 2 in. cover as shown, embedment length of 14 in..

EXAMPLE 8-11: Strength Method or Working Stress Method

Determination of anchorage for embedded item. Select embedment for a lifting eye.

Given:

Precast tee as shown

Grade 40 steel

f'_c = 3000 psi

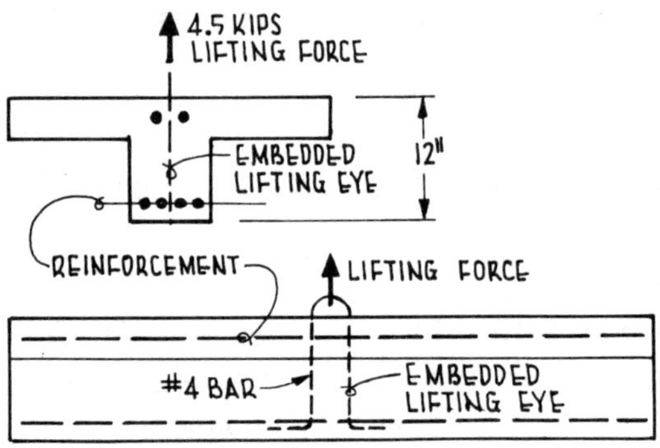

SOLUTION To develop full strength of the embedded No. 4 bars, a development length of 12 in. is required for a straight bar or 7 in. for a hooked bar. Since there is again not enough room for the straight bar, it must be hooked as shown. The capacity is then $2f_yA_s = 2 \times 40{,}000 \times 0.2 = 16$ kips.

In Example 8-11, it should be recognized that the solution did not include the actual load on the lifting eye. Nor do the tables include any provision for loads other than full capacity. The tables give the development length only to the full yield capacity of the bars.

For beams, code permits flexural reinforcement to be cut off when it is no longer needed. Such provisions apply only rarely to foundations, except that in grade beams some of the provisions may occasionally apply. In the next chapter, the code provisions for bond development are discussed as they apply to foundation design.

SUMMARY

The following major items were presented in the foregoing sections:

- Chloride salts and sulfur compounds in solution are two of the more common hazards to reinforced concrete foundations.

- Most of the desirable properties of concrete are related to strength; as strength increases, there are improvements in impermeability, hardness, durability, resistance to chemical attack, resistance to freeze–thaw, and most other desirable properties.

- Strength of concrete is strongly related to the water–cement ratio; as the water–cement ratio increases the strength decreases.

- The stress–strain curve for concrete is not ideal; an idealized but appropriate curve is substituted for the sake of stress analysis and design.

- The idealized yield point of concrete is taken at $0.85f'_c$, which is in substantial agreement with the results of extensive physical tests.

- The cover over reinforcement specified by code is a minimum; for foundations, the cover should be increased wherever unusually corrosive soil conditions warrant.

- In the strength method of design of reinforced concrete, loads are combined and then factored up to their theoretical ultimate value; the concrete section is then designed to sustain this ultimate load case with the section at its absolute maximum state of resistance, with all materials in full yield.

- Under service conditions, a section designed using the strength method will work back at some lower state of stress; the strength method does not require a check of the day-to-day stresses under service conditions.

- In the working stress method of design of reinforced concrete, loads are combined to produce the maximum working load which the member should ever be expected to sustain; the member itself is then designed to sustain this maximum load at the specified levels of working stresses in the steel and concrete.

- Reinforced concrete members may be designed similarly to other structural materials by solving for the required section modulus, using the flexure formula to solve for the section modulus.

- For design of reinforced concrete members at their ultimate state of flexure, the ultimate moment is used to solve for the plastic section modulus, with concrete stress at $0.85f'_c$ and steel in full yield.

- For design of reinforced concrete members at their working levels of flexure, the maximum working moment is used to solve for the elastic section modulus, with both concrete and steel stressed to their maximum specified levels of working stress.

• Reinforced concrete members are designed to sustain an average shear stress across the entire section rather than a peak shear stress at some particular point.

• Bond strength of concrete is one of the more variable values that must be used in the design of reinforced concrete; to reduce the dependence of the design on such an uncertain property, development lengths of reinforcement have been established by extensive tests.

REVIEW QUESTIONS

8-1. List the four basic ingredients used in making portland cement concrete.

8-2. Why is it necessary to reinforce concrete?

8-3. What kinds of chemical attack can be expected on concrete foundations subject to occasional submersion?

8-4. List the five types of portland cement and the distinguishing characteristics of each.

8-5. What is the usual age of concrete when its ultimate strength is measured?

8-6. How long does concrete continue to gain strength?

8-7. A concrete has a compressive strength of 5000 psi. Approximately what is its water–cement ratio?

8-8. Is water–cement ratio a ratio of volumes or weights?

8-9. A concrete has an ultimate strength of 3000 psi. What is its modulus of elasticity?

8-10. In the idealized stress–strain curve for concrete, what is its stress at the idealized yield point?

8-11. What is the diameter of a No. 7 reinforcing bar?

8-12. What percent elongation does a grade 40 steel undergo in its plastic range (Fig. 8-5)?

8-13. In structural design, a margin of strength is maintained above the known loading to allow for unanticipated loads. How is this margin provided in the ultimate strength method? In the working stress method?

8-14. What is the load factor for dead load? For live load?

8-15. What is the strength reduction factor ϕ, and why is it necessary?

8-16. An increase in stress is usually permitted in concrete members when they are subjected to transient lateral loads such as wind or earthquake. How much is this increase?

8-17. Define the flexural section modulus of a member in bending.

8-18. At about what percentage of ultimate stress should the service stress be held in a concrete beam?

8-19. What is the maximum strain that is assumed to occur in concrete at ultimate flexural load?

8-20. How is the factor β determined in the ACI analysis of concrete beams?

8-21. In the ACI design method, where does critical shear occur on a concrete beam?

8-22. What is the magnitude of the strength reduction factor ϕ for shear?

8-23. If a footing design is inadequate in shear, what must be done?

8-24. What is meant by the "bond strength" of concrete?

8-25. What is meant by "development length" of a reinforcing bar?

8-26. What is the primary advantage in using hooks for anchorage?

8-27. How is the development length of a reinforcing bar determined?

8-28. What is the efficiency of a lap splice in a concrete member?

8-29. In a compressive member such as a column, can reinforcement be spliced? Hooked?

8-30. What is the most common class of splice?

PROBLEMS

Moment (kip-ft)				Shear (kips)						
M_{DL}	M_{LL}	M_{EQ}	M_{WIND}	V_{DL}	V_{LL}	V_{EQ}	V_{WIND}	f'_c (psi)	Steel grade	Limits on d
58	41	39	42	6.5	4.5	4.0	4.8	3000	60	None
40	42	46	29	4.0	4.0	4.0	3.0	2000	60	None
39	46	31	22	3.5	4.0	2.5	3.5	3000	40	16 in.
68	54	19	38	5.8	4.5	3.0	2.0	2000	40	20 in.
72	76	39	58	8.0	8.2	4.5	5.5	3000	60	None
86	96	45	54	8.5	9.5	5.0	4.0	4000	40	None
90	62	36	51	8.0	6.0	4.0	5.0	4000	60	18 in.
104	59	63	22	8.9	5.0	2.0	3.0	3000	40	18 in.

Problems 8-1 through 8-8. Determine the maximum and minimum values for the nominal ultimate moment M_n and the nominal ultimate shear V_n acting on a reinforced concrete section for the loads listed above.

Problems 8-9 through 8-16. Determine the size and reinforcement of the concrete sections required to sustain the moments computed for Problems 8-1 through 8-8 using the strength method of design.

8-17. A rectangular cantilever beam is subject to a dead load moment of 64 kip-ft plus a live load moment of 72 kip-ft. Determine the required size and reinforcement for the section and sketch the chosen cross section. Use strength method, $f'_c = 4000$ psi, grade 60 steel.

8-18. A rectangular cantilever beam is subject to a dead load moment of 94 kip-ft, a live load moment of 61 kip-ft, and a wind load moment of 78 kip-ft. Select a suitable section and sketch your result. Use strength method, $f'_c = 3000$ psi, grade 60 steel.

Problems 8-19 through 8-26. Determine the size and reinforcement of the concrete sections required to sustain the moments of Problems 8-1 through 8-8 using the working stress method of design.

Problems 8-27 through 8-34. Review the results of Problems 8-9 through 8-16 for their adequacy in shear, using the strength method of design. Assume that the given shears are those at a distance d from face of support.

Problems 8-35 through 8-42. Review the results of Problems 8-19 through 8-26 for their adequacy in shear, using the working stress method of design. Assume that the given shears are those at a distance d from face of support.

Problems 8-43 through 8-51. Using the formulas for computing development lengths, determine the required development lengths for top bars for the given bar sizes. Check your results against the values given in Table A-5.

Problem	Bar size	f'_c (psi)	Steel grade
8-43	3	2000	60
8-44	4	3000	60
8-45	5	4000	60
8-46	6	2000	50
8-47	7	3000	50
8-48	8	4000	50
8-49	9	2000	40
8-50	10	3000	40
8-51	11	4000	40

9

Design of Concrete Foundations

The general theory for the design of reinforced concrete is developed in Chapter 8. Although brief, the development includes concepts of flexure, shear, and bond. The theory is applied to routine rectangular beams by means of simple examples.

In this chapter, the theory is applied specifically to shallow foundations. The three basic types of shallow foundations are included in the example problems: spread footings, strip footings, and grade beams. Code places some additional requirements on footings that do not apply to other structural elements; those requirements are discussed in detail in this chapter.

The design of foundations is heavily dominated by practice. Things are done in certain ways because "they have always been done that way." Although such stilted practices do little to encourage innovation, the consequences of failure in a foundation design are so serious that traditional proven methods will probably continue to be favored.

In defense of such traditional practices, it bears repeating that once built, foundations are generally covered and forgotten, never again to be seen or examined until the building is demolished (or until the foundation fails). The reliability of the design is obviously of paramount importance. Unproven systems, however innovative, can have little appeal under such circumstances.

DESIGN LOADS AND SOIL PRESSURES

The concrete in a footing must be designed to sustain the maximum soil pressure that could ever occur on the bottom of the footing. There is no concern whether this maximum pressure comes about through strength or settlement considerations in the soil. Whatever case of load produces, the maximum soil pressure also produces the maximum stress on the concrete section.

Even though a footing is proportional for settlements at DL + sustained LL, it still must be able to sustain the maximum load of DL + full LL whenever it occurs. Although this maximum load (and corresponding maximum pressure) will be sustained only for short periods, it still produces the maximum stress that the concrete in the footing will experience. Consequently, the design load on the concrete footing must always include the full live load.

In addition, it was shown in Chapter 6 that when a column or bearing wall is fixed to its footing, there can be as much as a 25% increase in the soil pressure at one side of the footing. It must be assumed that this increase could occur at any one of the four sides, although at only one side at a time. When a column or bearing wall is fixed to its footing, therefore, the concrete footing is designed for a magnified soil pressure of 1.25(DL + LL); it is assumed that this magnified pressure could occur at any one of the four sides.

There are thus two possible load cases for the design of the concrete section for a spread footing or a strip footing.

1. Where the column or wall is fixed to the footing, the soil pressures corresponding to a magnified load are used; the magnified design load is 1.25(DL + 100% LL).
2. Where the column or wall is hinged to the footing, the magnifier is not applicable; the soil pressures corresponding to a load of (DL + 100% LL) are used.

The foregoing two cases are used for design of the concrete sections regardless of whether the allowable soil pressure is governed by strength criteria or by settlement criteria.

DESIGN PRACTICES

It has been noted that concrete will undergo cracking on the tension side of the beam. A typical pattern of flexure/shear cracking in a uniformly loaded beam is shown in Fig. 9-1. Although these cracks must occur if the steel is to reach its design stress, the crack sizes are usually so small as to be barely visible. The size of the crack is secondary; the fact that it exists at all is the point of concern.

When a foundation experiences exposure to groundwater or to high levels of soil moisture, the flexural cracking provides a large number of points of relatively free access for the water and its dissolved impurities. As the crack size increases,

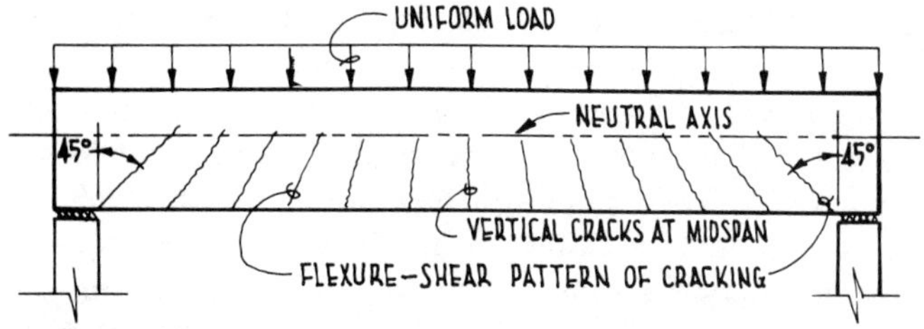

Figure 9-1 Typical flexure/shear cracking.

the freedom of entry of such impurities must also increase. The inevitable ion exchange can produce an increased risk of corrosion of reinforcement.

Many of the design practices in foundation design are aimed at controlling or reducing the crack sizes in order to reduce such risks. However, no computations of crack sizes need be performed, nor are elaborate precautions necessary under normal circumstances. The crack sizes may be controlled simply by limiting the tensile service stress in the reinforcement to relatively low levels; the strain in the adjacent concrete (and the resultant crack sizes) will then be correspondingly limited.

An expedient but effective means to limit the tensile strain in the steel is to use grade 40 steel for all reinforcement in foundations. This expedient is effective whether the strength method or working stress method is used. The design methods will automatically produce the desired limitations on strains.

It is a feature of the strength method of design, for example, that the tensile stress at service levels will be close to $0.6f_y$ [25]. Grade 40 steel will therefore be working at about 24,000 psi under day-to-day service loadings, but grade 60 steel will be working some 50% higher, at about 36,000 psi. Obviously, simply using grade 40 steel rather than grade 60 steel throughout the design of the foundation will automatically limit the service level stresses (and strains) to lower, more acceptable levels.

Even if grade 60 steel is the only steel available, however, the design can be performed as if it were grade 40, in which case the same limitation on the service stress will be achieved. A plastic section modulus chosen from Table A-4 for grade 40 steel will produce a stress under service loadings of about 24,000 psi regardless what grade of steel is finally used. The small additional cost of supplying grade 60 steel where only grade 40 is needed is justified under such conditions.

The same general level of service stresses can also be achieved by using the working stress method (rather than the strength method) for the foundation design. The allowable tensile stress in the reinforcement in the working stress method is limited to 20,000 psi for grade 40 steel and to 24,000 psi for grade 60 steel, as shown in Table A-1. The allowable stress might well be limited to even lower values, where corrosive groundwater presents unusually serious hazards.

On the other hand, where there is little danger of submersion in groundwater or where soil moisture is always low, such severe limits on crack sizes may not be justified. For such cases the designer may choose to use grade 60 steel and regular design procedures. The examples presented later in this chapter include such cases.

In the earlier discussions concerning bond strength and development length, it was noted that code allows flexural reinforcement to be cut off when it is no longer needed. Such points of discontinuity can produce "stress raisers" and points of potential cracking. Too, placing short sections of reinforcement accurately is troublesome, time consuming, and costly. Consequently, reinforcement in foundations is rarely cut off; it is usually made continuous from end to end of the footing or grade beam (although it may contain splices).

It has been noted that shear reinforcement is rarely used in foundation design. As a consequence, the size of the section will usually be governed by shear requirements rather than flexural requirements. In such circumstances the depth of footings can be expected to be somewhat larger than that of comparable members located in the superstructure above. As a general guide, the overall thickness of a spread footing can be expected to be about one-sixth the maximum lateral dimension, with a minimum thickness of about 16 in. being commonly observed.

When concrete is cast against soil, the minimum cover over the reinforcement is 3 in. The consequences of foundation failure are so serious, however, that this limit is usually observed as the barest minimum. In subsequent examples, the concrete cover is taken as 4 in. from the centerline of flexural reinforcement, which provides somewhat more than the minimum 3 in. clear cover.

SPREAD FOOTINGS

Spread footings are assumed to act as cantilevered beams in two directions. The critical section for bending in a spread footing that supports a concrete column is at the face of the column. For a footing that supports a steel column, the critical section for bending is taken halfway between the face of the steel column and the edge of the base plate. Critical sections are shown in Fig. 9-2.

It should be noted that the bending of the footing is assumed to occur uniformly along the full width of the footing as shown in Fig. 9-2, as if the column were actually a wall extending all the way across the footing. Although this assumption is admittedly a drastic simplification, its success over the years justifies its continued use. For rectangular footings, the procedure is repeated in the other direction to obtain the reinforcement in that direction.

The same simplifying assumption concerning bend lines also applies when designing the spread footing for beam shear. As for other types of beams, the critical section for beam shear is taken at a distance d from the face of the column, shown as lines AB and CD in Fig. 9-2. The ultimate shear stress on the concrete at the critical section is, as usual in beams,

$$v_c = 2\sqrt{f'_c} \tag{9-1}$$

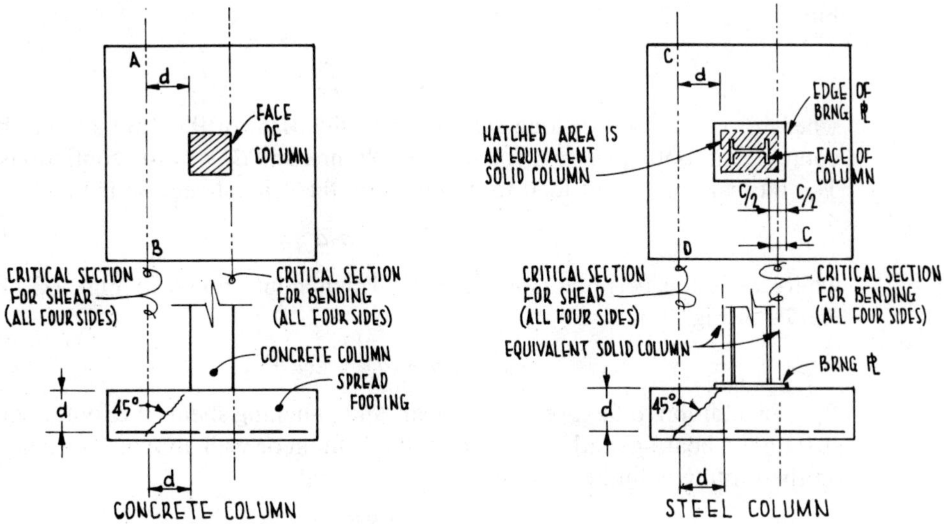

Figure 9-2 Critical sections in spread footings.

With no shear reinforcement, the ultimate shear force at the critical section may not be greater than V_c, where

$$V_c = v_c b_w d \tag{9-2}$$

and b_w is the total width of the footing at the critical section.

In addition to the foregoing design criteria for beam shear and flexure, Code imposes an additional limitation due to *punching shear*. Punching shear produces the pyramidal punchout shown in Fig. 9-3. Whereas critical beam shear is taken at a distance d from face of support, critical punching shear is taken at a distance $d/2$ from face of support, shown as rectangle $ABCD$ in Fig. 9-3.

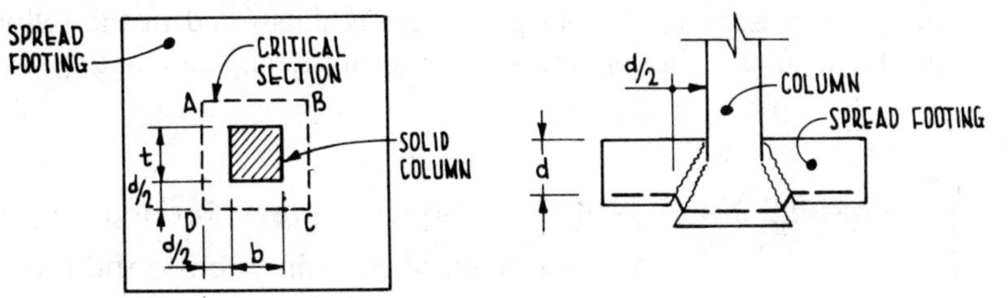

Figure 9-3 Punching shear.

Ultimate shear stress for punching shear in concrete without shear reinforcement is specified by code:

$$v_c = \left(\frac{1 + 2}{\beta_c}\right)(2\sqrt{f_c'}) \tag{9-3a}$$

but

$$v_c \leq 4\sqrt{f'_c} \tag{9-3b}$$

where β_c is the ratio of long side to short side (t/b) of the contact area between column and footing ($\beta_c = 1$ for square columns and $\beta_c = \infty$ for continuous walls). The ultimate punchout load to be taken by the critical section is then

$$V_c = v_c b_0 d \tag{9-4}$$

where b_0 is the perimeter of the critical punchout shown in Fig. 9-3. For line *ABCD* in Fig. 9-3,

$$b_0 = 2(b + d) + 2(t + d) \tag{9-5}$$

In addition to flexure, beam shear, and punching shear, the footing must also sustain the bearing load presented by the column or wall above. The bearing load on the footing is limited by code,

$$p_n = 0.85 f'_c A_g \tag{9-6}$$

where all symbols are those used earlier. The undercapacity factor ϕ is 0.7 for bearing stress between two structural elements such as these.

The allowable bearing load given by Eq. (9-6) may be increased by multiplying by the factor $\sqrt{A_2/A_1}$, but not by more than 2, where A_2 is the area of the top of the footing concentric with A_1 and, for footings, $A_1 = A_g$. In almost all cases where this code provision is applied to footings, the factor $\sqrt{A_2/A_1}$ will be much greater than 2, with the final result that bearing load will almost always be double that of Eq. (9-6).

For design of spread footings, there are thus four conditions to be met: bearing load on top face, punching shear, beam shear, and beam flexure. There are special design tables and graphs for use in designing spread footings; they are useful if one is frequently engaged in such designs. For infrequent use, it is far faster simply to guess the depth of footing and check the four conditions to see if the guess is adequate. That procedure is illustrated in the following examples, first for a square footing and second for a rectangular footing.

EXAMPLE 9-1: Design of a Square Spread Footing (Strength Method)

Assume base of column fixed to the footing.
Soil pressure limited by strength criteria.

Given:

$P_{DL} = 92$ kips
$P_{LL} = 76$ kips
Soil pressure $p_a = 3000$ psf
Column size 16 in. × 16 in.
$f'_c = 3000$ psi
Grade 40 steel

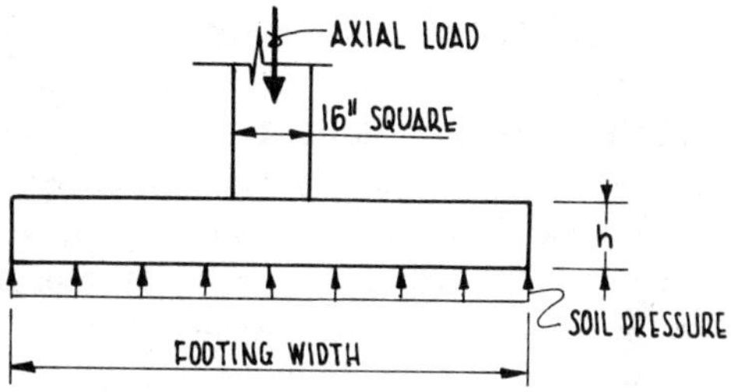

In foundation design, the difference in weight between concrete (145 pcf) and the displaced overburden (115 pcf) is commonly ignored, although some references [4] include it. In subsequent calculations here, it is ignored. If it were to be included, the increase in dead load would be estimated and added to the column dead load; at the conclusion of the calculations the accuracy of this initial estimate would be checked and verified.

SOLUTION The required contact area of the footing is the total load, $P_{DL} + P_{LL}$, multiplied by the magnifier 1.25 and divided by the allowable soil pressure, 3000 psf:

$$\text{area} = 1.25\left(\frac{P_{DL} + P_{LL}}{p_a}\right) = 1.25\left(\frac{92 + 76}{3}\right) = 70 \text{ ft}^2$$

Use square footing, 8 ft 6 in. or 102 in. The maximum soil pressures that can occur with this footing size are:

$$p_{DL} = \frac{1.25P_{DL}}{\text{area}} = \frac{1.25 \times 92}{8.5 \times 8.5} = 1.6 \text{ ksf}$$

$$p_{LL} = \frac{1.25P_{LL}}{\text{area}} = \frac{1.25 \times 76}{8.5 \times 8.5} = 1.3 \text{ ksf}$$

The first check is for bearing load on top of the footing. The total load P_n delivered to the top of the footing is, where $\phi = 0.7$ for bearing,

$$p_n = \frac{1.4P_{DL} + 1.7P_{LL}}{\phi} = \frac{1.4 \times 92 + 1.7 \times 76}{0.7}$$

$$= 369 \text{ kips}$$

The allowable bearing load on the footing is

$$P_n = 0.85f'_cA_g = 0.85 \times 30{,}000 \times 16 \times 16 = 653 \text{ kips}$$

This allowable load is multiplied by $\sqrt{A_2/A_1}$,

$$\sqrt{\frac{A_2}{A_1}} = \frac{102 \times 102}{16 \times 16} = 6.38 \text{ but} < 2$$

Hence

$$P_n = 2 \times 653 = 1306 \text{ kips} > 369 \text{ kips} \quad \text{(OK)}$$

The bearing capacity of the footing is more than adequate to carry the column load. It remains now to select the thickness of the footing.

A trial thickness h of about one-sixth of the footing width is assumed, or about 18 in. The effective depth d is then taken at 14 in. This trial thickness of 14 in. is first checked for punching shear. The average ultimate punching shear stress occurs at a distance $d/2$ from the face of the column on all four sides of the column as shown in the following sketch.

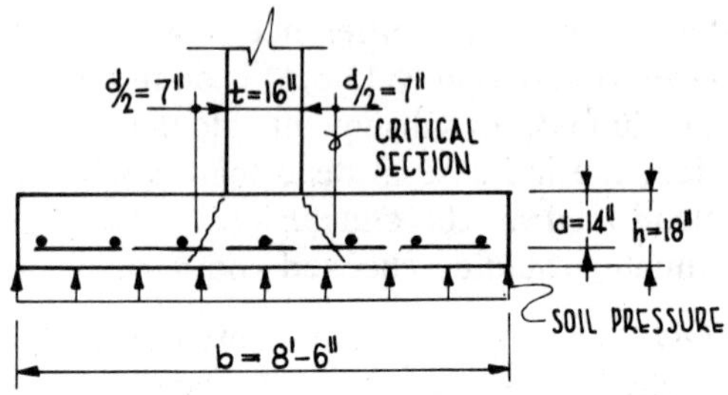

The total punching shear force acting on the section is found by statics:

$$P_{DL} = p_{DL}[b^2 - (t + d)^2] = 1.6(8.5^2 - 1.33^2) = 113 \text{ kips}$$

$$P_{LL} = p_{LL}[b^2 - (t + d)^2] = 1.3(8.5^2 - 1.33^2) = 92 \text{ kips}$$

Hence

$$P_n = \frac{1.4P_{DL} + 1.7P_{LL}}{\phi}$$

$$= \frac{1.4 \times 113 + 1.7 \times 92}{0.85} = 370 \text{ kips}$$

The average punching shear stress at ultimate load is given by code, where $\beta_c = 1$,

$$v_c = 2\sqrt{f'_c}\left(1 + \frac{2}{\beta_c}\right) = 6\sqrt{f'_c}$$

but v_c must not exceed $4\sqrt{f'_c}$. Hence $v_c = 4\sqrt{f'_c} = 220$ psi. The capacity of the footing to carry punching shear is then, where $b = 4(t + d) = 120$ in.,

$$V_c = v_c b_0 d = 220 \times 120 \times 14 = 370 > 370 \text{ kips} \quad \text{(OK)}$$

This trial depth of 14 in. is thus seen to be adequate to take the purchasing shear.

Next, the trial depth of 14 in. is checked for its capacity in beam shear. The critical shear force is found at a distance $d/2$ from the face of the column, as shown in the following sketch.

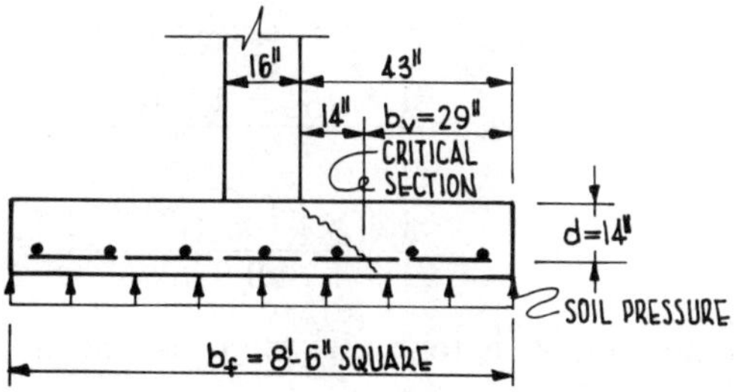

The shear acting on the footing beyond the line of the critical section is

$$V_{\mathrm{DL}} = p_{\mathrm{DL}} b_v b_f = 1.6 \left(\frac{29}{12}\right) 8.5 = 33 \text{ kips}$$

$$V_{\mathrm{LL}} = p_{\mathrm{LL}} b_v b_f = 1.3 \left(\frac{29}{12}\right) 8.5 = 27 \text{ kips}$$

Hence

$$V_n = \frac{1.4 V_{\mathrm{DL}} + 1.7 V_{\mathrm{LL}}}{\phi}$$

$$= \frac{1.4 \times 33 + 1.7 \times 27}{0.85} = 108 \text{ kips}$$

The capacity of the concrete in beam shear is

$$V_{\mathrm{cr}} = v_c b_w d = 2\sqrt{f_c'} \times 102 \times 14 = 156 > 108 \text{ kips} \quad \text{(OK)}$$

The trial depth of 14 in. is thus seen to be adequate to carry the beam shear on the section.

The trial depth of 14 in. is now checked for beam flexure, where $M = wL_c^2/2$ on the cantilever length L_c, as shown in the following sketch.

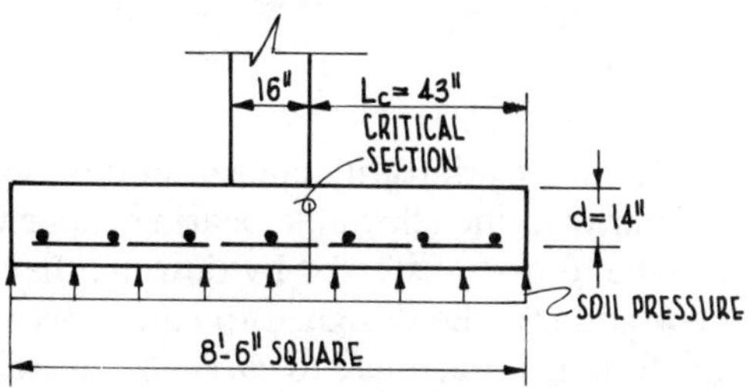

The moment acting on the critical section is

$$M_{\mathrm{DL}} = \frac{w_{\mathrm{DL}} L_c^2}{2} = \frac{1.6 \times 8.5(43/12)^2}{2} = 87 \text{ kip-ft}$$

$$M_{\mathrm{LL}} = \frac{w_{\mathrm{LL}} L_c^2}{2} = \frac{1.3 \times 8.5(43/12)^2}{2} = 71 \text{ kip-ft}$$

Hence

$$M_n = \frac{1.4M_{DL} + 1.7M_{LL}}{\phi}$$

$$= \frac{1.4 \times 87 + 1.7 \times 71}{0.9} = 269 \text{ kip-ft}$$

Section modulus Z_c is found by the usual means:

$$\frac{M_n}{0.85f'_c} = Z_c \qquad \frac{269{,}000 \times 12}{0.85 \times 3000} = \text{coefficient} \times 102 \times 14 \times 14$$

$$\text{coefficient} = 0.063$$

From Table A-5, select $\rho = 0.0050$ (ρ_{min} for grade 40). Solve for A_s,

$$A_s = \rho bd = 0.005 \times 102 \times 14 = 7.14 \text{ in.}^2$$

Use 12 No. 7 bars each way, evenly spaced. (A_s provided = 7.22 in.^{2}). The trial depth of 14 in. is thus seen to be adequate for flexure, beam shear, punching shear, and bearing; it is selected as the final depth of the footing.

The following sketch shows the final design, to include dowels from the footing to the column above.

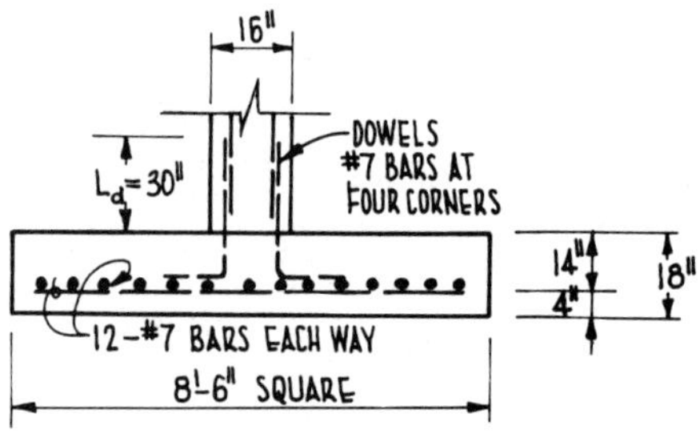

The dowels are rather arbitrarily chosen in this case. Had the applied column load been higher than the allowable bearing capacity of the footing, the excess load would have to be transferred by dowels; the size of the dowels and their embedment would then be designed to carry this excess load. In this footing, however, the footing is adequate to carry the entire load in bearing, so theoretically the dowels are not needed. For such cases, the practice is to design the dowels to carry about 10% of the flexural capacity of the column to account for any accidental moments on the footing. This moment is, where $Z_c = bh^2/6$ (approximately),

$$M_n = (0.85f'_cZ_c)(0.10) = \left(0.85 \times 3000 \times 16 \times \frac{16^2}{6}\right)(0.10)$$

$$= 174 \text{ kip-in.} = 14.5 \text{ kip-ft}$$

For dowels at the four corners,

$$\frac{M_n}{0.85f'_c} = Z_c \qquad \frac{174{,}000}{0.85 \times 3000} = \text{coefficient} \times 16 \times 16^2]$$

$$\text{coefficient} = 0.0167$$

From Table A-5, use minimum ρ; hence

$$A_s = \rho_{\min} b_w d = 0.005 \times 16 \times 14 = 1.12 \text{ in.}^2$$

Use No. 7 bars at the four corners.

As a point of interest, the difference between the dead load of the concrete and the weight of the soil in the foregoing example can now be checked to see how much error is involved in ignoring this difference. At a total depth of 18 in., the additional weight is, for a concrete weight of 145 pcf and a soil weight of 115 pcf,

$$\text{wt/ft}^2 = (145 - 115)\left(\frac{18}{12}\right) = 45 \text{ psf} \quad \text{(error)}$$

Of the total pressure of 2900 psf, this error is less than 2%.

The next example presents the design of a rectangular footing subjected to a column load. In rectangular footings, code requires that a major part of the shorter reinforcement must be located in a middle "band." The width of this middle band is equal to the narrower width of the footing as shown in Fig. 9-4. The portion of the reinforcement to be concentrated in the middle band width is given by code by the formula

$$\frac{\text{reinforcement in middle band width}}{\text{total reinforcement in short direction}} = \frac{2}{\beta + 1} \tag{9-7}$$

where β is the ratio of the long side to short side of the footing. For a square footing, $\beta = 1$ and the equation simply requires equal distribution in both directions. The use of this formula is demonstrated in the following example.

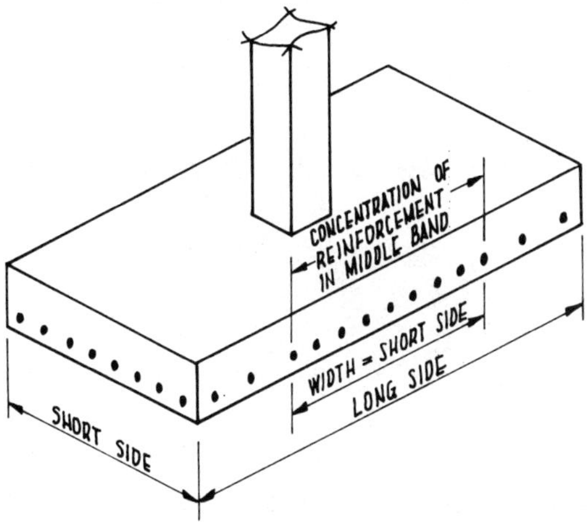

Figure 9-4 Middle band in rectangular footing.

EXAMPLE 9-2: Design of a Rectangular Footing (Strength Method)

Assume base of column fixed to footing.
Soil pressure limited by settlement criteria.

Given:

Property line as shown
$P_{DL} = 156$ kips
$P_{LL} = 88$ kips
Soil pressure $p_a = 2600$ psf
Column size 18 in. × 18 in.
$f'_c = 3000$ psi
Grade 40 steel

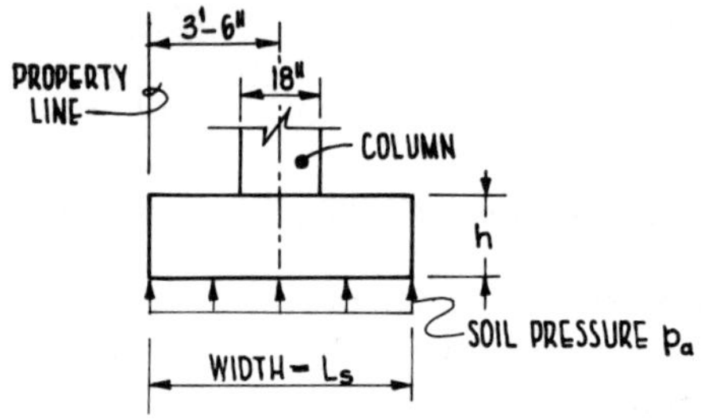

SOLUTION The required contact area of the footing is governed by settlement criteria:

$$\text{area} = \frac{P_{DL} + 50\% P_{LL}}{p_a} = \frac{156 + 0.5 \times 88}{2.6} = 76.9 \text{ ft}^2$$

Use footing 7 ft × 11 ft. The maximum soil pressures that can occur with this footing size are then

$$p_{DL} = \frac{1.25 P_{DL}}{\text{area}} = \frac{1.25 \times 156}{7 \times 11} = 2.53 \text{ ksf}$$

$$p_{LL} = \frac{1.25 P_{LL}}{\text{area}} = \frac{1.25 \times 88}{7 \times 11} = 1.43 \text{ ksf}$$

The bearing load on the footing is checked first. The load delivered to the top of the footing is

$$P_n = \frac{1.4 P_{DL} + 1.7 P_{LL}}{\phi}$$

$$= \frac{1.4 \times 156 + 1.7 \times 88}{0.7} = 526 \text{ kips}$$

The allowable load on the footing is

$$P_n = 0.85 f'_c A_g = 0.85 \times 3000 \times 18 \times 18 = 826 \text{ kips}$$

This allowable load is multiplied by the factor $\sqrt{A_2/A_1} = \sqrt{7 \times 11/1.5^2} = 5.85$, but may not be more than doubled. Hence the allowable load is

$$P_n = 2 \times 826 = 1652 > 526 \text{ kips} \quad \text{(OK)}$$

The footing is therefore adequate to receive the column load from above.

The footing is checked next for punching shear. Punching shear occurs along the pyramidal punchout shown in the following sketch.

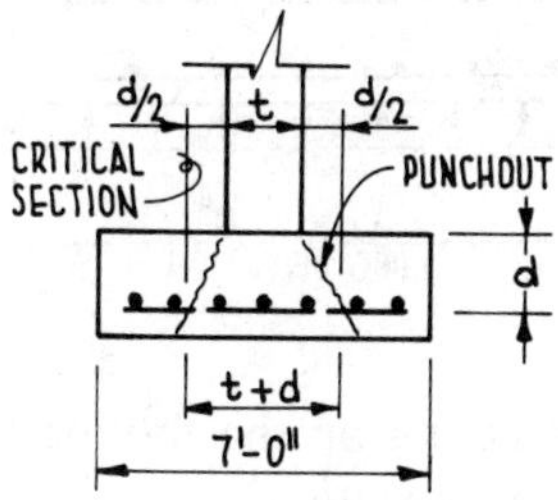

A trial thickness of about one-sixth of the longer side is assumed; hence $h = 11/6 = 2$ ft. The effective depth d is taken at 20 in. and $t + d = 38$ in. $= 3.17$ ft. The punching load is found by simple statics:

$$\begin{aligned} P_{\text{DL}} &= p_{\text{DL}}[L_L \times L_S - (t+d)^2] = 2.53(7 \times 11 - 3.17^2) \\ &= 169 \text{ kips} \end{aligned}$$

$$\begin{aligned} P_{\text{LL}} &= p_{\text{LL}}[L_L \times L_S - (t+d)^2] = 1.43(7 \times 11 - 3.17^2) \\ &= 96 \text{ kips} \end{aligned}$$

Hence

$$\begin{aligned} P_n &= \frac{1.4P_{\text{DL}} + 1.7P_{\text{LL}}}{\phi} \\ &= \frac{1.4 \times 169 + 1.7 \times 96}{0.85} = 470 \text{ kips} \end{aligned}$$

The punching shear stress in the concrete at ultimate load is

$$\begin{aligned} v_c &= 2\sqrt{f'_c}\left(1 + \frac{2}{\beta_c}\right) \qquad \text{where } \beta_c = 1 \\ &= 2\sqrt{f'_c}\,(3) = 6\sqrt{f'_c} \qquad \text{but} < 4\sqrt{f'_c} \end{aligned}$$

Hence

$$v_c = 4\sqrt{f'_c} = 4\sqrt{3000} = 219 \text{ psi}$$

The capacity of the footing in punching shear is then

$$V_c = v_c b_w d = 219(38 \times 4)20 = 666 > 470 \text{ kips} \quad \text{(OK)}$$

The trial depth of 20 in. is seen to be adequate for the punching shear.

The trial depth of 20 in. is next checked for beam shear at a distance d from the face of column, first in the long direction, then in the short direction. Locations of the critical sections for beam shear are shown in the sketch below.

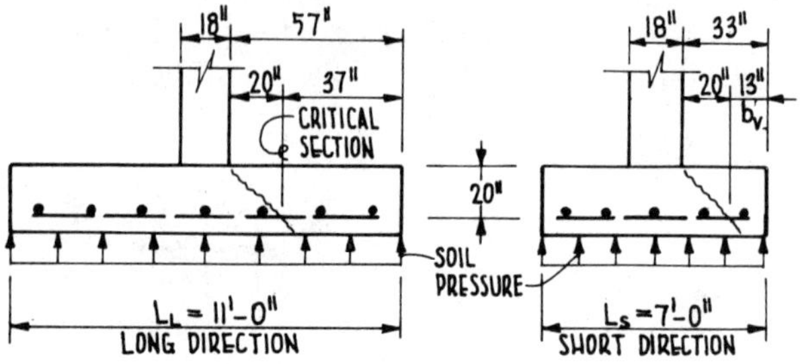

The shear force acting on the footing beyond the critical section in the long direction is found by statics:

$$V_{DL} = p_{DL} b_v L_S = 2.53 \left(\frac{37}{12}\right) 7.0 = 54.6 \text{ kips}$$

$$V_{LL} = p_{LL} b_v L_S = 1.43 \left(\frac{37}{12}\right) 7.0 = 30.9 \text{ kips}$$

Hence

$$V_n = \frac{1.4 V_{DL} + 1.7 V_{LL}}{\phi}$$

$$= \frac{1.4 \times 54.6 + 1.7 \times 30.9}{0.85} = 152 \text{ kips}$$

The capacity of the concrete in beam shear is

$$V_c = v_c b_w d = 2\sqrt{3000} \times 84 \times 20 = 184 > 152 \text{ kips} \quad \text{(OK)}$$

The section is thus seen to be adequate in the long direction. In the short direction, the shear force is

$$V_{DL} = p_{DL} b_v L_L = 2.53 \left(\frac{13}{12}\right) 11 = 30 \text{ kips}$$

$$V_{LL} = p_{LL} b_v L_L = 1.43 \left(\frac{13}{12}\right) 11 = 17 \text{ kips}$$

Hence

$$V_n = \frac{1.4 V_{DL} + 1.7 V_{LL}}{\phi}$$

$$= \frac{1.4 \times 30 + 1.7 \times 17}{0.85} = 83 \text{ kips}$$

The capacity of the concrete in beam shear is

$$V_c = v_c b_w d = 2\sqrt{3000} \times 132 \times 20 = 289 > 83 \quad \text{(OK)}$$

The trial depth of 20 in. is thus seen to be adequate for beam shear in both directions.

The trial depth of 20 in. is now used to design for beam flexure on the cantilever, first in the long direction, then in the short direction. The dimensions in the two directions are shown in the sketch below.

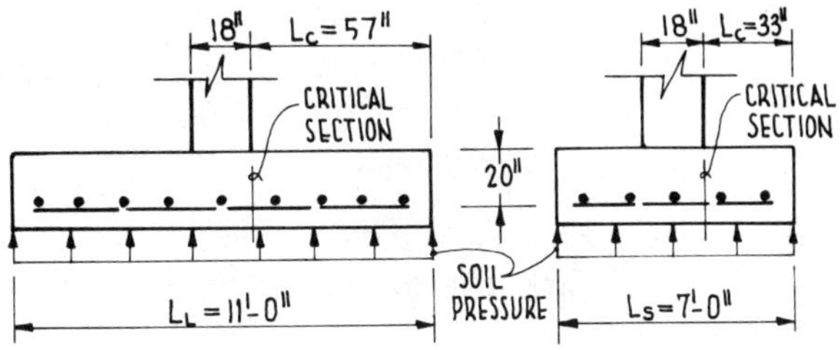

In the long direction, the moment at the critical section is

$$M_{\text{DL}} = \frac{w_{\text{DL}} L_c^2}{2} = \frac{2.53(57/12)^2}{2} = 28.5 \text{ kip-ft/ft}$$

$$M_{\text{LL}} = \frac{w_{\text{LL}} L_c^2}{2} = \frac{1.43(57/12)^2}{2} = 16.1 \text{ kip-ft/ft}$$

Hence

$$M_n = \frac{1.4 M_{\text{DL}} + 1.7 M_{\text{LL}}}{\phi}$$

$$= \frac{1.4 \times 28.5 + 1.7 \times 16.1}{0.9}$$

$$= 75 \text{ kip-ft/ft}$$

The plastic section modulus is computed:

$$\frac{M_n}{0.85 f'_c} = Z_c \qquad \frac{75{,}000 \times 12 \times 7}{0.85 \times 3000} = \text{coefficient} \times 84 \times 20^2$$

$$\text{coefficient} = 0.0735$$

From Table A-5, select $\rho = 0.005$ (minimum ρ). Compute $A_S = \rho bd = 0.005 \times 84 \times 20 = 8.4$ in.^{2} Use 12 No. 8 bars (long direction). In the short direction, the moment at the critical section is

$$M_{\text{DL}} = \frac{w_{\text{DL}} L_c^2}{2} = \frac{2.53(33/12)^2}{2} = 9.6 \text{ kip-ft/ft}$$

$$M_{\text{LL}} = \frac{w_{\text{LL}} L_c^2}{2} = \frac{1.43(33/12)^2}{2} = 5.4 \text{ kip-ft/ft}$$

Hence

$$M_n = \frac{1.4M_{\mathrm{DL}} + 1.7M_{\mathrm{LL}}}{\phi}$$

$$= \frac{1.4 \times 9.6 + 1.7 \times 5.4}{0.9}$$

$$= 25 \text{ kip-ft/ft}$$

The plastic section modulus is computed:

$$\frac{M_n}{0.85f'_c} = Z_c \qquad \frac{25{,}000 \times 12 \times 11}{0.85 \times 3000} = \text{coefficient} \times 132 \times 20^2$$

$$\text{coefficient} = 0.0245$$

From Table A-5, select $\rho = 0.005$ (minimum ρ). Compute $A_S = \rho bd = 0.005 \times 132 \times 20 = 13.2$ in.^{2} The portion of the total A_s that must be placed in the middle band, 7 ft wide,

$$\frac{A_S \text{ in middle band}}{\text{total } A_S} = \frac{2}{3.5/2 + 1} = 0.73$$

$$A_S \text{ in middle band} = 0.73 \times 13.2 = 9.64 \text{ in.}^2$$

Use 19 No. 8 bars, 13 in the middle 7 ft, with 3 additional bars in each end. The sketch below shows the final steel arrangement.

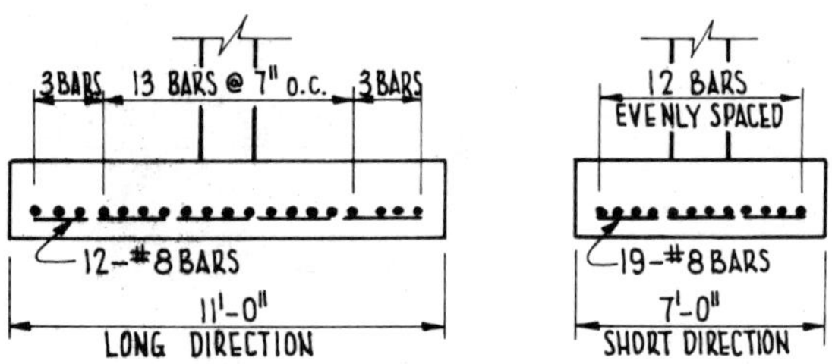

STRIP FOOTINGS

A strip footing is assumed to act as a cantilevered slab projecting from under the wall it supports. For strip footings supporting a concrete wall, the critical section for bending is at the face of the wall. If the wall is masonry, an equivalent concrete wall is used which is half the thickness of the masonry wall. The critical section for bending is then halfway between the middle of the wall and the edge of the wall, as shown in Fig. 9-5.

While code permits the critical section for beam shear to be measured from the actual face of the masonry wall, it is common practice to measure the critical shear from the face of the equivalent wall. In such an approach, the entire design

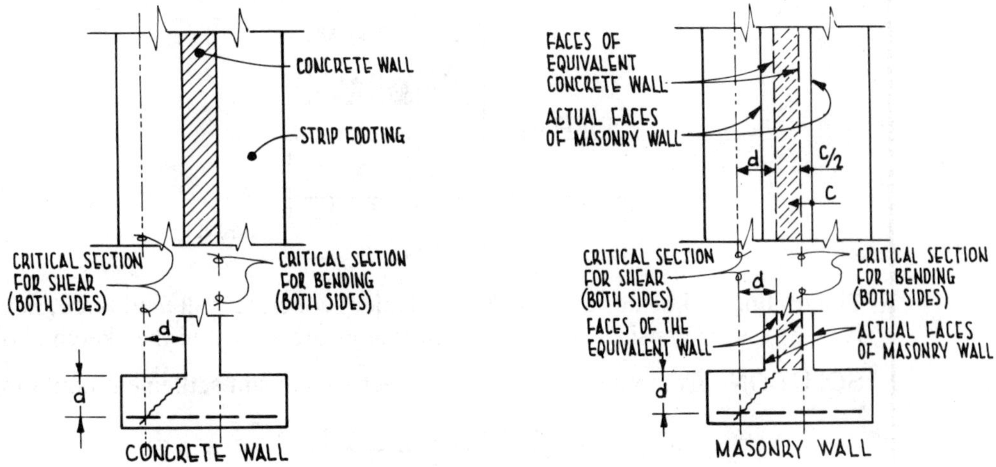

Figure 9-5 Critical sections in strip footings.

of the foundation is thus performed for an equivalent wall, affording a consistent treatment of the design. The ultimate strength of the equivalent wall in this approach is taken at twice the ultimate strength of the masonry, f'_m.

As with spread footings, shear reinforcement is rarely, if ever, used in strip footings. The strip footing is simply made thicker to take any excess shear. The thickness of strip footings may be expected to be somewhat less than the thickness of spread footings, but a thickness less than 12 in. is rarely seen.

There is no punching shear on a strip footing. A strip footing is designed only for beam shear and beam moment on the outstanding legs. There are no other special considerations in the design of a strip footing. It is simply a double cantilevered beam. A brief example will illustrate the procedure.

EXAMPLE 9-3: Design of Strip Footing for a Masonry Wall

Assume base of wall hinged to footing.
Soil pressure limited by settlement criteria.

Given:

P_{DL} = 8 kip/ft
P_{LL} = 5.5 kip/ft
Soil pressure p_a = 1600 psf
Wall thickness 12 in.
f'_m = 1350 psi
f'_c = 3000 psi
Grade 60 steel

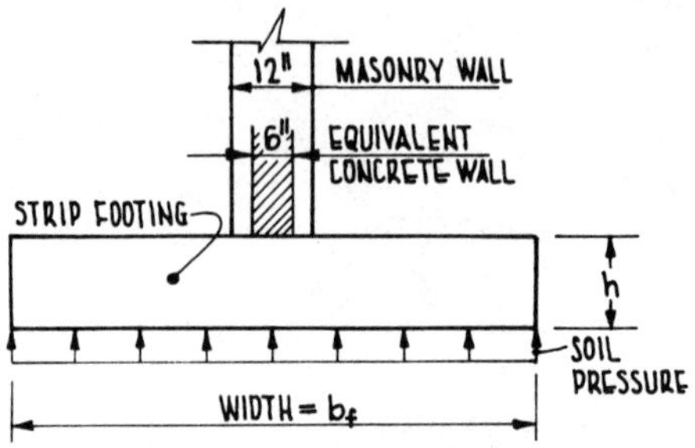

The footing will be designed for an equivalent concrete wall 6 in. thick, having $f'_c = 2f'_m = 2700$ psi. The relative dimensions are shown in the sketch above.

SOLUTION Compute the required contact area to suit settlement limitations:

$$\text{area} = \frac{P_{\text{DL}} + 50\% P_{\text{LL}}}{p_a} = \frac{8 + 0.5 \times 5.5}{1.60} = 6.72$$

Use a footing width of 7 ft. Compute the maximum soil pressures that can occur with this footing size:

$$p_{\text{DL}} = \frac{P_{\text{DL}}}{b_f} = \frac{8}{7} = 1.14 \text{ kips/ft}^2$$

$$p_{\text{LL}} = \frac{P_{\text{LL}}}{b_f} = \frac{5.5}{7} = 0.79 \text{ kips/ft}^2$$

Although not usually necessary for walls, the bearing load on the top of the footing will now be checked. The load delivered to the top of the footing is

$$p_n = \frac{1.4P_{\text{DL}} + 1.7P_{\text{LL}}}{\phi} = \frac{1.4 \times 8 + 1.7 \times 5.5}{0.7}$$

$$= 29.4 \text{ kips/ft}$$

The allowable bearing load on the footing is

$$P_n = 0.85 f'_c A_g = 0.85 \times 3000 \times 12 \times 6$$
$$= 184 \text{ kips/ft}$$

The allowable load is multiplied by the factor

$$\sqrt{A_2/A_1} = \frac{7 \times 1}{0.5 \times 1} = 4 \qquad \text{use 2}$$

$$P_n = 2 \times 184 = 368 \gg 29.4 \text{ kips/ft} \quad \text{(OK)}$$

The width of 7 ft is thus seen to be more than adequate for bearing capacity.

A trial thickness is now assumed for the footing. As for spread footings, a thickness of one-sixth the total width in flexure is a good start. For a total width of 7 ft, the estimated overall thickness h is then

$$h = \frac{\text{width}}{6} = 7 \times \frac{12}{6} = 14 \text{ in.}$$

The effective depth d is taken at 10 in. This trial depth of 10 in. for d is used to check the beam shear at a distance d from the face of the equivalent wall. Dimensions are shown in the sketch below.

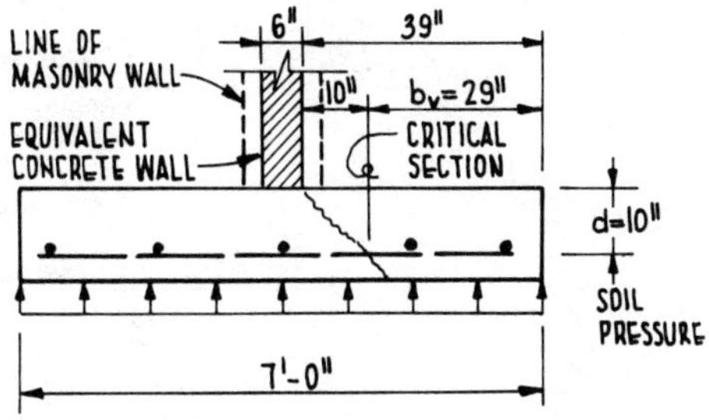

Shear on the section is, per foot of length,

$$V_{DL} = p_{DL} b_v L = 1.14 \left(\frac{29}{12}\right) \times 1 = 2.76 \text{ kips/ft}$$

$$V_{LL} = p_{LL} b_v L = 0.79 \left(\frac{29}{12}\right) \times 1 = 1.91 \text{ kips/ft}$$

Hence

$$V_n = \frac{1.4 V_{DL} + 1.7 V_{LL}}{\phi}$$

$$= \frac{1.4 \times 2.76 + 1.7 \times 1.91}{0.85}$$

$$= 8.37 \text{ kips/ft}$$

Capacity of the section is, per foot of length,

$$V_c = v_c b_w d = 2\sqrt{3000} \times 12 \times 10$$

$$= 13 > 8.37 \text{ kips/ft} \quad \text{(OK)}$$

The trial depth of 10 in. is thus seen to be adequate to take the beam shear.

The trial depth of 10 in. for d is now used for the design of the member in flexure. Critical section for flexure is at the face of the equivalent wall, as shown in the sketch below.

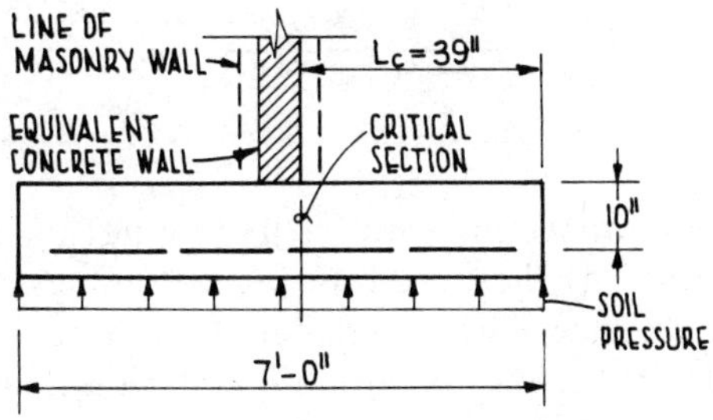

Moment acting on the critical section is

$$M_{DL} = \frac{p_{DL}L_C^2}{2} = \frac{1.14 \times (39/12)^2}{2} = 6.0 \text{ kip-ft/ft}$$

$$M_{LL} = \frac{p_{LL}L_C^2}{2} = \frac{0.79 \times (39/12)^2}{2} = 4.17 \text{ kip-ft/ft}$$

Hence

$$M_n = \frac{1.4M_{DL} + 1.7M_{LL}}{\phi}$$

$$= \frac{1.4 \times 6.02 + 1.7 \times 4.17}{0.9}$$

$$= 17.2 \text{ kip-ft/ft}$$

Section modulus Z is found by the usual means,

$$\frac{M_n}{0.85f_c'} = Z_c \qquad \frac{17{,}200 \times 12}{0.85 \times 3000} = \text{coefficient} \times 12 \times 10^2$$

$$\text{coefficient} = 0.0675$$

From Table A-5, select $\rho = 0.004$. Solve for $A_S = \rho bd = 0.004 \times 12 \times 10 = 0.480$ in.^{2}/ft. Use No. 6 bars at 11 in. o.c.

In the longitudinal direction, there is no computable load on the footing. The only requirement for steel is that for temperature and shrinkage, which in this case is $0.0018bt$. The required area of steel in the cross section is

$$A_S = 0.0018bt = 0.0018 \times 84 \times 14 = 2.12 \text{ in.}^2$$

Use 7 No. 5 bars, evenly spaced.

A sketch of the final design is shown below. The actual masonry wall is shown rather than the equivalent wall used throughout the foregoing calculations.

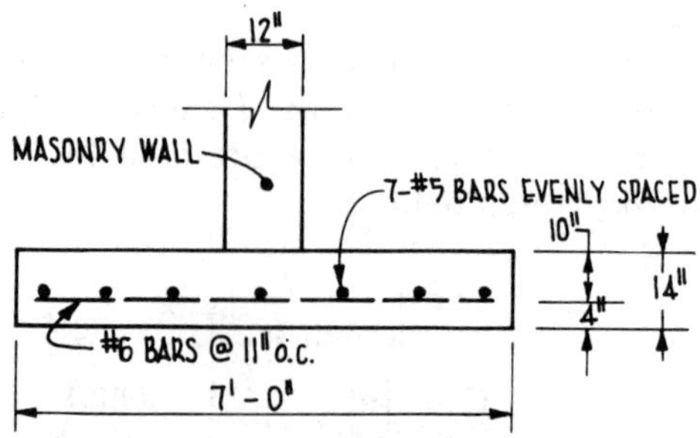

The dowels connecting a wall to its footing (if used) are as arbitrarily chosen for strip footings as they are for spread footings. If the wall is reinforced, it is common practice simply to match the wall reinforcement for size and location of dowels. If the wall is unreinforced, dowels may also be omitted.

GRADE BEAMS

Grade beams are one of the more economical foundations in common use for light buildings. They are used quite often for the foundations of lightweight prefabricated steel buildings as well as for one- or two-story concrete frames. They are not suited for heavy loads, however, and are rarely feasible for buildings having more than two stories.

As noted earlier, a grade beam is like any other continuous beam except that loads are upside down. It may be designed in the same way as any other continuous beam. Due to its continuity, however, it is necessary to design for both positive and negative moments along the length of the beam.

An accurate analysis of continuous beams is, of course, possible where the input data are accurate. Like so many other cases in foundations, however, the input data are subject to so much variation that a blanket approach is also used for the design of grade beams. Moments and shears throughout the length of the grade beam are established by the use of simple (but conservative) coefficients.

Typical spans of a continuous grade beam are shown in Fig. 9-6 along with its shear and moment diagrams. The maximum values of shear and moment are shown as formulas with simple coefficients. These (and other) coefficients are derived analytically and are given in the ACI code; they are called the ACI coefficients and are well known throughout the industry.

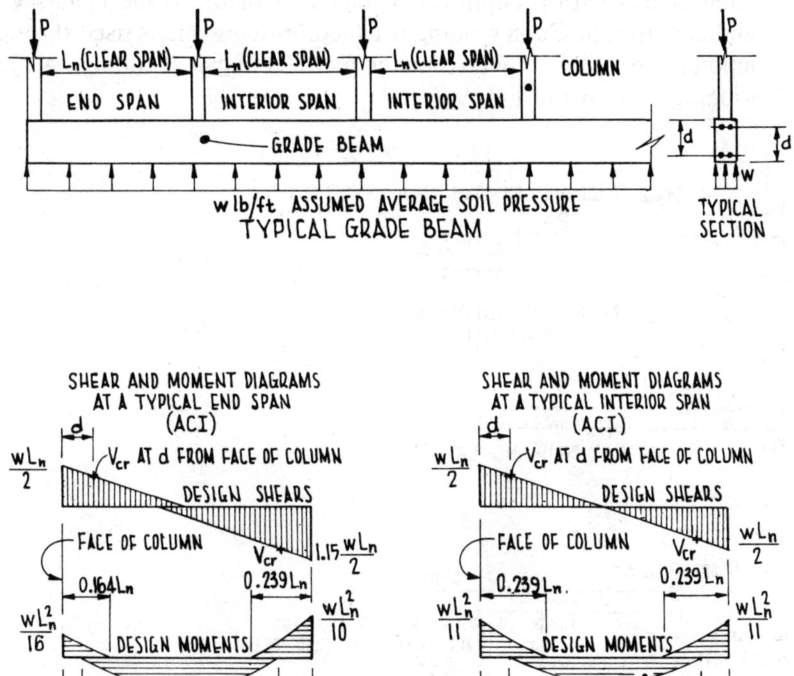

Figure 9-6 ACI coefficients for grade beams.

From Fig. 9-6 it can be seen that the maximum shear on a grade beam is $1.15wL_n$, where w is the contact pressure due either to dead load or to live load and L_n is the *clear* span between column faces. As noted earlier, the design value for shear is taken at a distance d from face of support as shown. The nominal ultimate shear force V_n includes the factored loads as well as the undercapacity factor ϕ:

$$V_n = \frac{1.4V_{DL} + 1.7V_{LL}}{\phi} \tag{9-8}$$

where $V_{DL} = 1.15w_{DL}L_n$ and $V_{LL} = 1.15w_{LL}L_n$.

Similarly, the maximum moment is given by $wL_n^2/10$, where w is the contact pressure due either to live load or dead load and L_n is the clear span between column faces. The nominal ultimate moment M_n is found as usual:

$$M_n = \frac{1.4M_{DL} + 1.7M_{LL}}{\phi} \tag{9-9}$$

where $M_{DL} = w_{DL}L_n^2/10$ and $M_{LL} = w_{LL}L_n^2/10$. It is important to note that the maximum value of moment is the same for both positive and negative moment; the same reinforcement will be required both at top and bottom of the grade beam.

A grade beam will probably be tall and slender. Without bracing, it will have a tendency to "roll" under load. Cross members placed between two lines of grade beams with a depth about half that of the grade beams will control this tendency to roll. Such bracing is called *bridging* and is used throughout concrete buildings to stabilize deep slender members against rolling. A typical layout of bridging is shown in Fig. 9-7.

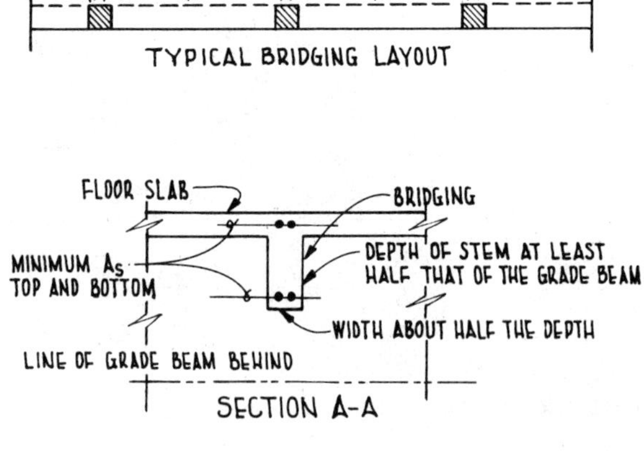

Figure 9-7 Typical bridging.

As spans increase, more lines of bridging are required. Although code does not specifically stipulate bridging requirements for grade beams, it is common practice to provide bridging (or some other positive support) at column lines and at midspan for spans 20 to 30 ft. For spans greater than 30 ft, bridging or some other means of positive support is usually provided at column lines and at third points of the span.

As with other types of footings, shear reinforcement is rarely used in grade beams. Since falsework and shoring are minimal, it is far more practical to increase the size of the beam slightly rather than to fabricate and place a large number of stirrups. As a consequence, the size of a grade beam is usually deeper than other concrete beams having a similar span.

In addition, there is usually no effect made to cut off the longitudinal reinforcement in a grade beam. The maximum required area of flexural steel, both top and bottom, is simply made continuous throughout the length of the grade beam. For such a design, only the highest numerical values of positive and negative moments are therefore needed when flexural reinforcement is selected.

Very often, the required width of the grade beam is so narrow that the column above would be wider than this required width; the columns would have to overhang the grade beam if it were to be centered on the grade beam. A simple solution is to make the grade beam as wide as the column; the column would then be properly seated on the grade beam. In such cases, the resulting soil pressure may be considerably less than the allowable pressure, but the overall suitability of the design justifies the configuration. The design used in the following example includes such circumstances.

EXAMPLE 9-4: Design of a Grade Beam Foundation (Strength Method)

Repetitive column loads at uniform spacing.
Strength method of design of concrete.

Given:

Load on the end column is roughly half the load on an interior column

Column size 12 in. × 12 in. for all columns

P_{DL} = 30 kips, P_{LL} = 24 kips (each column)

Allowable soil pressure is limited by strength criteria

P_a = 3000 psf

f'_c = 3000 psi

Grade 60 steel.

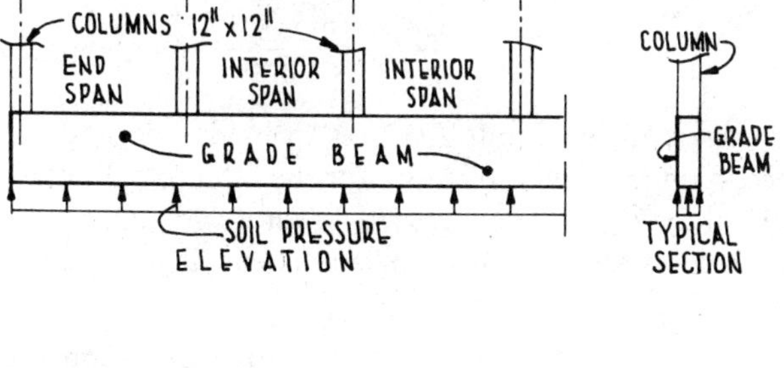

In this example, the difference in weight between the concrete grade beam and the soil it displaces are ignored.

SOLUTION The required contact area under the grade beam is computed by simple statics, assuming that the grade beam has a bearing area 26 ft long.

$$\text{area} = \frac{P_{DL} + P_{LL}}{P_a} = \frac{30 + 24}{3000}$$

$$= 18 \text{ ft}^2 \text{ per bay}$$

$$\text{width} = \frac{\text{area}}{\text{length}} = \frac{18}{26}$$

$$= 0.69 \text{ ft.} = 8.5 \text{ in.}$$

If the footing width were to be made 8.5 in. as indicated in the foregoing calculations, the column with its width at 12 in. would be eccentric on the grade beam. To keep the column load concentric on the grade beam, the grade beam is simply made 12 in. wide, recognizing that the soil pressure will be somewhat less than the allowable:

$$P_{DL} = \frac{P_{DL}}{\text{area}} = \frac{30}{1 \times 26} = 1.15 \text{ kips/ft}^2$$

$$P_{LL} = \frac{P_{LL}}{\text{area}} = \frac{24}{1 \times 26} = 0.92 \text{ kips/ft}^2$$

$$\text{Total } p = 1.15 + 0.92 = 2070 \text{ psf} < 3000 \text{ psf allowed}$$

The grade beam will now be checked for beam shear, using the ACI coefficients for maximum shear. Shear at any span:

$$V_{DL} = 1.15 p_{DL} L_n = \frac{1.15(1.15 \times 25)}{2} = 16.5 \text{ kips}$$

$$V_{DL} = 1.15 p_{LL} L_n = \frac{1.15(0.92 \times 25)}{2} = 13.2 \text{ kips}$$

Hence

$$V_n = \frac{1.4V_{DL} + 1.7V_{LL}}{\phi}$$

$$= \frac{1.4 \times 16.5 + 1.7 \times 13.2}{0.85} = 53.6 \text{ kips}$$

A trial depth of about $L/12$ or 26 in. is adopted for the grade beam, with a corresponding value of 22 in. for the effective depth d. This tall slender section of 12 in. × 26 in. is now reviewed for its adequacy in shear. The critical section for shear is shown on the partial shear diagram below.

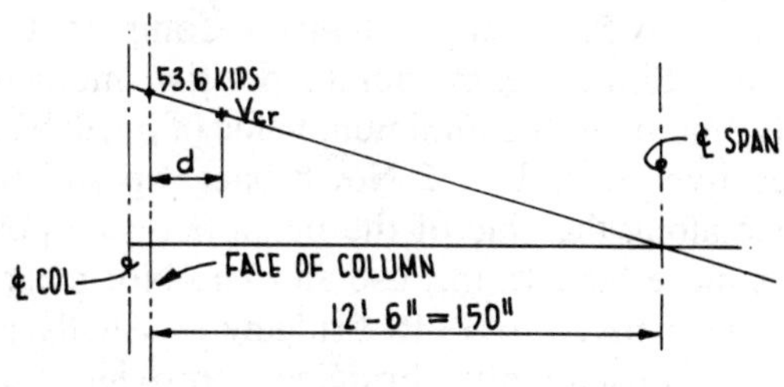

Critical shear on the section is found by ratios:

$$V_{cr} = \frac{150 - 22}{150} 53.6 = 45.7 \text{ kips}$$

The capacity of the section in shear is

$$\begin{aligned} V_c &= v_c b_w d = 2\sqrt{3000} \times 12 \times 22 \\ &= 29 \text{ kips} < 45.7 \text{ kips} \quad \text{(NG)} \end{aligned}$$

This section is not adequate for critical shear. A larger section, 38 in. deep, will be tried, $d = 34$ in.

Critical shear on the new section, 34 in. deep, is again computed by ratios:

$$V_{cr} = \frac{150 - 34}{150} 53.6 = 41.5 \text{ kips}$$

The capacity of this deeper section in shear is

$$V_c = v_c b_w d = 2\sqrt{3000} \times 12 \times 34 = 44.7 > 41.5 \quad \text{(OK)}$$

The trial depth of 34 in. is seen to be adequate for beam shear.

This trial depth of 34 in. is now reviewed for flexure. The maximum negative moment acting on the section is taken from the ACI coefficients:

$$M_{\text{DL}} = w_{\text{DL}} \times \frac{L_n^2}{10} = (1.15 \times 1) \times \frac{25^2}{10} = 71.9 \text{ kip-ft}$$

$$M_{\text{LL}} = w_{\text{LL}} \times \frac{L_n^2}{10} = (0.92 \times 1) \times \frac{25^2}{10} = 57.5 \text{ kip-ft}$$

Hence

$$\begin{aligned} M_n &= \frac{1.4M_{\text{DL}} + 1.7M_{\text{LL}}}{\phi} \\ &= \frac{1.4 \times 71.9 + 1.7 \times 57.5}{0.9} = 220 \text{ kip-ft} \end{aligned}$$

Required section modulus is computed as usual:

$$\frac{M_n}{0.85f'_c} = Z_c \qquad \frac{220{,}000 \times 12}{0.85 \times 3000} = \text{coefficient} \times 12 \times 34^2$$

$$\text{coefficient} = 0.075$$

From Table A-5, select $\rho = 0.0035$. Compute $A_s = \rho bd = 0.0035 \times 12 \times 34 = 1.428$ in.^{2}. Since the maximum positive moment is less than the maximum negative moment, the minimum value of ρ will be required both for positive and for negative steel. Use 2 No. 8 bars top and bottom continuous. Since the distance along the side of the beam between positive and negative reinforcement is more than 18 in., use an extra No. 8 bar at middepth.

For spans 20 to 30 ft, bridging is usually placed at column lines and at midspan. Consequently, bridging is provided for this grade beam at 4 m on center, at column lines and midspan. The stiffeners are selected as beams 12 in. wide $\times$ 20 in. deep (d = 16 in., with minimum ratio of steel, $\rho_{\min} = 0.0035$.

$$A_s = \rho_{\min} bd = 0.0035 \times 12 \times 16 = 0.672 \text{ in.}^2$$

Use 2 No. 6 bars top and bottom. The final design of the grade beam is shown in the following sketch.

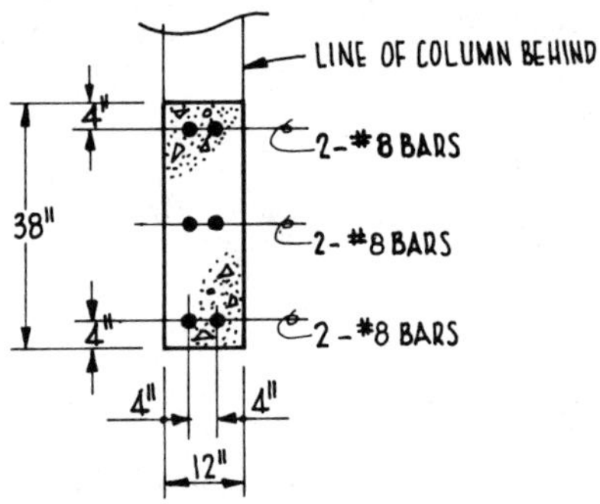

SUMMARY

The following practices in foundation design were presented in the foregoing discussions.

- Reinforced concrete footings are rarely reinforced for shear; the cross section is simply increased in size until it can sustain the shear without reinforcement.

- The crack sizes in reinforced concrete footings are usually controlled by limiting the service stress in the flexural reinforcement to a relatively low value.

- Spread footings and strip footings are designed as cantilevered concrete beams; grade beams are designed as continuous concrete beams.

- Spread footings are also designed for "punching" shear, a special case of shear in which the column penetrates the footing, punching out a block of concrete.

• Rectangular footings are designed with a heavier level of reinforcement in the "middle band" of the longer dimension.

• The ACI coefficients may be used to compute the design moments for continuous grade beams.

REVIEW QUESTIONS

9.1. Why are foundations designed using such staid conservative practices?

9.2. When settlements govern a design, the contact pressure is computed using a reduced, or sustained, live load, but when the concrete footing itself is designed, the highest attainable contact pressure must be used. Why?

9.3. Why is so much more concern shown for ordinary flexural cracking in the foundations than in the building above?

9.4. What does the assumption that the reinforcing steel in the foundation is grade 40 do for the design?

9.5. Why is negative flexural reinforcement in a grade beam usually made continuous, even through parts of the span where it is not actually needed?

9.6. What is the minimum cover over reinforcement in a foundation?

9.7. In a rectangular spread footing supporting a concrete column, where is the critical section for moment? For shear?

9.8. In a rectangular spread footing supporting a steel column on a bearing plate, where is the critical section for moment? For shear?

9.9. In what type of footing is "punching" shear a problem?

9.10. In a rectangular spread footing, how is the distribution of reinforcement made for that reinforcement perpendicular to the long axis?

9.11. In a strip footing supporting a concrete wall, where is the critical section for moment? For shear?

9.12. How is a footing designed for a masonry wall as compared to a concrete wall?

9.13. How is the design of a grade beam similar to that for a beam in the superstructure above? How is it different?

9.14. What addition to a foundation design can counter the tendency of a deep, narrow grade beam to "roll" under load?

9.15. How is the area of reinforcement in a bridging member established in the design of grade beams?

9.16. At what spacing would bridging be placed along a grade beam where columns occur at 36 ft on center?

9.17. Why is shear reinforcement not commonly used in foundations?

9.18. How are moments usually computed along the length of a grade beam?

9.19. Why is intermediate longitudinal reinforcement used at middepth of deep, narrow grade beams?

9.20. How does a bearing wall and its strip footing bridge a "soft spot" or a "hard spot" within the length of the foundation?

PROBLEMS

Problems 9-1 through 9-8. Design an isolated spread footing for the given columns under the given limitations.

Problem	Column size (in.)	Axial P_{DL} (kips)	Load P_{LL} (kips)	Soil pressure (psf)	f'_c (psi)	Steel grade	Remark
9-1	12 × 12	81	54	4000	3000	40	—
9-2	10 × 10	16	17	2000	3000	40	Property line 1.5 ft from centerline of column
9-3	16 × 16	122	99	3000	4000	50	—
9-4	10 × 10	37	28	2000	4000	50	Obstruction 2 ft from centerline of column
9-5	12 × 12	40	27	4000	3000	60	Obstruction 1.5 ft from centerline of column
9-6	14 × 14	79	37	3000	4000	60	—
9-7	12 × 12	50	34	3000	4000	40	—
9-8	12 × 12	69	67	2000	3000	60	Property line 3.5 ft from centerline of column

Problems 9-9 through 9-12. Design a strip footing for the given wall loads.

Problem	Wall size (in.)	Axial P_{DL} (kips/ft)	Load P_{LL} (kips/ft)	Soil pressure (psf)	f'_c (psi)	f'_m (psi)	Steel grade	Remark
9-9	8	14	12	4000	3000	1500	40	Masonry wal
9-10	12	7	9	2000	3000	—	50	Concrete wal
9-11	12	12	7	3000	4000	—	50	Concrete wal
9-12	16	15	12	3000	4000	1500	60	Masonry wal

Problems 9-13 through 9-16. Design a grade beam foundation for a repetitive column load as shown. The columns are square. The load on the end column may be assumed to be roughly half that on an interior column.

Problem	Column size (in.)	Bay length (ft)	Axial P_{DL} (kips)	Load P_{LL} (kips)	Soil pressure (psf)	f'_c (psi)	Steel grade
9-13	12	20	32	36	3000	3000	40
9-14	12	16	24	32	5000	4000	50
9-15	10	25	14	20	4000	4000	40
9-16	16	30	46	46	5000	3000	60

10

Other Topics in Shallow Foundations

In this chapter a few of the special circumstances that can affect the design of a foundation are discussed. Such circumstances are often completely unrelated, falling in no recognizable category. Or they may occur as an inescapable feature of the particular project, arising due to the type of soil, the location of the site, the severity of the weather, or the project design criteria.

One special circumstance, for example, is the design of buildings for which the service life is less than 10 years. For such buildings, the life is long enough to warrant a fixed foundation but not long enough to justify a permanent foundation. Some possibilities for reliable but expedient foundations for such buildings are suggested in this chapter.

Another such special topic is the design of buildings for remote locations. In such locations, the importation or production of aggregates for making concrete is often impractical. In these cases a foundation system of some nonstandard type may become feasible; a few such nonstandard foundations are discussed in this chapter.

Other topics in this chapter include foundations having irregular shapes, foundations for underpinning of buildings, foundations having high lateral loadings, exceptionally rigid foundations for stucco or plaster walls, unreinforced foundations, foundations of rubble or masonry, treated timber foundations, and foundations in expansive clays. Superfici... would seem to be of interest only rarely. They appear, however, ... regularity.

FOUNDATIONS HAVING IRREGULAR SHAPES

Circular, hexagonal, or octagonal spread footings are oft... for tanks, silos, towers, elevated water tanks, and other free... foundations Although the design of foundations for tall, free-standing struc... structures. scope of this book, the use of such foundation shapes for low free... yond the ...g struc-

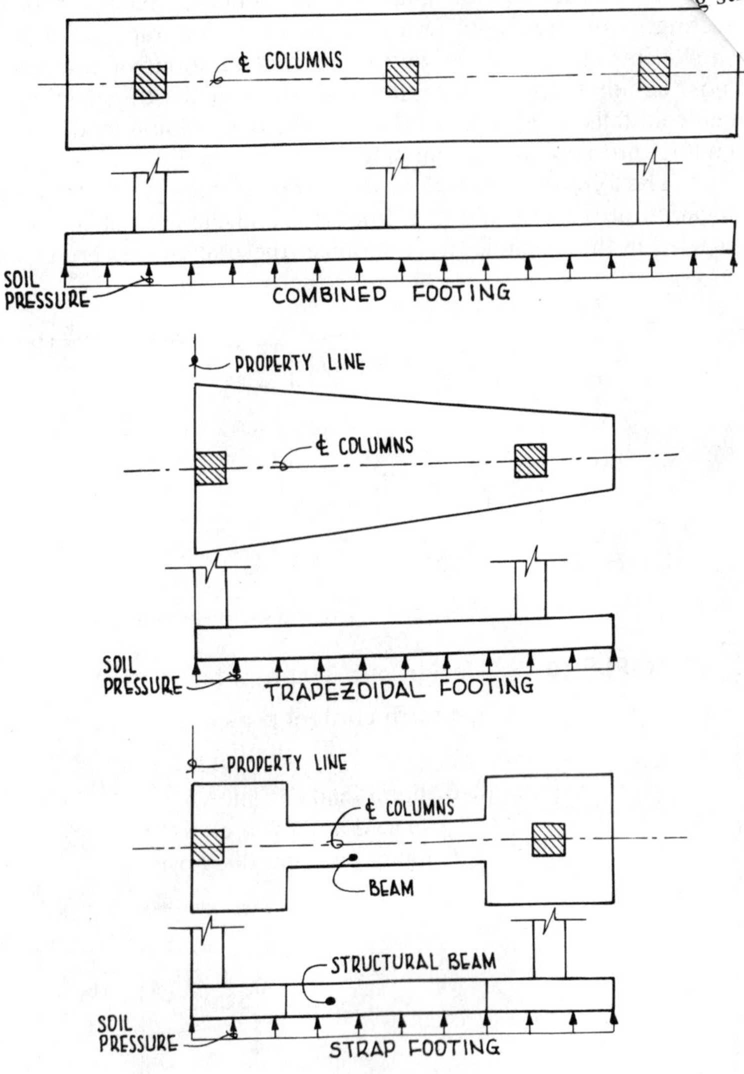

Figure 10-1 Common footing shapes.

ne procedures presented earlier, provided that the resultant of g falls within the middle ninth of the footprint. Foundations for ater tanks up to 10 ft in diameter and located at ground level are e same way as other shallow foundations; circular, hexagonal, or apes are frequently used in such applications.

such footings, the least width B for the design formulas is taken as the of circular footings or across the flats for polygonal footings. However, agnitude of settlement of free-standing structures is rarely limited to 1 in. primary limitation on settlement is usually controlled by the connecting pip-, which might tolerate up to $2\frac{1}{2}$ or 3 in. of settlement.

Another foundation shape that is commonly seen, especially as a combined footing, is the trapezoid. An example of such a trapezoidal combined footing is shown in Fig. 10-1 along with other configurations of combined footings. In its most common application, the dimensions of the trapezoid are set such that its centroid falls under the resultant of the two column loads, producing a uniform contact pressure at the founding line.

The following example will illustrate the design of a trapezoidal footing. The location of the centroid of a trapezoid and the area of the trapezoid, A, will be needed in the example; for reference, these quantities are given in Fig. 10-2.

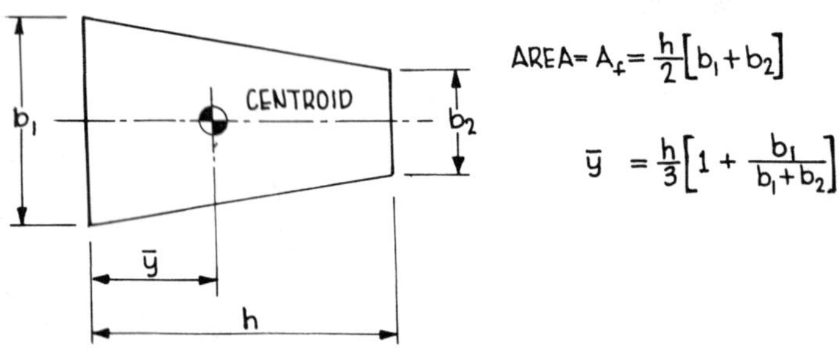

Figure 10-2 Area and centroid of a trapezoid.

EXAMPLE 10-1: Design of a Trapezoidal Footing

Uniform contact pressure.

Given:

Column loads as shown,
spaced at 20 ft o.c.
Contact pressure 3000 psf

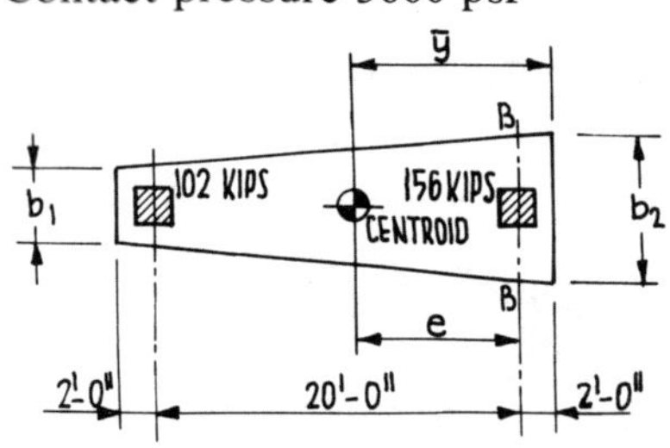

SOLUTION The overall length of the the trapezoid extends a reasonable distan... case, 2 ft was chosen. There are thus two un...sen arbitrarily such that and b_2, such that the resultant of loads falls ove... two columns. In this The magnitude of the resultant of loads is simply found, the widths b_1 ...id of the trapezoid. ...f the two column loads,

$$R = 102 + 156 = 258 \text{ kips}$$

The location of the resultant of loads is found by simple s... ...y summing moments about the line BB,

$$e = \frac{\sum Pe}{\sum P} = \frac{102 \times 20}{102 + 156} = 7.91 \text{ ft}$$

This resultant of loads is shown in the following sketch.

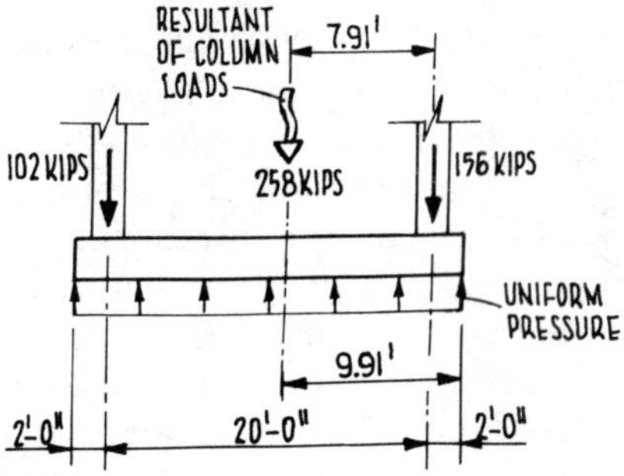

At a uniform contact pressure of 3000 psf, the required area is found routinely,

$$A_f = \frac{P}{p_a} = \frac{258{,}000}{3000} = 86 \text{ ft}^2$$

The values of A_f and $\bar{y}$ computed above are substituted into the equations given in Fig. 10-2, along with the overall length of 24 ft,

$$A_f = \frac{h}{2}(b_1 + b_2) \qquad \bar{y} = \frac{h}{3}\left(1 + \frac{b_1}{b_1 + b_2}\right)$$

After the substitution,

$$86 = \frac{24}{2}(b_1 + b_2) \qquad 9.91 = \frac{24}{3}\left(1 + \frac{b_1}{b_1 + b_2}\right)$$

These two equations are solved simultaneously, yielding

$$b_1 = 1.71 \qquad b_2 = 5.46 \text{ ft}$$

Use $b_1 = 1$ ft 9 in., $b_2 = 5$ ft 6 in.
These final results are shown in the following sketch, along with the final dimensions and pressures.

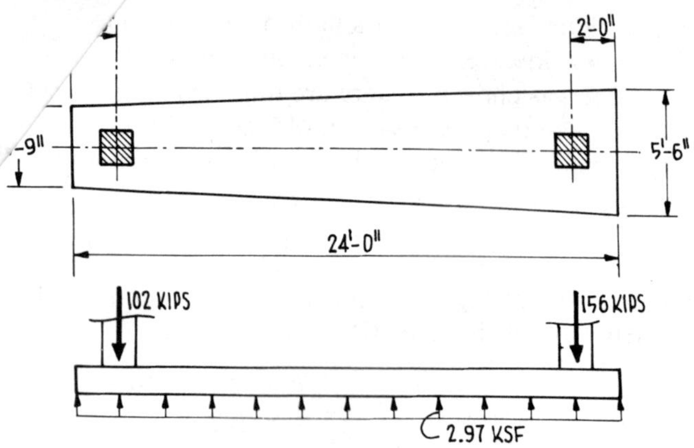

A moment's reflection will confirm that the least value that $\bar{y}$ can have in the foregoing example is $h/3$, or 8 ft, at which point the trapezoid becomes a triangle having $b_1 = 0$. Consequently, there is a very narrow range of column loads for which a trapezoid of reasonable dimensions can be selected. For any other values, the contact pressure will not be uniform.

Even when the contact pressure is uniform, however, the trapezoid does not settle uniformly, since at a uniform contact pressure, the larger end will settle more than the smaller end. Upon reflection, one must conclude that there are few circumstances where the trapezoidal combined footing offers any real advantage over the rectangular combined footing presented in Chapter 7. The trapezoidal footing has been popular for many years, however, and likely will remain popular for many more.

Another popular configuration for spread footings is the strap foundation, shown in Fig. 10-1 and with two of its variations in Fig. 10-3. The strap itself may lie in the plane of the footings or it may be a "strongback," located above or below the footings. Strap footings are also called cantilever footings or pumphandle footings.

It should be apparent that the load to be carried by each pad of the strap footing can be found by simple statics. From that point onward, the design of each pad is performed exactly like any other spread footing. The two pads can even be proportioned for equal settlements using the methods presented earlier, in the event their settlements are critical.

The cantilever footing arrangement shown in Fig. 10-3 may sometimes be useful in underpinning a building to make foundation repairs or to remove and replace an existing footing when a building is to be expanded. A simplified arrangement of such a case is shown in Fig. 10-4. Underpinning an existing structure is one of the more difficult problems in all of foundation engineering. It should be approached cautiously, and even then only with the help of expert and experienced counsel.

By far the most common use for strap footings is in placing a column in close proximity to the property boundary, as shown in Fig. 10-3. Such a case fre-

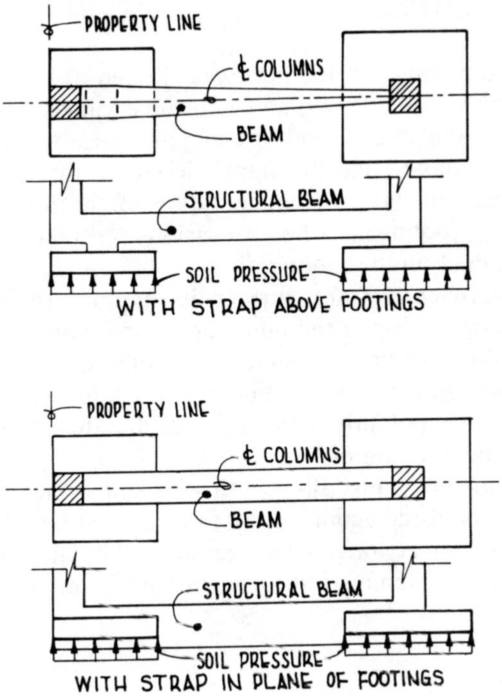

Figure 10-3 Strap footings.

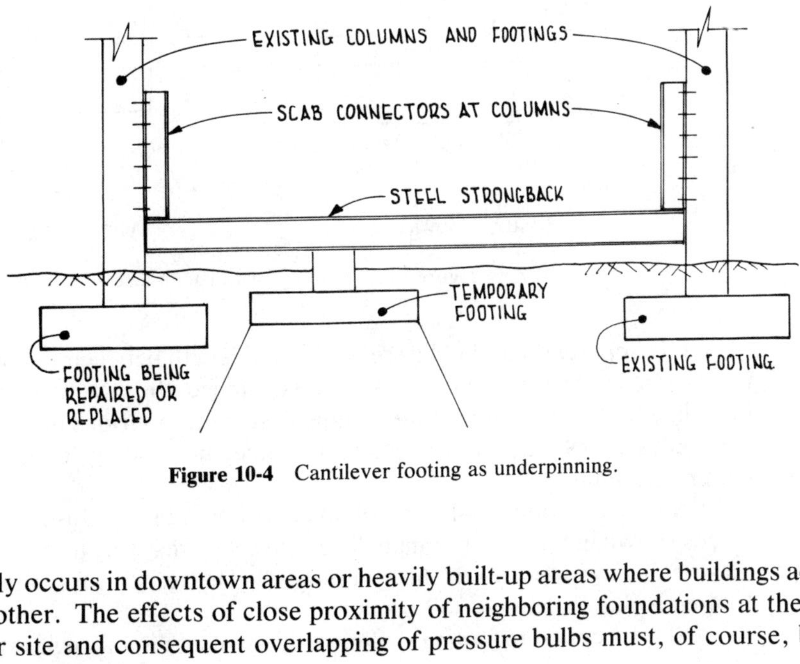

Figure 10-4 Cantilever footing as underpinning.

quently occurs in downtown areas or heavily built-up areas where buildings adjoin each other. The effects of close proximity of neighboring foundations at the particular site and consequent overlapping of pressure bulbs must, of course, be of prime concern.

LATERAL FRICTION LOADS ON FOOTINGS

The subject of lateral friction loads on footings was glossed over very lightly in earlier discussions. The building codes [9] give allowable values for friction between the bottom of the footing and the soil it contacts; these values of friction were used in earlier discussions to find the allowable lateral force on a footing. Regardless what the shape, function, or configuration of the footing was, the allowable lateral force on the footing in earlier discussions was taken at about 30% of the vertical force that acted on the footing.

Such a simplified approach warrants further discussion. Among the more notable omissions in the approach was the omission of any consideration of the shear strength of the soil along the line of contact. Also omitted was the potential benefits due to passive pressure of the confining soil against the sides of the footing. Also omitted was the potential benefits of deliberately roughening the bottom of the footing by the addition of "keys."

A typical footing is shown in Fig. 10-5, both with and without a shear key cast at its base. The shearing force acting on the footing must be taken by some combination of friction, soil shear, or passive pressure. Potential failure planes are shown under the footings, assuming a shear failure in the soil and subsequent sliding of the footing.

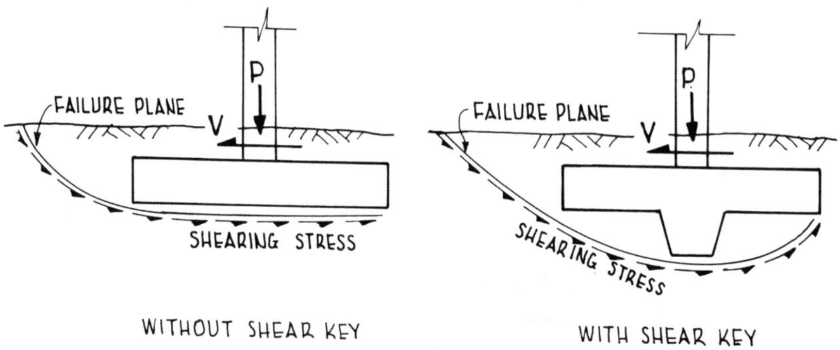

Figure 10-5 Footings with lateral load.

It is apparent that the difference in total length between the two shear planes for the two cases is quite small. As a consequence, the effectiveness of shear keys in adding to the resistance of the footing is subject to serious question. The use of such keys is a holdover from older practice; it is slowly dying out but is still commonly seen.

A typical footing under lateral load is shown in Fig. 10-6, with the resisting forces shown in their approximate locations. The resisting force under the footing may be due to friction or to shear in the soil. The resisting force on the side of the footing is due to passive soil pressure.

Since a concrete footing is almost always cast directly against the supporting

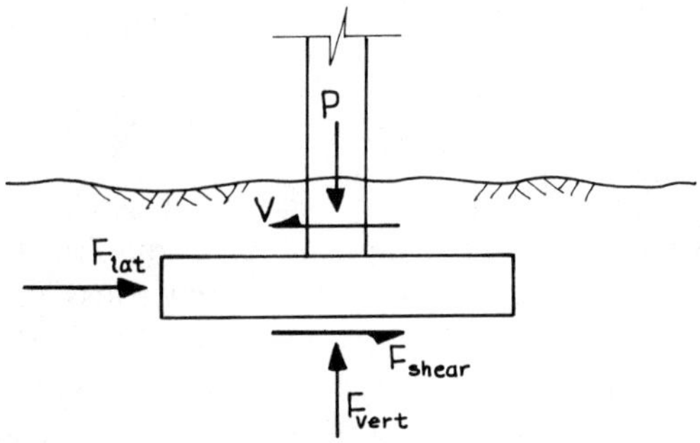

Figure 10-6 Forces on a footing.

soil, the line between the footing and the soil is never a smooth surface but a distinctly roughened surface, even in clayey soils. Further, some penetration into the soil by the water and mortar from the wet concrete is inevitable, producing a blurred contact surface. The coefficient of friction, therefore, can be expected to be quite high.

As friction continues to increase, the limiting value of the resisting force at the contact surface becomes the shear strength of the soil, s. This limiting value of resistance can be readily derived from the Coulomb equation,

$$s = c + \sigma \tan \phi \tag{10-1}$$

where c = cohesive strength of the soil
ϕ = angle of internal friction
σ = vertical pressure

For clays, $\phi = 0$ and the shear strength of the clay is equal to the cohesion c. The cohesion c in turn was shown in Chapter 2 to be half the unconfined compressive strength at failure, q_u; hence

$$s = \tfrac{1}{2} q_u \tag{10-2}$$

The unconfined compressive strength q_u was shown in Chapter 3 to be roughly three times the allowable bearing pressure p_a; hence with a factor of safety of 3,

$$s = \tfrac{1}{2} p_a = 0.5 p_a \tag{10-3}$$

The limiting value of shear strength for clays is thus seen to be far higher than the friction value f, where

$$f = 0.30 p_a \tag{10-4}$$

Hence sliding of the footing laterally will always govern in clays.

For sands, $c = 0$ and the shear strength of the sand is equal to the vertical pressure σ times $\tan \phi$. The total vertical pressure (including overburden) will always be slightly more than the allowable soil pressure p_a and the angle of internal friction will always be more than 26°, hence $\tan \phi > \frac{1}{2}$, and

$$s > \tfrac{1}{2} p_a \quad \text{or} \quad s > 0.5 p_a \tag{10-5}$$

Again, the limiting value of shear strength is seen to be far higher than the friction value, f, where

$$f = 0.30p_a \tag{10-6}$$

Hence sliding of the footing laterally will always govern in sands.

It is therefore concluded that for both sand and clay soils, the friction value for sliding of the footing, even though high, will always be less than the shear strength of the soil.

Refer again to Fig. 10-6. There are two resisting forces shown in the sketch, one due to the sliding friction force and one due to the passive resistance of the soil against the side of the footing. The total resistance was recognized earlier to be the sum of these two forces.

The passive resistance is rarely included as a reliable force in shallow footings; it is usually ignored. It is a viable force only where it can be assured that the overburden will be in place and undisturbed and that the concrete will be cast against undisturbed soil at the sides of the footing. Such ideal conditions are doubtful in foundation construction.

When these conditions are met, however, the design codes permit the forces to be summed to obtain the total resistance to lateral load. That, too, however, is subject to further doubt due to the inconsistency of deformations. For these forces to add, both forces would have to reach their computed maxima at the same time, a doubtful circumstance since one is a function of shear deformations and the other a function of compressive deformations.

In brief, the resistance of a footing to lateral loads can be taken safely at about 30% of the vertical load. Any refinements said to further enhance the capacity of a footing to resist lateral loads should be viewed rather skeptically.

FOUNDATIONS FOR STUCCO OR DECORATIVE MASONRY

Stucco, or exterior plaster, is a popular construction material throughout much of the world. Unfortunately, it is brittle and sometimes subject to cracking. Due largely to improper construction practices, it has become associated with cheap or shoddy construction and has fallen from favor in much of the developed world.

Stucco will crack for two reasons, either through improper composition and application or through foundation settlements. The composition and application of stucco is far outside the scope of a book on foundation design; such information may be found in standard references on construction materials [37]. The practices used for design of foundations suited to stucco walls, however, are well within the scope of this book as outlined in the following paragraphs.

The design procedures presented in earlier chapters assumed that differential settlement between any two structural supports would be limited to about $\frac{3}{4}$ in. As an angular displacement, this amount of settlement on a module of 20 ft comes out to be about 0.003 rad, or $\frac{3}{4}$ in. in 20 ft. For stucco and decorative masonry, the tolerable settlement is usually only about half this amount or even less [38]. A differential settlement of not more than $\frac{3}{8}$ in. in 20 ft (10 mm in 6 m) is recommended.

With such a severe limit on settlements, the use of stucco construction on clay soils would seem to be quite restricted. The surface of a clay stratum, for example, could shrink and swell more than $\frac{3}{8}$ in. in 20 ft just through the annual fluctuations in its moisture content. Even clusters of leafy vegetation and their transpiration could cause enough variation in moisture content in many clays to produce this much differential movement.

As may be supposed, however, the highest variations in moisture content (and resultant shrink and swell) will occur near the surface. Moisture content in clay soils becomes much more stable just a few feet below the surface, sharply reducing the potential for shrink/swell in such soils. It should not be inferred that differential movements would become negligible, but it is entirely possible that they could be reduced to half that at the higher levels.

Where stucco is being considered as a possible construction material, the limits on differential settlement should be stated at the time the soils investigation is begun. The soils report can then include estimates of the founding depths where differential settlements could be reduced to one-half the usual allowance or even less. With such information, a reliable foundation for supporting stucco or decorative masonry can be designed quite effectively.

For the sake of the preliminary design, the moisture content of clays having a plasticity index of 30 or less becomes reasonably stable at about 8 ft. The variation is not drastic, however, even at 5 ft and a depth of founding of 5 ft is suggested as a reasonable starting point for initial estimates. For heavier clays, a depth of founding up to 12 ft or even 15 ft may be necessary. These general figures apply to average years; effects of a 50-year drought or a 50-year flood could produce drastic differences in these estimates.

Stucco walls or decorative masonry at grade should be supported directly on deep grade beams rather than on strip footings. The grade beams should be continuous, with continuous reinforcement both at the top and bottom. Rigidity of the grade beam is of primary importance and rigidity comes with depth of the member; a grade beam whose depth is less than $L/16$ should not be used to support stucco, where L is the length between 90 turns in the grade beam.

The foregoing discussions apply to clay soils. The same concepts apply to sandy soils, but in general such problems are not as severe in sandy soils as in clay soils. For either soil, the determination of the minimum depth of founding for such walls should be made a part of the soils investigation; presumptive values should be avoided.

UNREINFORCED FOUNDATIONS

Over the past 60 years, the use of reinforcement in concrete structures has increased to the point where an unreinforced concrete member is today a rarity. Where a member can be configured such that no tension occurs in it, however, no purpose is served in reinforcing it. Concrete footings are one example of members which can be shaped such that no reinforcement is required.

A practical need for unreinforced concrete foundations was pointed out in Chapter 8. Where groundwater (or porewater) contains both sulfur and salt, the use of sulfur-resistant cement will often provide the necessary resistance to sulfate attack in the portland cement. The sulfur-resistant cement, however, has a lower threshhold of protection for the reinforcement, which is then subject to attack by the chlorides. The real risk to the concrete member would then be due to chloride attack on its reinforcement.

For such applications, the elimination of reinforcement provides a simple but effective remedy. Fortunately, spread footings and strip footings lend themselves readily to such a solution. For grade beams, however, the elimination of reinforcement is not usually practical.

Typical unreinforced spread footings are shown in Fig. 10-7. The first is shown with stepped sides, the second with sloped sides. The stepped sides are easier to form and cast but require slightly more concrete. With the 45° slope shown at the sides of the footing in Fig. 10-7, there is no tension in the concrete and no need for reinforcement except for occasional dowels. Although sulfur-resistant cement has a reduced level of protection for reinforcement, the level of protection for dowels can be improved by increasing the cover to 6 in. as shown in Fig. 10-7. The dowels to the column can be expected to be adequately protected where sufficient cover is maintained.

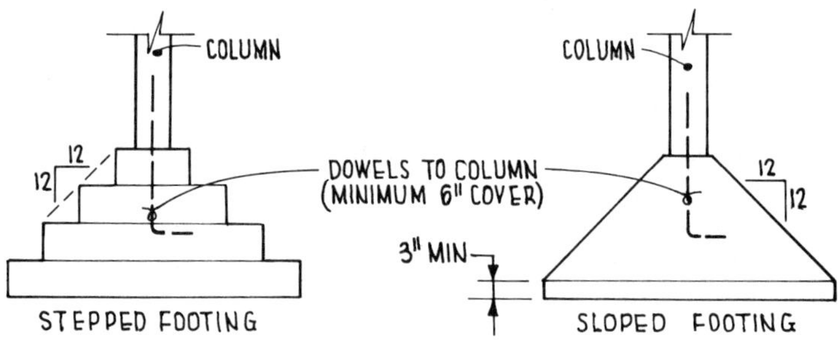

Figure 10-7 Unreinforced spread footings.

Except for requiring more bulk concrete, the configurations of Fig. 10-7 perform identically to the more familiar reinforced concrete pads. There is no difference at all in the way they are used. All the limitations and interactions in performance that occur in reinforced footings occur in unreinforced footings.

Unreinforced strip footings are designed in the same way as unreinforced spread footings. A typical example is shown in Fig. 10-8. Again, sides may be sloped or stepped as preferred.

Whether reinforced or unreinforced, a strip footing is not intended to carry any flexure along its length. The wall itself must therefore be designed to sustain any variations in load that occur along the length of the wall. The bond beams

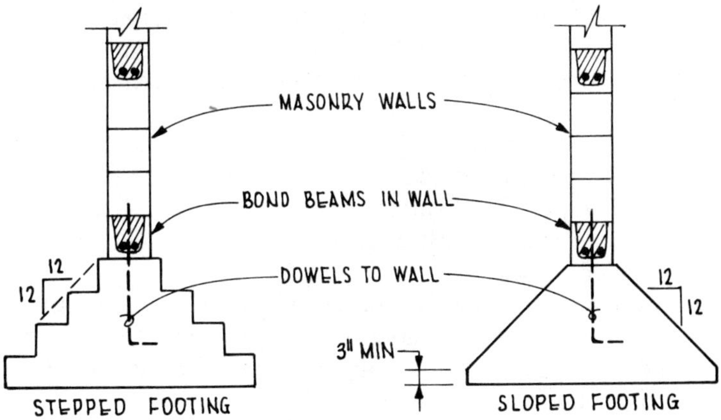

Figure 10-8 Unreinforced strip footings.

shown in Fig. 10-8 are the usual means to reinforce the wall for such variations along its length.

FOUNDATIONS OF RUBBLE OR MASONRY

There are many buildings in service today that are more than 100 years old. Almost all these older buildings that are built on shallow foundations are built on masonry or rubble foundations; portland cement concrete was simply not in such common use at that time. The use of masonry foundations can therefore be accepted as an established and proven practice.

Masonry foundations are built in the same way that unreinforced concrete foundations are built; the only difference is in the allowable compressive strength. Some typical examples are shown in Fig. 10-9. While such foundations have faded from use in recent times due to labor costs, they remain a completely workable and practical foundation.

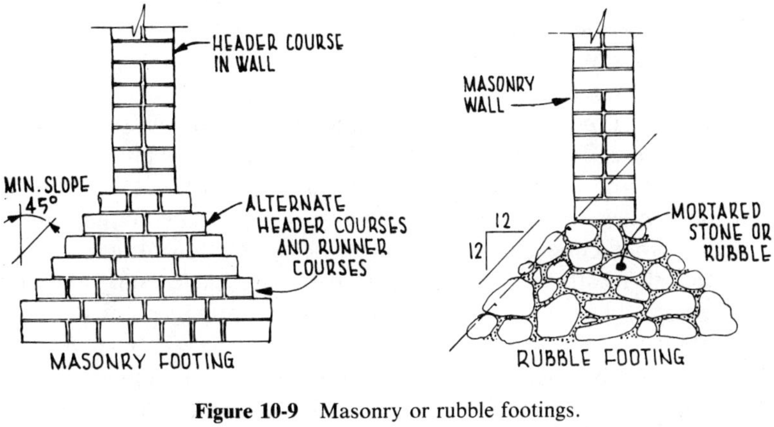

Figure 10-9 Masonry or rubble footings.

Rubble foundations are constructed from used or broken masonry or from fragments of stone. The rubble must of course be physically sound, even if stained and unsightly. Rubble foundations are no different structurally from other masonry foundations.

Very often, the rubble is not even mortared; the broken stones and masonry are manually placed and fitted without mortar to produce a tightly keyed masonry footing. The mortared bearing wall is then begun at the top of this unmortared keyed wall. The labor cost in building such a foundation is usually prohibitive in developed countries.

Masonry foundations built with lime mortar are still being built. Where concrete is expensive and labor is cheap, masonry foundations have traditionally been popular. They are also distinctly practical at remote locations, or where concrete is simply not available.

TREATED TIMBER FOUNDATIONS

Creosoted timber piles and creosoted railroad ties have a long and remarkable history as durable and successful foundations. Except for these two, there are few other historical examples that could be cited where timber foundations have been used. In earlier years, timber foundations simply could not endure the termites and decay organisms that accompany direct contact with earth.

In recent years, however, the development of effective chemical preservatives has made treated timber a potentially useful material for the construction of shallow foundations. To date, the use of treated timber foundations has largely been restricted to special circumstances where loads are light and alternatives are few. Nonetheless, the performance of such foundations has proven to be satisfactory and treated timber foundations can now be regarded as simply one more contemporary foundation system.

In its most common configuration as a shallow foundation, the treated timber foundation is a load-bearing stud wall sheathed with treated plywood on one side only; a typical section of such a wall is shown in Fig. 10-10. The wall is founded at some distance below grade on a treated timber plate. It is extended far enough above grade (about 12 in.) to permit untreated timber to be used throughout the building above it. Fasteners may be wrought iron, brass, or plastic, depending on the corrosiveness of the soil.

One of the more common applications of such a treated timber foundation is in the Great Lakes area, where the "all-weather wood foundation" [27] is promoted by the manufacturers of the preservatives. Its biggest advantage, as its name implies, is that construction can be undertaken at any time of year, even in winter when concrete foundation work is difficult or impossible. The acceptance of the all-weather wood foundation by mortgage companies (and federal loan agencies) for ordinary 20-year or 30-year mortgages attests to its durability and reliability.

Municipal building codes [9] do not permit masonry to be supported by timber. Consequently, the treated timber foundation is at present suitable only for

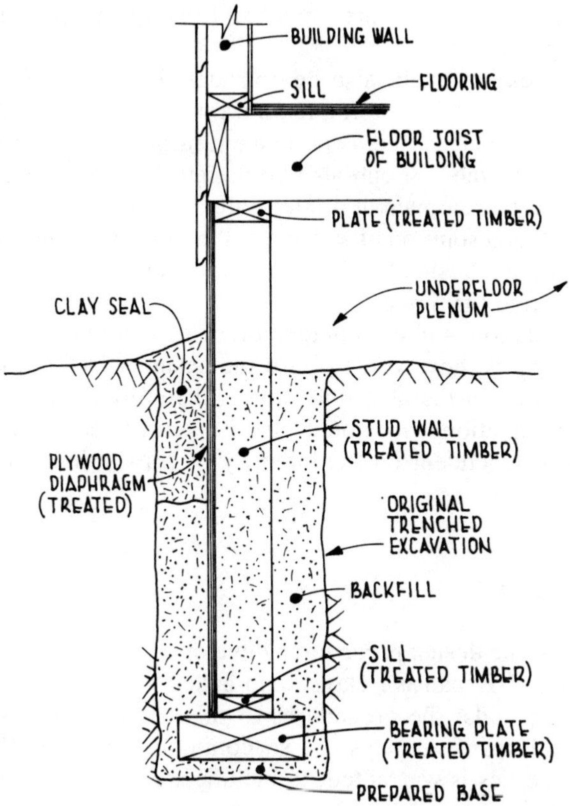

Figure 10-10 Treated timber foundation.

wood frame construction. Where the foundation wall can be made rigid enough, however, it could also be suitable for stucco or exterior plaster.

The rigidity of the foundation wall is of course dependent on its depth. The depth can be as much as 8 ft, since the treated plywood sheathing comes in lengths of 8 ft; with proper blocking the depth could be made even deeper. With such depths, the foundation wall can be made extremely rigid at a relatively nominal cost, even to the point of being rigid enough to support stucco or exterior plaster.

The structural design of the load-bearing stud wall is no different from the design of any other load-bearing stud wall. The design might even include basement walls, where the studs are designed to take lateral earth pressure as well as vertical loads. Such routine exercises in timber design are not included here.

Excavation of the foundation trench may be accomplished by simple trenching machines where the soil is suitable for such trenching. Once leveled and backfilled, the foundation wall is unaffected by weather. The modern preservatives used for treating timber are not leached out due to immersion in groundwater.

Under examination, the treated timber foundation is seen to be little more than a variation of the ordinary strip footing. As such, it may be considered for any light frame construction where a strip footing is a feasible foundation system.

Its possibilities for use in remote locations are obvious, as well as in other locations where concrete is simply not available.

The treated timber foundation might also be considered as an easily constructed expedient foundation, suitable as a foundation for temporary construction. The fact that the foundation is classed as permanent rather than temporary might be a factor in its favor at times. It can also be designed for easy removal should conditions require it to be removed at a later date.

It has been pointed out that a solution of salt and sulfur is a serious hazard to reinforced concrete foundations. It should also be pointed out that this same solution is a natural preservative for timber.

The treated timber foundation is thus a contemporary innovation in foundations. It has no history of failures, however, to warn the designer of those practices which are to be avoided. Considerable reflection is therefore advisable before using treated timber foundations; it should also be remembered that the integrity of the foundation is dependent on the integrity of conventional timber fasteners.

FOUNDATIONS ON EXPANSIVE CLAYS

In Chapter 1 it was stated that the design procedures developed in this text would be applicable to ordinary soils. Expansive clays, also called active clays, were excluded from those ordinary soils; the procedures in this book should not be applied to expansive clays. Expansive clays are so common, however, that a brief discussion of their properties is warranted, to include some general discussions on the design of foundations in such soils.

Expansive clays have the distinct characteristic of shrinking and swelling as their water content decreases and increases. Magnitudes of vertical movement at the ground surface can be quite large, as much as 2 or 3 ft annually in more severe cases with several inches being relatively common in less extreme cases. The shrink–swell phenomenon is a constant process, responding almost constantly to rainfall, leaky sewer lines, water-line breaks, dry periods, heat waves, and other common and frequent events that can affect the moisture content in the subgrade around and under a building.

Movements of such magnitudes can be highly destructive to buildings, in which deflections are typically considered to be large if they are even as much as $\frac{1}{2}$ in. There are few natural phenomena that can produce deflections in buildings of the magnitude that an expansive clay can produce. In addition, the forces generated by a confined clay can be tremendous, far more than routine buildings could ever sustain; the existence of such clays on a building site obviously represent a serious hazard.

The occurrence of expansive clays is not rare; in fact, such clays are quite common. In Mississippi, for example, almost 75% of the state is covered by such clays, with activities ranging from mild to extremely severe [28]. The overconsolidated permian clays of Oklahoma and the montmorillonite soils of Wyoming are also well known as problem clays. Many other states show similar streaks of expansive clays, with varying degrees of activity and associated problems.

As in so many other areas of soils, there is little agreement on what constitutes an expansive clay. In very general terms, the activity of such clays is indicated by the plasticity index, PI. A soil having a PI greater than about 25 is usually considered to be an active soil; this rather arbitrary but simple indicator is accepted here. As the PI increases above 25, the activity and shrink–swell potential increases markedly.

In an earlier section, the concept was developed that placing a foundation at a lower elevation where the moisture content is more stable will help to reduce deflections and settlements. That concept is also valid for foundations in expansive clays. The drawback in using this approach in expansive clays is that the moisture content may undergo rather sharp and rapid changes at depths up to 12 to 15 ft or even more. The cost of placing footings below such depths can often be prohibitive; some means to use less expensive shallow foundations adapted to these expansive clays therefore becomes a highly desirable alternative.

The reason for the rapid variations in moisture content at such depth lies in the fissures and cracks that exist in weathered clays, as shown in Fig. 10-11. These fissures occur due to the repeated wetting, drying, shrinking, and swelling of the clay as it weathers. The heavier or more plastic these clays are, the deeper, the more numerous, and more widely disseminated these fissures become.

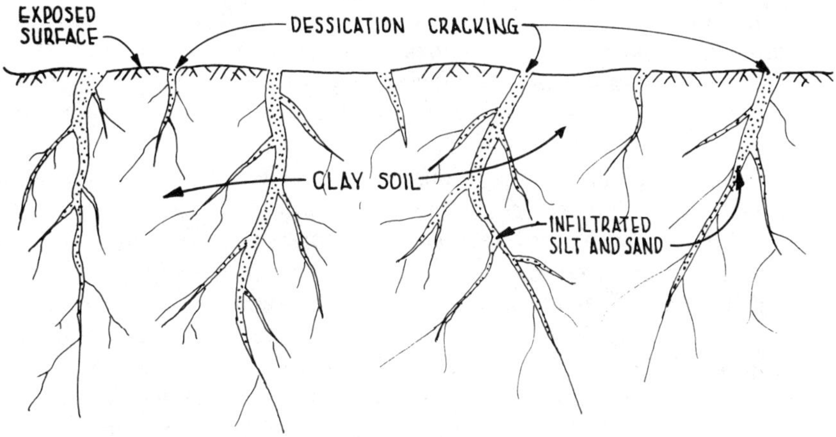

Figure 10-11 Desiccation cracking.

The extreme drying (and fissuring) of clay soils is termed desiccation. When the clay is dry and the fissures stand open, they gradually fill with silt and sand, thereafter providing a permanent but nonuniform network of access routes for later penetration by surface water. With repeated cycles of wetting and drying, the fissures become larger and deeper, the silt deposits become thicker and deeper, and the dissemination of surface water becomes faster, deeper, and more thoroughly dispersed.

For many years there was a belief that when a desiccated clay was covered by a building or a pavement, the evaporation of the porewater at the surface would be drastically reduced. Due to this reduced evaporation rate, the groundwater from below, rising by capillarity into the covered soil, would not evaporate as quickly as before. The water content in the clay would therefore increase sharply, contributing to the swelling of the clay. It was considered essential by some designers at the time to maintain an open space under the ground floor of a building where air could be circulated; the evaporation of the water could then continue, thus preventing or at least reducing the swelling.

More recent research has shown that there is no such continuous rise of groundwater in a desiccated clay [29]; the existence of the desiccation fissures thoroughly and effectively disrupts any such phenomenon. Once the water content becomes stabilized, there is no further need for concern about intrusion of water from below. The water content in a desiccated clay may therefore be maintained at a stable level simply by controlling the intrusion of surface water from above.

In any heavy clay located where shallow foundations would normally be founded, it should always be presumed that the clay will be desiccated. The problem of rapid intrusion of surface water should therefore always be presumed to be a design condition. While the rapidity of the intrusion may vary, the susceptibility of these clays to shrink–swell phenomena should never be doubted.

There are undoubtedly many ways to seal the surface of an area to prevent water intrusion. Bituminous coatings, plastic films, stabilized clay toppings, or other impermeable clay toppings spring immediately to mind. In designing any such seals or toppings, a key factor to its success would be the provision of *rapid* drainage of surface water away from the protected area (slopes of $2\frac{1}{2}\%$ or even more).

In designing an impervious clay topping over the building site, it should be remembered that reworking and remixing (remolding) a desiccated clay will destroy the fissure pattern, restoring its natural impermeability. A thin surface layer of clay (10 in. or so) that has been picked up, reworked, remolded, and replaced will therefore provide a relatively impermeable barrier to surface water. With adequate drainage of this remolded layer, the undisturbed clay below this level will be relatively safe from the rapid intrusion of surface water.

Although simply remolding the clay will produce the desired result, the resulting remolded layer would again be subject to the same cycle of shrinking, cracking, and fissuring. In a short while, this thin layer will have its own pattern of fissures and its watertightness will again be diminished. If the layer were to be lime stabilized and remolded, however, its PI would be reduced markedly and its susceptibility to such fissuring would be reduced significantly.

Lime stabilization has been used successfully in such applications in highway construction [29]. It is worth considering in other applications where the clay is suitable, such as sealing a building foundation. As indicated in Fig. 10-12, the layer of remolded and stabilized clay may be placed some distance below the surface, permitting the usual landscaping and grading to be carried out.

It should not be inferred from the foregoing discussions that no water will

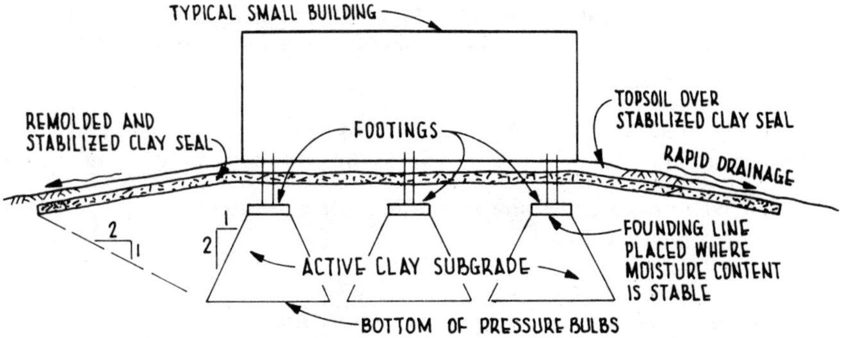

Figure 10-12 Typical surface seal.

penetrate the remolded stratum and percolate downward. Such weathering is of course inevitable. It will be slowed, however, and the clay at the founding line will maintain a more stable water content which was the initial purpose for providing the seal. The rapid drying and wetting cycle with its destructive nonuniform heaving can thus be reduced to manageable levels.

Even with the seal at the surface, however, the founding line must still be placed down at some lower level where the water content is stable. But with the remolded layer protecting the stratum below from rapid changes in the available water supply, the water content can be expected to be stable at much higher elevations than would otherwise be possible. Consequently, the cost of the foundation may be reduced significantly through the use of such features in the design.

The foundation design cannot anticipate later events that might provide an uncontrolled source of water. Typical of such events are sewer leaks, water pipe breaks, spring flooding, deliberate penetrations of the seal to plant trees or shrubs, as well as any number of other very real hazards. Such possibilities can only be weighed, judged, and countered by the individual designer as they may apply to a particular project.

SUMMARY

The following points summarize the conclusions drawn in the discussions of this chapter.

- Foundations having irregular shapes may be designed using the principles outlined in earlier chapters; the theory is not restricted to familiar shapes.

- Irregular shapes may produce quirks in settlement characteristics and should be selected with the full awareness that settlement is proportional to pressure times width, not to pressure alone.

- Joining two spread footings with a strap or a strongback is sometimes done to permit carrying a load off center from its foundation.

• The allowable lateral load on a footing may be taken at 30% of the vertical load; features in the design intended to increase this capacity should be fully verified before they are adopted.

• Foundation settlements may be reduced in some cases by placing the founding line at a lower level where the soil moisture content is more stable.

• Unreinforced foundations may be worth considering where corrosion of reinforcement is a problem.

• Foundations of rubble or masonry, unreinforced, are a completely workable and proven foundation; in some circumstances they may be far less expensive than trying to produce concrete for conventional foundations.

• Treated timber foundations have become a reality in contemporary construction.

• The variation in the water content in a desiccated expansive clay may be controlled by controlling the intrusion of surface water; once the water content is stabilized, the subsurface water makes no further contribution.

• The use of an impermeable seal at a level near the surface of an expansive clay can serve to stabilize the water content below that level; a much shallower foundation might then be considered.

REVIEW QUESTIONS

10-1. Where might a trapezoidal footing be useful?

10-2. What is the difference between using a pumphandle foundation and using two ordinary spread footings?

10-3. What effect does a "key" have on footings that are subjected to lateral loads?

10-4. A coefficient of friction of 0.3 is comparatively high. How is such a high coefficient justified in a design procedure that is as conservative as foundation design?

10-5. Why is the passive pressure of the soil against the sides of a footing usually ignored when calculating the resistance of a footing to lateral loads?

10-6. What is the prime consideration to be met in designing a foundation for stucco or decorative masonry or any other brittle continuous material?

10-7. Why is clay such a problem soil when designing for stucco or decorative masonry?

10-8. Why would unreinforced concrete be more suited to a saltwater environment than reinforced concrete?

10-9. Why is a solution containing both salt and sulfur so much more damaging than a solution containing only one of these chemicals?

10-10. What is the difference in the response of the soil to an unreinforced footing and a reinforced footing?

10-11. When an unreinforced footing is used to support a bearing wall, how are soft spots or hard spots in the underlying soil bridged by the wall and footing?

10-12. How are rubble foundations different from unreinforced foundations?

10-13. Why are timber foundations now accepted in a practice as conservative as foundation design?

10-14. Why should plastic timber fasteners even be considered when designing a treated timber foundation?

10-15. At what PI should a clay be suspected of being expansive?

10-16. Why is desiccation such a major contributor to rapid expansion in expansive clays?

10-17. How is it possible to control expansion in a desiccated clay simply by controlling the intrusion of water from above, with no regard for intrusion of water from below?

10-18. How does remolding a clay restore its impermeability?

10-19. What does lime stabilization do to clay that reduces its susceptibility to swell?

10-20. Why would a slow leak in a water pipe be more hazardous than an actual pipebreak in an expansive clay?

Appendix: Design Tables

The tables given in this appendix are adequate for the design problems introduced in this book. For more detailed analysis or for larger projects, the use of more comprehensive methods with their correspondingly more comprehensive design aids is recommended. The tables herein have been reduced as much as possible where such reduction will not incur an undue amount of interpolation.

The coefficients for beam constants given in Table A-4 are dimensionless. The coefficients are therefore the same for both SI units and Imperial units. The remaining tables are in Imperial units, but their counterparts in SI units may be found in numerous other handbooks and pamphlets.

The tables of development lengths, Tables A-5 and A-6, are sharply limited. For larger sizes of bars, for wire fabric reinforcement, for unusual conditions of cover or embedment, or under any circumstances where these tables are not exactly applicable, a more comprehensive set of tables should be consulted.

The author's earlier textbook, *Design of Reinforced Concrete* (Prentice-Hall, 1988) contains a full development of the simplified approach to concrete presented herein, using the section modulus of the section rather than the more detailed engineering equations. The tables in that book are more extensive, and include compression steel for flexural members.

TABLE A-1 ALLOWABLE STRESSES

a. In Concrete

Type of stress (normal weight concrete: 145 pcf)	ACI318-83	Strength (psi)			
		2000	3000	4000	5000
Modulus of elasticity, E_c	$57{,}000\sqrt{f'_c}$	2.5×10^6	3.1×10^6	3.6×10^6	4.0×10^6
Modular ratio, $n = E_s/E_c$	$\frac{509}{\sqrt{f'_c}}$	11.4	9.29	8.04	7.20
Flexure stress, f_c, extreme compression fiber	$0.45f'_c$	900	1350	1800	2250
Shear stress, v					
Beams, walls, one-way slabs	$1.1\sqrt{f'_c}$	49	60	70	78
Two-way slabs and footings[a]	$(1 + 2/\beta_c)\sqrt{f'_c}$ but $< 2\sqrt{f'_c}$	89	110	126	141
Bearing stress, f_c					
Full area	$0.3f'_c$	600	900	1200	1500
Less than full area (all sides)[b]	$0.3f'_c\sqrt{A_2/A_1}$				

[a] β_c is the ratio of long side to short side of a concentrated load or reaction area.
[b] A_1 is the loaded area; A_2 is the maximum area of the portion of the supporting surface that is geometrically similar to and concentric with the loaded area.

b. In Reinforcement

Modulus of elasticity of all steel: $E_s = 29{,}000{,}000$ psi

Tensile and compressive stresses

Grade 40 steel: $f_y = 40{,}000$ psi, $f_s = 20{,}000$ psi
Grade 50 steel: $f_y = 50{,}000$ psi, $f_s = 20{,}000$ psi
Grade 60 steel: $f_y = 60{,}000$ psi, $f_s = 24{,}000$ psi

TABLE A-2

a. Area of Steel in a Section 1 ft Wide

Spacing (in.)	Bar size								
	3	4	5	6	7	8	9	10	11
2	0.66	1.18	1.84	2.65	3.61	4.71			
$2\frac{1}{2}$	0.53	0.94	1.47	2.12	2.89	3.77	4.80		
3	0.44	0.79	1.23	1.77	2.41	3.14	4.00	5.07	6.25
$3\frac{1}{2}$	0.38	0.67	1.05	1.51	2.06	2.69	3.43	4.34	5.35
4	0.33	0.59	0.92	1.33	1.80	2.36	3.00	3.80	4.68
$4\frac{1}{2}$	0.29	0.52	0.82	1.18	1.60	2.09	2.66	3.38	4.16
5	0.27	0.47	0.74	1.06	1.44	1.88	2.40	3.04	3.75
$5\frac{1}{2}$	0.24	0.43	0.67	0.96	1.31	1.71	2.18	2.76	3.41
6	0.22	0.39	0.61	0.88	1.20	1.57	2.00	2.53	3.12
$6\frac{1}{2}$	0.20	0.36	0.57	0.82	1.11	1.45	1.84	2.34	2.88
7	0.19	0.34	0.53	0.76	1.03	1.35	1.71	2.17	2.68
$7\frac{1}{2}$	0.18	0.31	0.49	0.71	0.96	1.26	1.60	2.03	2.50
8	0.17	0.29	0.46	0.66	0.90	1.18	1.50	1.90	2.34
9	0.15	0.26	0.41	0.59	0.80	1.05	1.33	1.69	2.08
10	0.13	0.24	0.37	0.53	0.72	0.94	1.20	1.52	1.87
11	0.12	0.21	0.33	0.48	0.66	0.86	1.09	1.38	1.70
12	0.11	0.20	0.31	0.44	0.60	0.79	1.00	1.27	1.56
13	0.10	0.18	0.28	0.41	0.56	0.72	0.92	1.17	1.44
14	0.09	0.17	0.26	0.38	0.52	0.67	0.86	1.09	1.34
15	0.09	0.16	0.25	0.35	0.48	0.63	0.80	1.01	1.25
16	0.08	0.15	0.23	0.33	0.45	0.59	0.75	0.95	1.17
17	0.08	0.14	0.22	0.31	0.42	0.55	0.71	0.89	1.10
18	0.07	0.13	0.20	0.29	0.40	0.52	0.67	0.84	1.04

b. Properties of Reinforcing Bars

	Bar size								
	3	4	5	6	7	8	9	10	11
Diameter (in.)	0.375	0.500	0.625	0.750	0.875	1.00	1.13	1.27	1.41
Area (in.^{2})	0.11	0.20	0.31	0.44	0.60	0.79	1.00	1.27	1.56
Perimeter (in.)	1.18	1.57	1.96	2.36	2.75	3.14	3.55	3.99	4.43
Weight (lb/ft)	0.38	0.67	1.04	1.50	2.04	2.67	3.40	4.30	5.31

TABLE A-3 STEEL AREA IN SQUARE INCHES FOR COMBINATIONS OF BAR SIZES

		Number of bars				Number of bars				Number of bars			
No.	Area	1	2	3	4	1	2	3	4	1	2	3	4
No. 3													
1	0.11												
2	0.22												
3	0.33												
4	0.44												
5	0.55												
6	0.66												
No. 4													
1	0.20												
2	0.39												
3	0.59												
4	0.79												
5	0.98												
6	1.18												
No. 5		No. 5 plus No. 4											
1	0.31	0.50	0.70	0.90	1.09								
2	0.61	0.81	1.01	1.20	1.40								
3	0.92	1.12	1.31	1.51	1.71								
4	1.23	1.42	1.62	1.82	2.01								
5	1.53	1.73	1.93	2.12	2.32								
6	1.84	2.04	2.23	2.43	2.63								
No. 6		No. 6 plus No. 5				No. 6 plus No. 4							
1	0.44	0.75	1.06	1.36	1.67	0.64	0.83	1.03	1.23				
2	0.88	1.19	1.50	1.80	2.11	1.08	1.28	1.47	1.67				
3	1.33	1.63	1.94	2.25	2.55	1.52	1.72	1.91	2.11				
4	1.77	2.07	2.38	2.69	2.99	1.96	2.16	2.36	2.55				
5	2.21	2.52	2.82	3.13	3.44	2.41	2.60	2.80	2.99				
6	2.65	2.96	3.26	3.57	3.88	2.85	3.04	3.24	3.44				
No. 7		No. 7 plus No. 6				No. 7 plus No. 5				No. 7 plus No. 4			
1	0.60	1.04	1.48	1.93	2.37	0.91	1.21	1.52	1.83	0.80	0.99	1.19	1.39
2	1.20	1.64	2.09	2.53	2.97	1.51	1.82	2.12	2.43	1.40	1.60	1.79	1.99
3	1.80	2.25	2.69	3.13	3.57	2.11	2.42	2.72	3.03	2.00	2.20	2.39	2.59
4	2.41	2.85	3.29	3.73	4.17	2.71	3.02	3.33	3.63	2.60	2.80	2.99	3.19
5	3.01	3.45	3.89	4.33	4.77	3.31	3.62	3.93	4.23	3.20	3.40	3.60	3.79
6	3.61	4.05	4.49	4.93	5.38	3.91	4.22	4.53	4.84	3.80	4.00	4.20	4.39
No. 8		No. 8 plus No. 7				No. 8 plus No. 6				No. 8 plus No. 5			
1	0.79	1.39	1.99	2.59	3.19	1.23	1.67	2.11	2.55	1.09	1.40	1.71	2.01
2	1.57	2.17	2.77	3.37	3.98	2.01	2.45	2.90	3.34	1.88	2.18	2.49	2.80
3	2.36	2.96	3.56	4.16	4.76	2.80	3.24	3.68	4.12	2.66	2.97	3.28	3.58
4	3.14	3.74	4.34	4.95	5.55	3.58	4.03	4.47	4.91	3.45	3.76	4.06	4.37
5	3.93	4.53	5.13	5.73	6.33	4.37	4.81	5.25	5.69	4.23	4.54	4.85	5.15
6	4.71	5.31	5.92	6.52	7.12	5.15	5.60	6.04	6.48	5.02	5.33	5.63	5.94

No. 9		No. 9 plus No. 8				No. 9 plus No. 7				No. 9 plus No. 6			
1	1.00	1.79	2.57	3.36	4.14	1.60	2.20	2.80	3.41	1.44	1.88	2.33	2.77
2	2.00	2.79	3.57	4.36	5.14	2.60	3.20	3.80	4.41	2.44	2.88	3.33	3.77
3	3.00	3.79	4.57	5.36	6.14	3.60	4.20	4.80	5.41	3.44	3.88	4.33	4.77
4	4.00	4.79	5.57	6.36	7.14	4.60	5.20	5.80	6.41	4.44	4.88	5.33	5.77
5	5.00	5.79	6.57	7.36	8.14	5.60	6.20	6.80	7.41	5.44	5.88	6.33	6.77
6	6.00	6.79	7.57	8.36	9.14	6.60	7.20	7.80	8.41	6.44	6.88	7.33	7.77
No. 10		No. 10 plus No. 9				No. 10 plus No. 8				No. 10 plus No. 7			
1	1.27	2.27	3.27	4.27	5.27	2.05	2.84	3.62	4.41	1.87	2.47	3.07	3.67
2	2.53	3.53	4.53	5.53	6.53	3.32	4.10	4.89	5.68	3.13	3.74	4.34	4.94
3	3.80	4.80	5.80	6.80	7.80	4.59	5.37	6.16	6.94	4.40	5.00	5.60	6.21
4	5.07	6.07	7.07	8.07	9.07	5.85	6.64	7.42	8.21	5.67	6.27	6.87	7.47
5	6.33	7.33	8.33	9.33	10.3	7.12	7.90	8.69	9.48	6.94	7.54	8.14	8.74
6	7.60	8.60	9.60	10.6	11.6	8.39	9.17	9.96	10.7	8.20	8.80	9.40	10.0
No. 11		No. 11 plus No. 10				No. 11 plus No. 9				No. 11 plus No. 8			
1	1.56	2.83	4.09	5.36	6.63	2.56	3.56	4.56	5.56	2.35	3.13	3.92	4.70
2	3.12	4.39	5.66	6.92	8.19	4.12	5.12	6.12	7.12	3.91	4.69	5.48	6.26
3	4.68	5.95	7.22	8.48	9.75	5.68	6.68	7.68	8.68	5.47	6.26	7.04	7.83
4	6.25	7.51	8.78	10.1	11.3	7.25	8.25	9.25	10.3	7.03	7.82	8.60	9.39
5	7.81	9.07	10.3	11.6	12.9	8.81	9.81	10.8	11.8	8.59	9.38	10.2	11.0
6	9.37	10.6	11.9	13.2	14.4	10.4	11.4	12.4	13.4	10.2	10.9	11.7	12.5

TABLE A-4 COEFFICIENTS FOR SECTION CONSTANTS[a]

a. $f'_c = 2000$ psi (13.8 N/mm^2), $\beta = 0.85$, $n = 11.5$

	At ultimate loads				At service levels		
	Grade 40 steel		Grade 60 steel		Elastic coeffs. for all steel		
A_s/bd	Z_c	svcf_c	Z_c	svcf_c	S_c	I_0	f_s/f_c
0.0034			0.112	7:1000	0.111	0.027	35.84
0.0040			0.131	8:1100	0.119	0.031	32.55
0.0050	0.111	6: 900	0.161	9:1200	0.129	0.037	28.60
0.0060	0.131	7: 900	0.189	9:1400	0.138	0.043	25.70
0.0070	0.151	7:1000	0.217	10:1500	0.146	0.048	23.45
0.0080	0.171	8:1100	0.243	11:1600	0.153	0.053	21.64
0.0090	0.189	8:1200	0.267	12:1700	0.159	0.058	20.14
0.0100	0.208	9:1300	0.291	12:1800	0.165	0.062	18.88
0.0110	0.225	9:1300			0.170	0.067	17.80
0.0120	0.243	10:1400			0.175	0.071	16.86
0.0130	0.259	10:1400			0.179	0.075	16.03
0.0140	0.275	10:1500			0.184	0.079	15.30
0.0150	0.291	11:1600			0.187	0.082	14.64
0.0160	0.306	11:1600			0.191	0.086	14.04
0.0170	0.320	11:1600			0.194	0.089	13.50
0.0180	0.334	12:1700			0.198	0.093	13.01

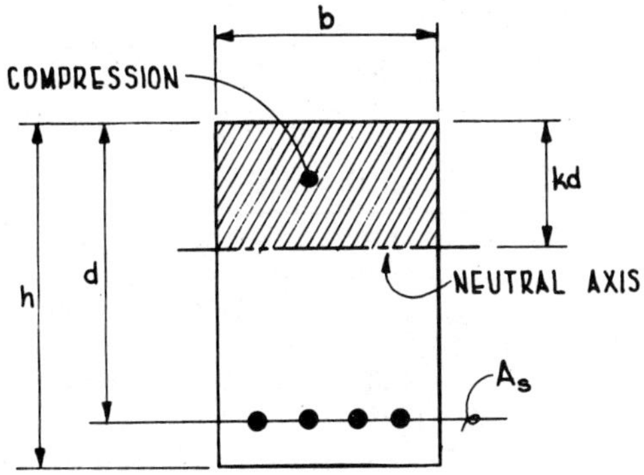

b. f'_c = 3000 psi (20.7 N/mm²), β = 0.85, n = 9.8

	At ultimate loads				At service levels		
	Grade 40 steel		Grade 60 steel		Elastic coeffs. for all steel		
A_s/bd	Z_c	svcf_c	Z_c	svcf_c	S_c	I_0	f_s/f_c
0.0034			0.076	8:1100	0.103	0.023	32.82
0.0040			0.090	8:1200	0.110	0.026	29.84
0.0050	0.075	7: 900	0.111	10:1400	0.120	0.031	26.27
0.0060	0.090	7:1000	0.131	11:1500	0.128	0.036	23.63
0.0070	0.104	8:1100	0.151	12:1700	0.136	0.041	21.59
0.0080	0.118	9:1200	0.171	12:1800	0.143	0.046	19.95
0.0090	0.131	9:1300	0.189	13:1900	0.149	0.050	18.60
0.0100	0.145	10:1400	0.208	14:2000	0.154	0.054	17.45
0.0110	0.158	10:1500	0.225	15:2100	0.159	0.058	16.47
0.0120	0.171	11:1600	0.243	15:2200	0.164	0.061	15.61
0.0130	0.183	11:1600	0.259	16:2300	0.168	0.065	14.86
0.0140	0.196	12:1700	0.275	17:2400	0.172	0.068	14.19
0.0150	0.208	12:1800	0.291	17:2500	0.176	0.072	13.59
0.0160	0.220	13:1800	0.306	18:2600	0.180	0.075	13.05
0.0170	0.231	13:1900			0.183	0.078	12.56
0.0180	0.243	13:2000			0.186	0.081	12.11
0.0190	0.254	14:2000			0.189	0.084	11.70
0.0200	0.265	14:2100			0.192	0.087	11.32
0.0210	0.275	15:2100			0.195	0.090	10.96
0.0220	0.286	15:2200			0.197	0.092	10.64
0.0230	0.296	15:2200			0.200	0.095	10.33
0.0240	0.306	16:2300			0.202	0.098	10.05
0.0250	0.315	16:2300			0.205	0.100	9.78
0.0260	0.325	16:2400			0.207	0.102	9.53
0.0270	0.334	17:2400			0.209	0.105	9.29
0.0280	0.343	17:2400			0.211	0.107	9.07

c. f'_c = 4000 psi (27.6 N/mm²), β = 0.85, n = 8.1

	At ultimate loads				At service levels		
	Grade 40 steel		Grade 60 steel		Elastic coeffs. for all steel		
A_s/bd	Z_c	svcf_c	Z_c	svcf_c	S_c	I_0	f_s/f_c
0.0034			0.058	8:1200	0.097	0.020	30.81
0.0040			0.068	9:1300	0.104	0.023	28.03
0.0050	0.057	7:1000	0.084	10:1500	0.113	0.028	24.70
0.0060	0.068	8:1100	0.100	11:1600	0.122	0.032	22.25
0.0070	0.079	8:1200	0.116	12:1800	0.129	0.037	20.35
0.0080	0.090	9:1300	0.131	13:1900	0.135	0.041	18.81
0.0090	0.100	10:1400	0.146	14:2100	0.141	0.045	17.55
0.0100	0.111	10:1500	0.161	15:2200	0.147	0.048	16.48
0.0110	0.121	11:1600	0.175	16:2300	0.152	0.052	15.56
0.0120	0.131	12:1700	0.189	17:2400	0.156	0.055	14.76
0.0130	0.141	12:1800	0.203	17:2500	0.161	0.059	14.06
0.0140	0.151	13:1800	0.217	18:2600	0.165	0.062	13.44
0.0150	0.161	13:1900	0.230	19:2700	0.168	0.065	12.88
0.0160	0.171	14:2000	0.243	19:2800	0.172	0.068	12.37
0.0170	0.180	14:2100	0.255	20:2900	0.175	0.071	11.91
0.0180	0.189	15:2100	0.267	21:3000	0.178	0.074	11.49
0.0190	0.199	15:2200	0.279	21:3100	0.181	0.076	11.10
0.0200	0.208	16:2300	0.291	22:3200	0.184	0.079	10.75
0.0210	0.217	16:2300	0.302	22:3200	0.187	0.082	10.42
0.0220	0.225	16:2400			0.189	0.084	10.11
0.0230	0.234	17:2400			0.192	0.087	9.83
0.0240	0.243	17:2500			0.194	0.089	9.56
0.0250	0.251	18:2600			0.197	0.091	9.31
0.0260	0.259	18:2600			0.199	0.094	9.07
0.0270	0.267	18:2700			0.201	0.096	8.85
0.0280	0.275	19:2700			0.203	0.098	8.64
0.0290	0.283	19:2800			0.205	0.100	8.44
0.0300	0.291	19:2800			0.207	0.102	8.26
0.0320	0.306	20:2900			0.210	0.106	7.91
0.0340	0.320	21:3000			0.214	0.110	7.59
0.0360	0.334	21:3100			0.217	0.114	7.30
0.0380	0.347	22:3200			0.220	0.118	7.04

d. f'_c = 5000 psi (34.5 N/mm²), β = 0.83, n = 7.5

	At ultimate loads				At service levels		
	Grade 40 steel		Grade 60 steel		Elastic coeffs. for all steel		
A_s/bd	Z_c	svcf_c	Z_c	svcf_c	S_c	I_0	f_s/f_c
0.0034			0.047	9:1300	0.093	0.018	29.32
0.0040			0.055	10:1400	0.099	0.021	26.69
0.0050	0.046	7:1100	0.068	11:1600	0.108	0.026	23.54
0.0060	0.055	8:1200	0.081	12:1700	0.116	0.030	21.22
0.0070	0.064	9:1300	0.094	13:1900	0.124	0.034	19.42
0.0080	0.073	10:1400	0.107	14:2100	0.130	0.037	17.97
0.0090	0.081	10:1500	0.119	15:2200	0.136	0.041	16.77
0.0100	0.090	11:1600	0.131	16:2300	0.141	0.044	15.75
0.0110	0.098	12:1700	0.143	17:2500	0.146	0.048	14.88
0.0120	0.107	12:1800	0.155	18:2600	0.150	0.051	14.13
0.0130	0.115	13:1900	0.167	19:2700	0.155	0.054	13.46
0.0140	0.123	13:1900	0.178	19:2800	0.159	0.057	12.87
0.0150	0.131	14:2000	0.189	20:2900	0.162	0.060	12.34
0.0160	0.139	14:2100	0.200	21:3000	0.166	0.063	11.86
0.0170	0.147	15:2200	0.211	22:3100	0.169	0.066	11.42
0.0180	0.155	16:2300	0.222	22:3200	0.172	0.068	11.02
0.0190	0.163	16:2300	0.232	23:3300	0.175	0.071	10.65
0.0200	0.171	17:2400	0.243	24:3400	0.178	0.073	10.32
0.0210	0.178	17:2500	0.253	24:3500	0.181	0.076	10.00
0.0220	0.186	17:2500	0.263	25:3600	0.183	0.078	9.71
0.0230	0.193	18:2600	0.272	25:3700	0.186	0.081	9.44
0.0240	0.200	18:2700	0.282	26:3700	0.188	0.083	9.19
0.0250	0.208	19:2700	0.291	26:3800	0.190	0.085	8.95
0.0260	0.215	19:2800			0.193	0.087	8.73
0.0270	0.222	20:2900			0.195	0.090	8.51
0.0280	0.229	20:2900			0.197	0.092	8.32
0.0290	0.236	20:3000			0.199	0.094	8.13
0.0300	0.243	21:3000			0.201	0.096	7.95
0.0320	0.256	22:3100			0.204	0.100	7.62
0.0340	0.269	22:3200			0.208	0.103	7.32
0.0360	0.282	23:3300			0.211	0.107	7.04
0.0380	0.294	24:3400			0.214	0.110	6.79
0.0400	0.306	24:3500			0.217	0.114	6.56
0.0420	0.317	25:3600			0.219	0.117	6.35
0.0440	0.329	26:3700			0.222	0.120	6.15
0.0460	0.339	26:3800			0.224	0.123	5.96

[a] Values are coefficients for section constants: Z_c = coeff. × bd^2; S_c = coeff. × bd^2; I_0 = coeff. × bd^3. Units for 'svcf_c' are given first in newtons per square millimeter, then in pounds per square inch. Value of I_0 is only for short-term live loads. For long-term loads, use $\frac{1}{2}E_c$ with I_0.

TABLE A-5 DEVELOPMENT LENGTH OF STRAIGHT DEFORMED REINFORCING BARS

			Required Embedment in Inches to Develop Full Strength										
			Development length			Splice lengths in tension						Splice length
Ultimate concrete strength	Bar size	Grade of steel	Tension		Compression, all bars	Class A		Class B		Class C		Compression, all bars
			Top	Other		Top	Other	Top	Other	Top	Other	
2000 psi	3	40	12	12	8	12	12	12	12	14	12	12
		50	12	12	8	12	12	14	12	18	13	12
		60	13	12	10	13	12	16	12	21	15	12
	4	40	12	12	9	12	12	15	12	19	14	12
		50	14	12	11	14	12	18	13	24	17	13
		60	17	12	13	17	12	22	16	29	20	15
	5	40	15	12	11	15	12	20	14	26	19	13
		50	19	14	14	19	14	25	18	33	23	16
		60	23	16	17	23	16	30	21	39	28	19
	6	40	22	16	13	22	16	29	21	38	27	15
		50	28	20	17	28	20	36	26	47	34	19
		60	33	24	20	33	24	43	31	56	40	23
	7	40	30	22	16	30	22	39	28	51	37	18
		50	38	27	20	38	27	49	35	64	46	22
		60	45	32	23	45	32	59	42	77	55	26
	8	40	39	28	18	39	28	51	37	67	48	20
		50	49	35	22	49	35	64	46	84	60	25
		60	59	42	27	59	42	77	55	100	72	30
	9	40	50	36	20	50	36	65	46	85	61	23
		50	62	45	25	62	45	81	58	106	76	28
		60	75	54	30	75	54	97	70	127	91	34
	10	40	63	45	23	63	45	82	59	108	77	25
		50	79	57	28	79	57	103	74	135	96	32
		60	95	68	34	95	68	124	88	162	116	38
	11	40	78	56	25	78	56	102	73	133	95	28
		50	98	70	32	98	70	127	91	166	119	35
		60	117	84	38	117	84	153	109	199	142	42

TABLE A-5 *(continued)*

Required Embedment in Inches to Develop Full Strength												
			Development length			Splice lengths in tension						Splice length
Ultimate concrete strength	Bar size	Grade of steel	Tension		Compression, all bars	Class A		Class B		Class C		Compression, all bars
			Top	Other		Top	Other	Top	Other	Top	Other	
3000 psi	3	40	12	12	8	12	12	12	12	14	12	12
		50	12	12	8	12	12	14	12	18	13	12
		60	13	12	8	13	12	16	12	21	15	12
	4	40	12	12	8	12	12	15	12	19	14	12
		50	14	12	9	14	12	18	13	24	17	13
		60	17	12	11	17	12	22	16	29	20	15
	5	40	14	12	9	14	12	18	13	24	17	13
		50	18	13	11	18	13	23	16	30	21	16
		60	21	15	14	21	15	27	20	36	26	19
	6	40	18	13	11	18	13	23	17	31	22	15
		50	23	16	14	23	16	29	21	38	27	19
		60	27	19	16	27	19	35	25	46	33	23
	7	40	25	18	13	25	18	32	23	42	30	18
		50	31	22	16	31	22	40	29	52	37	22
		60	37	26	19	37	26	48	34	63	45	26
	8	40	32	23	15	32	23	42	30	55	39	20
		50	40	29	18	40	29	52	37	68	49	25
		60	48	34	22	48	34	63	45	82	59	30
	9	40	41	29	16	41	29	53	38	69	50	23
		50	51	36	21	51	36	66	47	87	62	28
		60	61	44	25	61	44	80	57	104	74	34
	10	40	52	37	19	52	37	67	48	88	63	25
		50	65	46	23	65	46	84	60	110	79	32
		60	78	56	28	78	56	101	72	132	94	38
	11	40	64	46	21	64	46	83	59	109	78	28
		50	80	57	26	80	57	104	74	136	97	35
		60	96	68	31	96	68	125	89	163	116	42

TABLE A-5 *(continued)*

Ultimate concrete strength	Bar size	Grade of steel	Required Embedment in Inches to Develop Full Strength									
			Development length			Splice lengths in tension						Splice length
			Tension		Compression, all bars	Class A		Class B		Class C		Compression, all bars
			Top	Other		Top	Other	Top	Other	Top	Other	
4000 psi	3	40	12	12	8	12	12	12	12	14	12	12
		50	12	12	8	12	12	14	12	18	13	12
		60	13	12	8	13	12	16	12	21	15	12
	4	40	12	12	8	12	12	15	12	19	14	12
		50	14	12	8	14	12	18	13	24	17	13
		60	17	12	9	17	12	22	16	29	20	15
	5	40	14	12	8	14	12	18	13	24	17	13
		50	18	13	10	18	13	23	16	30	21	16
		60	21	15	12	21	15	27	20	36	26	19
	6	40	17	12	9	17	12	22	16	29	20	15
		50	21	15	12	21	15	27	20	36	26	19
		60	25	18	14	25	18	33	23	43	31	23
	7	40	21	15	11	21	15	28	20	36	26	18
		50	27	19	14	27	19	35	25	45	32	22
		60	32	23	17	32	23	42	30	54	39	26
	8	40	28	20	13	28	20	36	26	47	34	20
		50	35	25	16	35	25	45	32	59	42	25
		60	42	30	19	42	30	54	39	71	51	30
	9	40	35	25	14	35	25	46	33	60	43	23
		50	44	32	18	44	32	57	41	75	54	28
		60	53	38	21	53	38	69	49	90	64	34
	10	40	45	32	16	45	32	58	42	76	54	25
		50	56	40	20	56	40	73	52	95	68	32
		60	67	48	24	67	48	87	62	114	82	38
	11	40	55	40	18	55	40	72	51	94	67	28
		50	69	49	22	69	49	90	64	118	84	35
		60	83	59	27	83	59	108	77	141	101	42

TABLE A-5 *(continued)*

Required Embedment in Inches to Develop Full Strength												
			Development length			Splice lengths in tension						Splice length
			Tension			Class A		Class B		Class C		
Ultimate concrete strength	Bar size	Grade of steel	Top	Other	Compression, all bars	Top	Other	Top	Other	Top	Other	Compression, all bars
	3	40	12	12	8	12	12	12	12	14	12	12
		50	12	12	8	12	12	14	12	18	13	12
		60	13	12	8	13	12	16	12	21	15	12
	4	40	12	12	8	12	12	15	12	19	14	12
		50	14	12	8	14	12	18	13	24	17	13
		60	17	12	9	17	12	22	16	29	20	15
	5	40	14	12	8	14	12	18	13	24	17	13
		50	18	13	9	18	13	23	16	30	21	16
		60	21	15	11	21	15	27	20	36	26	19
	6	40	17	12	9	17	12	22	16	29	20	15
		50	21	15	11	21	15	27	20	36	26	19
		60	25	18	14	25	18	33	23	43	31	23
	7	40	20	14	11	20	14	25	18	33	24	18
5000 psi		50	25	18	13	25	18	32	23	42	30	22
		60	29	21	16	29	21	38	27	50	36	26
	8	40	25	18	12	25	18	32	23	42	30	20
		50	31	22	15	31	22	40	29	53	38	25
		60	37	27	18	37	27	49	35	63	45	30
	9	40	32	23	14	32	23	41	29	54	38	23
		50	40	28	17	40	28	51	37	67	48	28
		60	47	34	20	47	34	62	44	81	58	34
	10	40	40	29	15	40	29	52	37	68	49	25
		50	50	36	19	50	36	65	47	85	61	32
		60	60	43	23	60	43	78	56	102	73	38
	11	40	49	35	17	49	35	64	46	84	60	28
		50	62	44	21	62	44	80	57	105	75	35
		60	74	53	25	74	53	96	69	126	90	42

TABLE A-6 DEVELOPMENT LENGTH OF HOOKED BARS[a]

Concrete strength	Steel grade	Bar size 3	4	5	6	7	8	9	10	11
2000 psi	40	7	9	11	13	16	18	20	22	25
	50	8	11	14	17	20	22	25	28	31
	60	10	13	17	20	23	27	30	34	37
3000 psi	40	6	7	9	11	13	15	16	18	20
	50	7	9	11	14	16	18	21	23	25
	60	8	11	14	16	19	22	25	27	30
4000 psi	40	6	6	8	9	11	13	14	16	17
	50	6	8	10	12	14	16	18	20	22
	60	7	9	12	14	17	19	21	24	26
5000 psi	40	6	6	7	8	10	11	13	14	16
	50	6	7	9	11	12	14	16	18	19
	60	6	8	11	13	15	17	19	21	23

[a] To back of hook (in.).

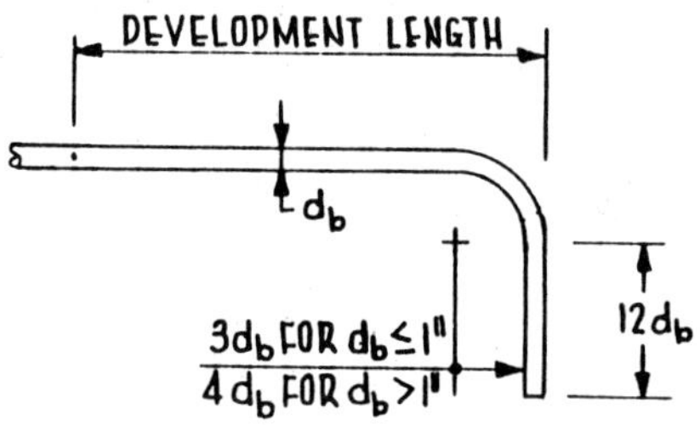

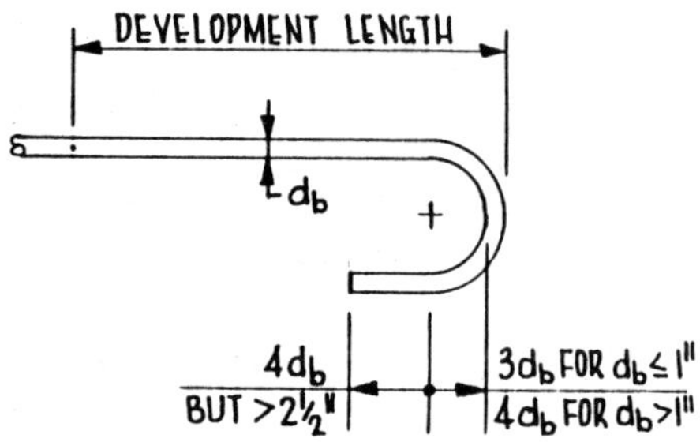

References

1. Atterberg, A., Die Plastizität der Tone, *Internationale Mitteilungen für Bodenkunde,* Vol. I (1911), pp. 10–43
2. Casagrande, A., Classification and Identification of Soils, *Transactions of the American Society of Civil Engineers,* (1948), pp. 901–902.
3. Jumikis, A. R., *Introduction to Soil Mechanics,* D. Van Nostrand, Princeton, N.J., 1967.
4. Peck, R. B., Hansen, W. E., and Thornburn, T. H., *Foundation Engineering,* 2nd ed., Wiley, New York, 1974.
5. Schmertmann, J. H., Static Cone to Compute Static Settlement over Sand, *Journal of the Soil Mechanics and Foundations Division, ASCE,* Vol. 96, No. SM3 (1970).
6. Boussinesq, M. J., *Application des Potentiels, a l'Étude de l'Équilibre et du Mouvement des Solides Élastiques,* Gauthier-Villars, Paris, 1885.
7. Winterkorn, H. F., and Fang, H. Y., ed., *Foundation Engineering Handbook,* Van Nostrand Reinhold, New York, 1975, Chapter 3.
8. Meyerhof, G. G., The Bearing Capacity of Foundations under Eccentric and Inclined Loads, *Proceedings, 3rd International Conference on Soil Mechanics and Foundation Engineering,* Zurich, 19 Vol. I, pp. 440–445.
9. International Conference of Building Officials, *Uniform Building Code,* ICBO, Whittier, Calif., 1983.

10. Sowers, G. F., *Introductory Soil Mechanics and Foundations: Geotechnical Engineering,* 2nd ed., Macmillan, New York, 1979.
11. Teng, W. C., *Foundation Design,* Prentice-Hall, Englewood Cliffs, N.J., 1962.
12. Skempton, A. W., Notes on the Compressibility of Clays, *Quarterly Journal of the Geological Society of London,* Vol. C (1944), pp. 119–135.
13. Hough, B. K., *Basic Soils Engineering,* 2nd ed., Ronald Press, New York, 1957.
14. Schmertmann, J. H., Improved Strain Influence Factor Diagrams, *Journal of the Soil Mechanics and Foundations Division, ASCE,* Vol. 104, No. GT8 (1978).
15. Foster, C. R. and Fergus, S. M., *Stress Distribution in a Homogeneous Soil,* Research Report 12-F, Highway Research Board, 1951.
16. Morgan, J. R., and Scala, A. J., Flexible Pavement Behaviour and Application of Elastic Theory—A Review, *Proceedings of the 4th Conference of the Australian Road Research Board,* Melbourne, 1968, Vol. 4, Part 2, p. 1201.
17. Bozozuk, M., The Gloucester Test Fill, Ph.D. thesis, Purdue University, 1972.
18. Skempton, A. W., Peck, R. B., and MacDonald, D. H., Settlement Analysis of Six Structures in Chicago and London, *Proceedings of the Institution of Civil Engineers,* Part 1, Vol. 4 (1955).
19. Schleicher, F., Zur Theorie des Baugrundes, *Der Bauingenieur,* No. 48 (1926), p. 49.
20. Perloff, W. H., Pressure Distribution and Settlement, in *Foundation Engineering Handbook* (H. F. Winterkorn and H. Y. Fang, eds.), Van Nostrand Reinhold, New York, 1975, Chapter 4, pp. 150–151.
21. American Association of State Highway and Transportation Officials, *Standard Specifications for Highway Materials and Methods of Sampling and Testing, AASHTO,*
22. Proctor, R. R., Fundamental Principles of Soil Compaction, *Engineering News-Record,* August 31, September 9, September 21, September 28, 1933.
23. ACI Committee 318-83, *Building Code Requirements for Reinforced Concrete,* American Concrete Institute, Detroit, Mich., 1983.
24. Wang, C., and Salmon, C. G., *Reinforced Concrete Design,* 3rd ed., Harper & Row, New York, 1979.
25. French, S. E., *Design of Reinforced Concrete,* Prentice-Hall, Englewood Cliffs, N.J., 1988.
26. Fitzgerald, R. W., *Mechanics of Materials,* 2nd ed., Addison-Wesley, Reading, Mass., 1982.
27. National Forest Products Association, *All-Weather Wood Foundation System; Design, Fabrication, Installation Manual,* NFPA, Washington, D.C., 1982.
28. Mississippi State Highway Department, *A Study of Active Clays as Related to Highway Design,* MSHD, Jackson, Miss., 1972.
29. Teng, T. C., and Clisby, M. B., Experimental Work for Active Clays in Mississippi, *Transportation Engineering Journal, ASCE, Proceedings,* Vol. 101, No. TE1 (1975).
30. Meyerhof, G. G., The Influence of Roughness of Base and Ground Water on the Ultimate Bearing Capacity of Foundations, *Geotechnique,* Vol. 5, No. 3 (1955), p. 3, 227.
31. K. Terzaghi, *Theoretical Soil Mechanics,* Wiley, New York, 1943.
32. Ahlvin, R. G., and Ulery, H. H., *Tabulated Values for Determining the Complete Pattern of Stresses, Strains and Deflections beneath a Uniform Circular Load on a Homogeneous Half-Space,* Bulletin 342, Highway Research Board, 1962.

33. Eggested, A., Deformation Measurements below a Model Footing on the Surface of Dry Sand, *Proceedings of the European Conference on Soil Mechanics and Foundation Engineering,* Wiesbaden, Germany, 1963, Vol. 1, p. 233.
34. American Society of Civil Engineers, *Subsurface Investigation for Design and Construction of Foundations of Buildings,* ASCE Manual of Engineering Practice 56, ASCE, New York, 1962.
35. General Services Administration, Structural Engineering, in *GSA Handbook,* PBS-P-3475.1B, GSA, Washington, D.C., March 15, 1971.
36. American Institute of Steel Construction, *Manual of Steel Construction,* 8th ed., AISC, New York, 1980.
37. Huntington, W. C., and Mickadeit, E. R., *Building Construction,* 5th ed., Wiley, New York, 1981.
38. NavFac DM-2, *Structural Engineering,* Design Manual 2, Naval Facilities Engineering Command, Alexandria, Va., 1983.

Index

C

D

E

F

G